MW00443006

THE SUGAR MASTERS

THE UGAR MASTERS

*Planters and Slaves in Louisiana's
Cane World, 1820–1860*

RICHARD FOLLETT

LOUISIANA STATE UNIVERSITY PRESS
BATON ROUGE

Copyright © 2005 by Louisiana State University Press
All rights reserved
Manufactured in the United States of America
Louisiana Paperback Edition, 2007
First printing

Designer: Amanda McDonald Scallan
Typeface: Whitman
Typesetter: G & S Typesetters, Inc.
Printer and binder: Edwards Brothers, Inc.

Library of Congress Cataloging-in-Publication Data

Follett, Richard J., 1968–
  The sugar masters : planters and slaves in Louisiana's cane world, 1820–1860 / Richard Follett.
      p. cm.
  Includes bibliographical references and index.
  ISBN 0-8071-3038-9 (cloth : alk. paper)
  1. Plantation life—Louisiana—History—19th century. 2. Slavery—Louisiana—History—18th century. 3. Sugar grow-
ing—Louisiana—History—19th century. 4. Plantation owners—Louisiana—Social conditions—19th century. 5. Slaves—
Louisiana—Social conditions—19th century. 6. African Americans—Louisiana—Social conditions—19th century.
7. Louisiana—History—1803–1865. 8. Louisiana—Race relations. I. Title.
E445.L8F65 2005
338.1′7361′0976309034—dc22
                                                                                  2004022416
ISBN-13: 978-0-8071-3247-0 (paper : alk. paper)

The paper in this book meets the guidelines for permanence and durability of the Committee on Production Guidelines
for Book Longevity of the Council on Library Resources. ∞

CONTENTS

ACKNOWLEDGMENTS

This book was researched and written in Baton Rouge, Louisiana, Galway, Ireland, and Brighton, England. It is a pleasure to thank both friends and colleagues, who have assisted me along the way, and my family, who have lived with the sugar masters for years. My sincere thanks to you all.

I owe a particular debt of gratitude to the staff of Hill Memorial Library at Louisiana State University. A thousand thanks to Judy Bolton, who helped me at every stage of this project. Judy not only shared her expertise of the Louisiana collections with me, but her fellow archivists made the Hill Reading Room a happy, enjoyable, and cheerful place to work. Although I would like to thank every person who paged and repaged the collections I used, special thanks to Faye Phillips, Christina Riquelmy, Merle Suhayda, Tara Zachary, and Emily Robison. Flying back and forth across the Atlantic to conduct research trips or deliver conference papers is costly, and I would like to express my gratitude to the organizations and institutions that have provided me with financial support during this project: the Fulbright Commission, Louisiana State University, the Royal Irish Academy, and the British Academy. The National University of Ireland, Galway, and the University of Sussex generously supported me with fellowships, travel bursaries, and academic leave. Many thanks to Nicholas Canny in Galway and Pete Nicholls, Steve Burman, John Whitley, and Maria Lauret at Sussex. Back at LSU, many thanks to the university for the Josephine A. Roberts Award and to the Rural Life Museum for their invitation to share my research with members of the public at the Ione Burden symposium.

At Louisiana State University Press, Sylvia Frank Rodrigue has helped me in countless ways during the editorial process. I could not have asked for a more professional and understanding editor, and I warmly thank her for all her assistance. Many thanks to Lee Sioles and Alisa Plant for their work during the final stages of the project. Fellow academics on both sides of the Atlantic have read and commented on various drafts of chapters, conference papers, and departmental research seminars. I thank colleagues for their suggestions, comments, and hospitality, whether at the biggest conference or the smallest works-in-progress seminar. Particular thanks to conference commentators Bruce Levine, Alex Lichtenstein, Martin Klein, Madhavi Kale, T. Stephen Whitman, Billy Smith, Francille Wilson, and Christine Daniels for their valuable suggestions and remarks. Others who have generously shared their research with me or provided vital, formal and informal, comments include John Bezís-Selfa, Moon Ho-Jung, Cindy Hahamovitch, Stephen Fender, Richard Kilbourne, Paul Lovejoy, Rebecca Scott, William Dusinberre, and the late Peter Parish.

I owe a deep debt of gratitude to colleagues who have given so freely of their time to read substantial portions and draft chapters of the manuscript. Many thanks to Peter Kolchin, James Roark, Jonathan Prude, Christopher Clark, John Heitmann, Gavin Wright, Michael Les Benedict, Maria Lauret, Michael Tadman, Jay Kleinberg, Joan Cashin, Catherine Clinton, Martin Crawford, Liese Perrin, Clive Webb, and Vivien Hart. Peter Coclanis read the manuscript in its entirety on two separate occasions and his suggestions, along with those of an anonymous reader for LSU Press, proved especially valuable.

Four colleagues, however, merit special comment. Rick Halpern and I have been working on race and labor in the Louisiana cane world for much of the past decade. We have shared research, conference panels, grants, and our homes on so many occasions—thanks to Rick for all the help. My former colleague at Sussex, Peter Way, is a steadfast friend and searching critic; his impact on my research and writing has been second to none. Paul Paskoff's enthusiasm and resolute support for this project, from its inception to completion, has been a source of tremendous encouragement to me. Long may our antebellum discussions in Highland Coffees continue. My greatest debt of gratitude lies with William J. Cooper Jr., mentor and friend.

None of the excellent advice and support I have received would have meant very much without the encouragement of my family. My parents have been unwavering in their support and have helped in innumerable ways through thick and thin. It is good to say thanks for everything. Thanks also to all my family in Spain and Canada. Marina has lived with every episode in the sugar masters' lives and has endured the blood, sweat, and tears of writing and publishing this book. I cannot begin to thank her and our daughter Carla enough.

ABBREVIATIONS

AH	*Agricultural History*
AHR	*American Historical Review*
JAH	*Journal of American History*
JEH	*Journal of Economic History*
JIH	*Journal of Interdisciplinary History*
JSH	*Journal of Southern History*
LH	*Louisiana History*
LHQ	*Louisiana Historical Quarterly*
LSM	Louisiana State Museum, New Orleans. Microfilm.
LSU	Louisiana and Lower Mississippi Valley Collections, Hill Memorial Library, Louisiana State University, Baton Rouge
TEX	Barker Texas History Center, University of Texas, Austin. Microfilm.
TUL	Howard-Tilton Library, Tulane University, New Orleans. Microfilm.
UNC	Southern Historical Collection, Manuscripts Department, Wilson Library, University of North Carolina, Chapel Hill. Microfilm.

THE SUGAR MASTERS

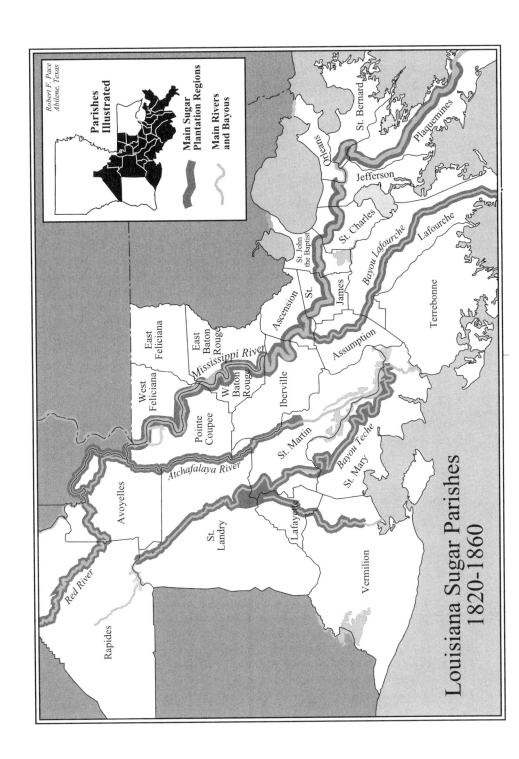

Louisiana Sugar Parishes
1820–1860

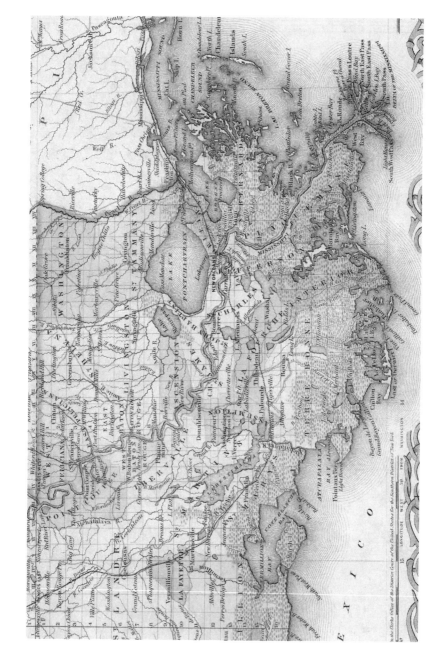

PROLOGUE

The Sugar Masters

ITH UNION AND CONFEDERATE troops massing in northern Virginia, the London *Times* correspondent William Howard Russell hurried upstream after his brief sojourn in New Orleans. Anxious to visit the plantations of the Louisiana sugar country, Russell arrived at John Burnside's expansive sugar holdings some thirty miles south of the state capitol, Baton Rouge. Climbing the bell tower of the plantation house, Russell surveyed a vast agricultural kingdom: "The view from the belvedere . . . was one of the most striking of its kind in the world. If an English agriculturist could see six thousand acres of the finest land in one field, unbroken by hedge or boundary, and covered with the most magnificent crops of . . . sprouting sugar-cane . . . he would surely doubt his senses. But there is literally such a sight—six thousand acres, better tilled than the finest patch in all the Lothians, green as Meath pastures, which can be turned up for a hundred years to come." Russell left the region both impressed and horrified by the superior slave workmanship and industrial productivity of the late antebellum sugar estates. The slaves conducted "all the work of skilled labourers," he noted, and the regimented field gangs moved through the rich greensward with almost military discipline. In the slave quarters, the Englishman observed the "deep dejection" among the enslaved, even as they "stood by, shy, curtsying, and silent, as I broke in upon their family circle." In the plantation hospital, five patients lay "listlessly" on their beds, suffering from pneumonia, fevers, and swelling. Appalled by the destructive impact of the sugar regime, Russell concluded that "nothing but 'involuntary servitude' could go through the toil and suffering required to produce sugar." Russell's week-long stay in the sugar country revealed a merciless industry in which cane farming and mechanized sugar production fused on industrial-scale plantations. Alarmingly modern in their total and structured exploitation of Louisiana's cane world, the sugar planters he met nevertheless seemed divorced from the social currents of nineteenth-century progress. They exercised a "patriarchal sway" over those in their household and shared, Russell concluded, a "particular turn of mind."[1]

This study examines the planters' "turn of mind" in Louisiana's antebellum sugar country. Shrewdly capitalist in their business affairs, the sugar masters defended slavery as an organic institution, an idealized hierarchy in which the slaveholder stood at the apex of society. This fantasy, however, did not preclude them from viewing their slaves as proficient and capable sugar workers who could be readily exploited. As aggressive promoters of the most industrialized sector of southern agriculture, the sugar masters found few discrepancies between their personae as slaveholders and as thoroughly modern businessmen who transformed the swampy landscape of the lower Mississippi Valley into a rich and productive plantation belt. Upon introducing steam-powered sugar mills in the 1840s, Louisiana's sugar elite proudly declared that they had increased sugar production fifteen-fold since the turn of the century. The use of steam-powered sugar mills, drilled gang work, modern assembly-line techniques, and disciplined clock-ordered management enabled them to maximize production and confront the meteorological constraints to cane sugar farming in Louisiana.[2]

1. William H. Russell, *My Diary North and South* (Boston: T.O.H.P. Burnham, 1863), 257–85.

2. Although the wage relationship was central to the capitalist mode of production, historians who limit their definitions of capitalism prioritize wage relations over other variables of capitalist behavior. These include: economic rationality through profit-maximizing; wealth accumulation; market responsiveness and the degree of commercialization; economic specialization, rationality of spirit, and capital accounting; security-seeking risk reduction; and the use of technical and managerial improvements to enhance production. The absence of wage relations ensured that the American South was not ipso facto "capitalist," but that marketplace values guided the thoughts and actions of dynamic slaveholders who were, as Carville Earle contends, "technically precocious, creative, and receptive to innovation." Peter A. Coclanis, *The Shadow of a Dream: Economic Life and Death in the South Carolina Low Country, 1670–1920* (New York, 1989), 59–60; James Oakes, *The Ruling Race: A History of the American Slaveholders* (New York, 1982), 25; Carville Earle, "The Price of Precocity: Technical Choice and Ecological Constraint in the Cotton South, 1840–1890," *AH* 66 (summer 1992): 26 (quote); Gavin Wright, *The Political Economy of the Cotton South: Households, Markets, and Wealth in the Nineteenth Century* (New York, 1978), 3, 43–88; William Dusinberre, *Them Dark Days: Slavery in the American Rice Swamps* (New York, 1996), 6–7.

The antebellum sugar elite achieved these rising production levels on modern, factory-like plantations with regimented and skilled slave labor. Purchasing thousands of young men and women in the New Orleans slave markets, Louisiana planters strove to create an ideal workforce who would toil at the metered cadence of the steam age and produce the next generation of enslaved labor. The potential for industrial action and sabotage—from cutting the cane to shipping the sugar—proved endless, yet few estates experienced systemic resistance to mechanization and industrialization. In fact, the relative success of the sugar masters depended to a large degree on the slaves' acquiescence to the new machines and the agro-industrial discipline of the sugarhouse. To encourage slaves to act in the planters' interest and to put their advanced technology into practice, slaveholders both menaced their bondspeople with the threat of the whip and promised them wages and payment in kind. In return for stability and diligent labor over the year, the slaves obtained Christmas bonuses, extra food, and payment for the produce of their garden plots and overtime work. The incentive and overtime work schemes, however, proved highly calculating, as planters stretched out the working day and exploited every moment for profit. As we will see, the sugar masters fused a manipulative management style with an idealized notion of the master-slave relationship. By orchestrating payment and bestowing incentives, planters could appear before their slaves as paternalistic stewards who shared estate profits with their bondspeople. Unsurprisingly, African Americans viewed overtime work and compensation in a wholly different light. Nonetheless, the compromises hammered out between master and slave ultimately aided productivity as the enslaved advanced their own interests by accommodating the machine and the planters' agenda. For a few hundred dollars—a pittance in the annual cost of making sugar—planters could advance toward the sugar-grinding season confident that their crews would labor for long hours without bringing the entire production process to a halt.[3]

3. On master-slave compromises that ultimately enhanced productivity, see Peter A. Coclanis, "How the Low Country Was Taken to Task: Slave-Labor Organization in Coastal South Carolina and Georgia," in *Slavery, Secession, and Southern History*, ed. Paquette and Ferleger (Charlottesville, Va., 2000), 66, 67; Christopher Morris, "The Articulation of Two Worlds: The Master-Slave Relationship Reconsidered," *JAH* 85 (December 1998): 1007; David Brion Davis, "Looking at Slavery from Broader Perspectives," *AHR* 105 (April 2000): 465; O. Nigel Bolland, "Proto-Proletarians? Slave Wages in the Americas between Slave Labour and Free Labour," in *From Chattel Slaves to Wage Slaves*, ed. Turner (London, 1995), 123–47.

In the past decade, historians of the slave South have increasingly focused on the ways in which planters combined modern and premodern impulses, embracing both the capitalist market and a social ideology based on hierarchy, honor, and paternalism. The brilliant though controversial work of Robert Fogel, Stanley Engerman, James Oakes, and Eugene Genovese has suggested that southern planters were either shrewd entrepreneurs whose raison d'être lay in business success or paternalistic lords who defended their hierarchical society from capitalism and socially destabilizing liberal egalitarianism. Genovese, in particular, has argued that the master-slave relationship and daily acts of slave resistance encouraged an ambivalence to progress among southern planters. The nonmarket, prebourgeois nature of southern social relations, he concluded, engendered a distinct culture and economy that was not opposed to capitalism, but that was alien to the democratic, liberal impulse of northern free-labor ideologues. In contrast, those such as William Dusinberre, who favor an essentially capitalist formulation of the southern planter, have maintained that slaveholders were profit-centered business-men who valued gentlemanly honors, but whose factory-like plantations "resembled in complexity and uncertainty the most advanced operations of a northern capitalist."[4]

The intellectual gulf that once separated these distinct schools of thought has significantly narrowed, though it has not been entirely eliminated. Both Oakes and Genovese now argue that southern slaveholders vigorously sought to profit

---

4. For historiographical overviews, see Mark D. Smith, *Debating Slavery: Economy and Society in the Antebellum American South* (Cambridge, 1998). The seminal works in Eugene D. Genovese's oeuvre include: *The Political Economy of Slavery: Studies in the Economy and Society of the Slave South* (New York, 1965); *The World the Slaveholders Made: Two Essays in Interpretation* (New York, 1969); *Roll, Jordan, Roll: The World the Slaves Made* (New York, 1972); with Elizabeth Fox-Genovese, *Fruits of Merchant Capital: Slavery and Bourgeois Property in the Rise and Expansion of Capitalism* (New York, 1983); *The Slaveholder's Dilemma: Freedom and Progress in Southern Conservative Thought, 1820–1860* (Columbia, S.C., 1992); *A Consuming Fire: The Fall of the Confederacy in the Mind of the White Christian South* (Athens, Ga., 1998). By contrast, Robert Fogel, Stanley Engerman, James Oakes, and William Dusinberre have stressed the profit orientation of slaveholders and the capitalist nature of slavery. In later works, however, Oakes, Dusinberre, and Shearer Davis Bowman argued that southern planters were intensely capitalist but not democratic and sought to maintain strict social hierarchies at the expense of liberal capitalism. See Robert W. Fogel and Stanley L. Engerman, *Time on the Cross: The Economics of American Negro Slavery* (New York, 1974); Robert W. Fogel, *Without Consent or Contract: The Rise and Fall of American Slavery* (New York, 1989); Oakes, *Ruling Race*; James Oakes, *Slavery and Freedom: An Interpretation of the Old South* (New York, 1990); Shearer Davis Bowman, *Masters and Lords: Mid-Nineteenth-Century U.S. Planters and Prussian Junkers* (New York, 1993); Dusinberre, *Them Dark Days*, 6 (quote); William Dusinberre, *Slavemaster President: The Double Career of James Polk* (New York, 2003).

from slavery while casting about for an "alternate route to modernity" where the social impulse of liberal capitalism made few inroads into their organically structured, nonmarket, and hierarchical society. Balancing both the capitalist and precapitalist impulses of antebellum society, most scholars now underscore the hybridity of planter culture, planters' selective adaptation of entrepreneurial values and practices, and the semicapitalist nature of the Old South.[5]

Sugar planters and northern industrialists might have shared a similar mercantilist outlook, but the social values of the slave plantation sharply distinguished them. Culturally if not commercially distinct from their northern peers, southern planters assimilated aspects of individualistic, capitalist, and market-oriented thought into a social ethic that emphasized personal integrity, social standing, human mastery, and above all, independence of thought and action. As Peter Kolchin has observed, the dualistic tendencies within the planter's mind between profit orientation and paternalistic values created a hybrid culture that remained socially, if not economically, at odds with the American North.[6]

5. Recent interpretations of southern slavery as a complex hybrid of capitalist and precapitalist values, where slave owners demonstrated capitalist pretensions with a labor system and social values that espoused paternalistic and patriarchal values, include: Genovese, *The Slaveholder's Dilemma*, 13 (quote); Joyce E. Chaplin, *An Anxious Pursuit: Agricultural Innovation and Modernity in the Lower South, 1730–1815* (Chapel Hill, N.C., 1993); Christopher Morris, *Becoming Southern: The Evolution of a Way of Life, Warren County and Vicksburg, Mississippi, 1770–1860* (New York, 1995); Robert Olwell, *Masters, Slaves, and Subjects: The Culture of Power in the South Carolina Low Country, 1740–1790* (Ithaca, N.Y., 1998); Mark M. Smith, *Mastered by the Clock: Time, Slavery, and Freedom in the American South* (Chapel Hill, N.C., 1997); Jeffrey R. Young, *Domesticating Slavery: The Master Class in Georgia and South Carolina, 1670–1837* (Chapel Hill, N.C., 1999); Edward E. Baptist, *Creating an Old South: Middle Florida's Plantation Frontier before the Civil War* (Chapel Hill, N.C., 2002); James David Miller, *South by Southwest: Planter Emigration and Identity in the Slave South* (Charlottesville, Va., 2002); Richard Follett, "On the Edge of Modernity: Louisiana's Landed Elites in the Nineteenth-Century Sugar Country," in *The American South and the Italian Mezzogiorno,* ed. Dal Lago and Halpern (Basingstoke, 2002), 73–94; William Kauffman Scarborough, *Masters of the Big House: Elite Slaveholders of the Mid-Nineteenth-Century South* (Baton Rouge, La., 2003).

6. Peter Kolchin, *American Slavery, 1619–1877* (New York, 1993), 173. On the planters' social ethic, see, among others: Bertram Wyatt-Brown, *Southern Honor: Ethics and Behavior in the Old South* (New York, 1982); Kenneth S. Greenberg, *Masters and Statesmen: The Political Culture of American Slavery* (Baltimore, 1985); Steven M. Stowe, *Intimacy and Power in the Old South: Ritual in the Lives of the Planters* (Baltimore, 1987); Bertram Wyatt-Brown, *The Shaping of Southern Culture: Honor, Grace, and War, 1760s–1880s* (Chapel Hill, N.C., 2001). For the debate over southern distinctiveness, see Peter Kolchin, *A Sphinx on the Land: The Nineteenth-Century South in Comparative Perspective* (Baton Rouge,

This study contributes to these debates by arguing that antebellum sugar planters effectively balanced their capitalist pretensions with a social ethic that celebrated their independent mastery of an economic, social, and racial world. To be sure, planters occasionally idealized the master-slave relationship and intermittently presented themselves as paternalistic stewards, but they also celebrated their "enterprise and energy," their "intelligence and skill" in slave management, and the efficacy of their plantation regimes. By examining the ways in which these elements interrelate, *Sugar Masters* explores the manipulative thrust of Louisiana slaveholding and reveals the often depressing dynamics of plantation capitalism in the American South's most industrialized and perhaps most exploitative "agricultural business."[7]

My attention focuses predominantly on the three decades before the Civil War, a period in which the sugar industry expanded rapidly and during which plantation slavery reached its apogee as an economic institution and its nadir for those enslaved in the cane world. Although aspects of the plantation slavery in south Louisiana will appear familiar to those who study the cotton and rice states, the peculiarities of sugar made a direct impact upon the slaveholding regime. In fact, and as we shall see, Louisiana's sugar country evolved as a hybrid region that combined aspects of Caribbean and American slavery. The plantation elite drew upon the slaveholding culture of the American South and northern business practices, and it matched the cold-blooded exploitation of the West Indian sugar lords. Thus Louisiana's historical experience should be framed not solely in the context of U.S. slavery, but also from the perspective of the Caribbean sugar-producing world. Geographically located on the northern rim of that cane-growing basin, Louisiana evolved as the last of the New World's cane sugar colonies, and its industry combined the collected experience and suffering of plantation sugar production with lessons drawn from American slavery and industrialization.

Predominantly a social history of plantation slavery in Louisiana's cane world, this book builds upon a rich legacy of scholarship on sugar production to explore

La., 2003); Immanuel Wallerstein, "What Can One Mean by Southern Culture?" in *The Evolution of Southern Culture,* ed. Bartley (Athens, Ga., 1988), 1–13.

7. Frederick L. Olmsted, *A Journey in the Seaboard Slave States,* 2 vols. (1856; reprint, New York, 1904), 2 : 320; James Robertson, *A Few Months in America* (London, 1855), 90 (first quote); *De Bow's Review* 15 (December 1853): 648 (second quote); Walter Brashear to Francis C. Brashear, 16 September 1833, Brashear and Lawrence Family Papers, UNC (third quote).

how masters and slaves combined modern and premodern values in a singular and profitable manner. No study of sugar can ignore the influential works of J. Carlyle Sitterson, whose landmark *Sugar Country* and sugar-related articles set the benchmark for the field in the 1940s and 1950s. Publishing prior to the historiographical and methodological revolution that transformed the study of slavery in the 1960s and 1970s, Sitterson empirically chronicled the history of the sugar industry in tremendous detail. Later studies by David Whitten, Mark Schmitz, John Heitmann, and John Rehder added considerable nuance to the economic history of the antebellum sugar industry, while Craig Bauer and Michael Wade, among others, proffered stimulating studies of individual planters. Aspects of Louisiana slavery and Creole identity have recently come under scrutiny, adding further precision to our historical portrait of the region and our understanding of the forbidding sugar and cotton regimes of the state. This book builds upon and engages with previous scholarship to examine the sugar regime at the height of the slave era, to explore the master-slave relationship, and finally to consider how planters reconciled their profit-minded entrepreneurialism with their collective self-image as paternalistic lords. As historian Bruce Levine observes, by understanding slavery as a social relationship that combined capitalist and precapitalist elements, we can "determine the specific way in which these elements interacted with each other, thereby laying bare the system's actual dynamics." Chapter 1 accordingly provides a brief overview of the Louisiana cane sugar industry, while chapter 2 examines slaveholders' demographic management of the slave labor force. Chapters 3 and 4 explore slaveholding management, supervision, and the use of pseudo-market rewards and harsh punishments within the industry. Chapters 5 and 6 address how the paternalistic character of the master-slave relationship coexisted with both industrial capitalist practices and the slaves' quest for greater independence.[8]

8. Bruce Levine, "Modernity, Backwardness, and Capitalism in the Two Souths," in *The American South and the Italian Mezzogiorno,* ed. Dal Lago and Halpern (Basingstoke, 2002), 235 (quote). On the sugar industry in antebellum Louisiana, see J. Carlyle Sitterson, *Sugar Country: The Cane Sugar Industry in the South, 1753–1950* (Lexington, Ky., 1953), 1–227; David O. Whitten, "Sugar Slavery: A Profitability Model for Slave Investments in the Antebellum Louisiana Sugar Industry," *Louisiana Studies* 12 (summer 1973): 423–42; Whitten, "Tariff and Profit in the Antebellum Louisiana Sugar Industry," *Business History Review* 44 (summer 1970): 226–33; Mark D. Schmitz, "The Economic Analysis of Antebellum Sugar Plantations in Louisiana" (Ph.D. diss., University of North Carolina, 1974); Schmitz, "Economies of Scale and Farm Size in the Antebellum Sugar Sector," *JEH* 37 (December 1977): 959–80; John A. Heitmann, *The Modernization of the Louisiana Sugar Industry, 1830–1910* (Baton Rouge, La., 1987); John B. Rehder, *Delta Sugar: Louisiana's Vanishing Plantation Landscape* (Baltimore, 1999). On

Although the history of Louisiana's sugar masters partly mirrored that of other slaveholding societies, the specificities of sugar production directly affected the nature of the slave regime. To delineate its impact, it is necessary briefly to explain how sugar was made. From the earliest experimentation with cane sugar farming in the 1750s until the innovation of modern frost-resistant canes, sugar planting was a forced crop in Louisiana. Predominantly grown in the tropics, nineteenth-century varieties of sugarcane required good, well-drained soil, ample moisture, and a long, frost-free growing season of at least 250 days. While southern portions of Louisiana met this requirement, central parts of the state struggled with a growing season that remained unconducive to commercial sugar production. Furthermore, most planters in Louisiana (save those in the most southerly parishes) faced the perennial risk of frost damage. Icy winds swept down the Mississippi Valley each autumn, freezing the cane buds and destroying the cane's commercial value. Often, frost damage was followed swiftly by a warming front, which triggered oxidization; as a result, the freshly extracted cane juice became so excessively thick and viscous that it failed to crystallize upon evaporation. With the danger of annual frosts and warming post-freeze breezes, the effective growing season for antebellum cane farmers in Louisiana remained approximately nine months. In the frost-free Caribbean, by contrast, canes matured after eighteen to twenty months and produced 14 to 15 percent sucrose by volume. No planter in antebellum

individual planters, see Michael G. Wade, *Sugar Dynasty: M. A. Patout & Son, Ltd., 1791–1993* (Lafayette, La., 1995); Craig. A. Bauer, *A Leader among Peers: The Life and Times of Duncan Farrar Kenner* (Lafayette, La., 1993); David O. Whitten, *Andrew Durnford: A Black Sugar Planter in Antebellum Louisiana* (Natchitoches, La., 1981). For a succinct overview of the antebellum sugar industry, see John Rodrigue, *Reconstruction in the Cane Fields: From Slavery to Free Labor in Louisiana's Sugar Parishes, 1862–1880* (Baton Rouge, La., 2001), 9–32, and Ira Berlin, *Generations of Captivity: A History of African-American Slaves* (Cambridge, Mass., 2003), 146–49, 179–81, 198. On the complexity and hybridity of Louisiana's slave community, see V. Alton Moody's now dated *Slavery on Louisiana Sugar Plantations* (1924; reprint, New York, 1976), 35–42, 64–66, 72–95; and Joe Gray Taylor, *Negro Slavery in Louisiana* (Baton Rouge, La., 1963). More recent methodological, empirical, and analytical approaches include Gwendolyn Midlo Hall, *Africans in Colonial Louisiana: The Development of Afro-Creole Culture in the Eighteenth Century* (Baton Rouge, La., 1992); Ann Patton Malone, *Sweet Chariot: Slave Family and Household Structure in Nineteenth-Century Louisiana* (Chapel Hill, N.C., 1992); Roderick A. McDonald, *The Economy and Material Culture of Slaves: Goods and Chattels on the Sugar Plantations of Jamaica and Louisiana* (Baton Rouge, La., 1993); Walter Johnson, *Soul by Soul: Life inside the Antebellum Slave Market* (Cambridge, Mass., 1999) and Judith Kelleher Schafer, *Becoming Free, Remaining Free: Manumission and Enslavement in New Orleans, 1846–1862* (Baton Rouge, La., 2003).

Louisiana could mature their canes to such an extent, nor could they produce such sucrose-rich cane juice. The risk of frost damage alone imposed a rigid time constraint on sugar production. Slaves planted the canes in January and February and harvested them from mid-October to December. Some planters tarried longer and pitted their fortunes against the early frosts, but all faced a potentially costly gamble between harvesting their immature cane or risking their luck against the volatility of the Gulf climate.[9]

Meteorological and environmental pressures not only defined the seasonal pattern of labor; it also defined the work culture of the sugar industry. As former slave Hunton Love recalled, "Sugar wuz king in those days" and, like an absolute monarch, all suffered under its influence. African American Ellen Betts remembered that because of the punishing, relentless tempo of agro-industrial sugar production, "To them what work hour in, hour out, them sugar cane fields sure stretch from one end of the earth to the other." And stretch they surely did. While the grueling labor cycle peaked in the harvest season, plantation work continued almost unchecked throughout the year. Immediately after the cane was processed in mid- to late December, slaves began seeding and planting the next year's cane crop—a task that seldom matched the intensity of harvest labor, but one that nonetheless involved extensive and arduous work. Drainage canals required constant maintenance before plow hands furrowed the soil into perpendicular rows, which were planted with seed cane. Until mechanization in the twentieth century, planting cane required diligent, prolonged, and backbreaking labor, which extended into the late winter as planters hurried to prepare the land and their cultivation ridges for planting. Not every field, however, required replanting. A perennial, sugarcane yielded up to six crops in the Caribbean, though most planters in Louisiana allowed it to ratoon for only two years before replanting. In the spring and early summer, slaves hoed and chopped away the weeds that grew among the sprouting canes. By mid-June or early July, the cane shoots were robust enough to survive without constant attention. During this lay-by, when canes required no additional cultivation, slaves turned to a myriad of maintenance tasks—the cultivation of corn, the collection of wood, and the ever-constant maintenance of the levees—which kept the destructive power of the Mississippi and its tributaries in check. Draining and

9. On cane botany, see Frank Blackburn, *Sugar-Cane* (London, 1984); A. C. Barnes, *The Sugar Cane* (New York, 1974); W. J. Evans, *The Sugar Planter's Manual* (Philadelphia, 1848), 82. On temperatures, see *De Bow's Review* 27 (November 1857): 519; Fred B. Kniffen, *Louisiana: Its Land and People* (Baton Rouge, La., 1968), 21.

ditching enabled planters to reclaim some of the land toward the backswamp; during the late summer and early autumn, planters either hired white workers to conduct this unhealthy task or they ordered their own slaves to build canals and ditches, which drew water from the fields.[10]

The overlapping tasks placed continuous pressure on slave crews, who toiled in the early autumn to complete all harvest preparations before the grinding season began in mid- to late October or, at the latest, early November. During this season, operations continued around the clock as cane cutters advanced over the fields, slicing the canes at the root, stripping them of their leaves, and transporting 80 percent of the crop to the mill for processing. The remaining portion was stored as seed-cane for the following planting season. Slaves fed the canes through the mill and extracted the sugar juice. The evaporation process took place over four open kettles, each of which varied in size before reaching the smallest kettle, or battery, where granulation began. As a roaring furnace kept the kettles at the correct temperatures, skilled sugar makers added lime to the boiling juice to remove impurities (an inexact though critical stage). Slaves then skimmed the clarified juice before ladling it into the next kettle for evaporation. In the final kettle, the sugar maker thrust a copper spoon into the battery and examined the thick, grainy appearance of the semi-molten juice before ordering it to be struck. Slaves transferred the clarified sugar into wooden vats, known as coolers, where the sugar would crystallize over the course of twenty-four hours. It was then packed into large wooden hogsheads (containing between 1,000 and 1,200 pounds of sugar) with holes in their bases and allowed to drain until all the molasses separated from the brown crystalline sugar. Planters then either shipped the sealed hogshead to

10. Interview with Hunton Love, date unknown, WPA Ex-Slave Narrative Collection, LSU (first quote); B. A. Botkin, ed., *Lay My Burden Down: A Folk History of Slavery* (Chicago, 1945), 127 (second quote). On cultivation, harvesting, and processing, see Sitterson, *Sugar Country,* 112–56; Walter Prichard, "Routine on a Louisiana Sugar Plantation under the Slavery Regime," *Mississippi Valley Historical Review* 14 (September 1927): 168–78; Glen R. Conrad and Ray F. Lucas, *White Gold: A Brief History of the Louisiana Sugar Industry, 1795–1995* (Lafayette, La., 1995), 14–22. On the spatial limitations to sugar farming, see Sam B. Hilliard, "Site Characteristics and Spatial Stability of the Louisiana Sugarcane Industry," *AH* 53 (January 1979): 254–69. For postbellum agricultural developments, see Louis Ferleger, "Farm Mechanization in the Southern Sugar Sector after the Civil War," *LH* 23 (winter 1982): 21–34. For recent scholarship on the interaction of environment and human agency, see esp. Judith A. Carney, *Black Rice: The African Origins of Rice Cultivation in the Americas* (Cambridge, Mass., 2001); Mart A. Stewart, *"What Nature Suffers to Groe": Life, Labor, and Landscape on the Georgia Coast, 1680–1920* (Athens, Ga., 2002).

market to be consumed raw or transported it for refining in one of several northern cities. Upon completion of the harvest in mid- to late December, the slaves enjoyed a brief respite from farmwork before returning to the fields and yet another
sugar crop.

The seasonal nature of sugar production imposed a forbidding regime on the
slaves. It impelled speed through the entire production process and placed a premium on discipline, organization, and labor stability during the grinding season.
The exigencies of cane farming obliged planters to mechanize, and this in turn
shaped management and the work culture of the slaves. The sugar regime defined
the type of bondspeople planters sought to buy, it gravely affected slave women's
capacity to bear children, and it left an appalling legacy of death in its wake. Yet
above all, the onset of agro-industrial sugar production forced planters and slaves
to negotiate the terms of bondage and to eke out a set of compromises that ultimately enabled the mills to run from morning to night. Such settlements, however, hardly proved triumphant for the slaves. By acceding to the industrial order,
they ultimately tightened the slaveholders' bonds of exploitation and found themselves trapped in the planters' web of duties and obligations. Inadvertently,
African Americans contributed to the system that enslaved them. Although they
struggled to improve their lives and extend their rights, slaves in the sugar country encountered a manipulative and wholly oppressive regime.

one

"A SHARE OF AMBITION"

*T*HE MODERN TRAVELER who pulls off U.S. 90 and turns towards Weeks—some twenty miles south of Lafayette—passes towering cane fields and saline domes that dot the landscape. Named for its land- and slaveholding family, Weeks lies in the heart of the old Attakapas district, just south of McIlhenny's Tabasco factories, which export the produce of Iberia Parish to grocery stores across the globe. In the 1840s and 1850s, just as now, the Teche country left an indelible imprint on the dietary map of the United States, as sugar—not hot pepper sauce—flowed to consumers in New York, Philadelphia, and rural Illinois. Coastal cutters and steam-powered riverboats foreshadowed the diesel fumes of today's eighteen-wheelers, though both Bayou Teche and its tarmac replacement carried Attakapas commerce beyond Louisiana and to grocers' shelves across the nation.

Weeks lay at the western axis of the antebellum sugar industry, and the vast family estate that overlooked Vermillion Bay stood as an exemplary model of plantation growth in Louisiana's cane world. Establishing his plantation on Grand Cote Island in 1820, William F. Weeks masterminded an immense agricultural enterprise that by the 1850s included two hundred slaves, two thousand acres of land, and a "superior" sugar mill that incorporated a steam-granulating pan. Like many of the Anglo-American settlers in the Attakapas district, the Weeks family had migrated to the rich sugar lands of St. Mary and St. Martin Parish in search of wealth and fortune in sugar. The relatively affordable lands and welcome tariff support in the wake of the War of 1812 proffered rich opportunities for sugar planting. By 1814, David Weeks had acquired Grand Cote Plantation; he later bought Parc Perdu in the western reaches of the Teche country. A significant sugar magnate in his own right, he expanded operations at Grand Cote still further by the close of the decade, when his young wife reported, "We are as busy as bees. We have forty-seven hogsheads of beautiful sugar made and we are not half done yet." The fertile soils at Grand Cote yielded almost four hundred hogsheads of sugar in 1850, and that figure was significantly bettered a short eight years later, when Weeks recorded that his plantation had produced 711 hogsheads of sugar. Fully aware that his success rested both on personal ambition and agro-industrial progress, William Weeks matched his father David's exploits, victoriously announcing, "I am very much gratified at my crop on C[ypre] Mort, which, taken in connection with the building of the sugar house and other improvements, is extraordinary—and what is more remarkable—I shall beat all my neighbors with superior forces."[1]

Like his contemporaries, William Weeks stood at the vanguard of an industry that combined technology, science, expert management, and slave labor. But Weeks and his fellow antebellum sugar masters also required a tough, driving temperament and a rational eye for profit and innovation. Francis DuBose Richardson was exemplary in his possession of these attributes. Candidly analyzing Richardson's capacity for the sugar trade, Moses Liddell drew a striking portrait of the "energy, activity, and ingeniousness" required for cane cultivation. "I think FDR may succeed pretty well," Liddell observed. "He has industry and management, and some experience, and has had a little success, he is dependant, and has a share of ambition to press on." Liddell shrewdly added that Richardson "is willing and desirous to go ahead" and that "he will be very useful to himself, his neighborhood, country, and to me in the management of my interests." But above all, Liddell continued, success rested on "strength and capital [combined] with remarkable energy and unbounded perseverance to succeed well." Liddell's son-in-law, John Hampden Randolph of Nottoway Plantation, also possessed such qualities, recognizing that the true key to prosperity in the sugar bowl lay in "perseverance" and, above all, "good management." With their entrepreneurial talents channeled into sugar making, Richardson and Weeks shared discipline, drive, and a perceptive capacity to measure risk and financial success. In mid-1858, Weeks aptly revealed these attributes, ruing that "I see now that we have been dancing

---

1. Description of Plantation and Lands To Be Placed upon the Inventory of the Property Belonging to the Estate of Mary Clara Conrad, October, n.d., 1865; Mary C. Moore to Alfred Conrad, 29 November 1802 [1820] (first quote); William F. Weeks to John Moore, 24 December 1858 (second quote), all in David Weeks and Family Papers, LSU; P. A. Champomier, *Statement of the Sugar Crop Made in Louisiana, 1858–1859*, 30.

too fast for the music—or rather that we have, contrary to our custom, expended more than our receipts." As he explained, "The purchase of slaves—and the amount expended toward the establishment of a new place and machinery necessary for the advantageous working of Grand Cote" drew heavily upon operating funds. Despite these temporary impediments, Weeks remained optimistic, noting that "not one dollar of what has gone through my hands has been expended uselessly [as] we have a fair prospect for a crop if no bad luck befalls us—we will pay all off next year and be ahead." Like many of his sugar-master brethren, Weeks combined personal industriousness with the capital required for modernization and technical improvement.[2]

Commentators who traveled in the sugar region frequently remarked on the agricultural and industrial transformation that the planters orchestrated throughout south Louisiana. In his annual report to the U.S. Commissioner of Patents, Charles Fleischmann concluded that "there is no sugar growing country, where all the modern improvements have been more fairly tested and adopted than in Louisiana." Attributing the success of these advances to the "enterprise and high intelligence of the Louisiana planters, who spare no expense to carry this important branch of agriculture and manufacture to its highest perfection," Fleischmann—like other observers—noted that despite the climatic limitations to cane cultivation in south Louisiana, the planters achieved a "proud triumph" in adopting the latest boiling apparatus and in "fulfilling all the conditions that science and experience have pointed out . . . for obtaining a pure and perfect crystalline sugar." One anonymous contributor to the *Baton Rouge Gazette* similarly lauded his fellow sugar masters for their skillful mastery of the "mechanical and chemical sciences which now become so apparent in this country." Acquainted with several "going-a-head" planters, the correspondent announced that by introducing improvements in agriculture and machinery, the sugar master "will reap his harvest in half the time, and with half the labor and expense" than he had done with primitive agronomy and animal-powered sugar mills. Ever partisan, the *Franklin Planters' Banner* declared that the sugar masters displayed both "intelligence and skill" in their planting operations, combined with "good management on the improved principle adopted in Louisiana." This blend of management and skill

2. Moses Liddell to John R. Liddell, 28 July 1845 (first quote); Moses Liddell to John R. Liddell, 25 August 1845 (second quote); John H. Randolph to John R. Liddell, 22 March 1846 (third quote), all in Moses and St. John Richardson Liddell Family Papers, LSU; William F. Weeks to John Moore, 14 July 1858, David Weeks and Family Papers, LSU (fourth quote).

not only assured the relative economic success of the U.S. sugar industry, the St. Mary Parish paper concluded; it also gave Louisiana planters a marked advantage over their competitors in Mexico, Cuba, and the West Indies. New Orleans editorialist James De Bow could hardly contain his optimism about the cane industry, announcing, "We congratulate our country on the spirit of enterprise which prevails. The competition evinced in the improvement of the manufacture of sugar shows energetic feelings among our planters." De Bow's conclusions paralleled those of his contemporary, Pierre Rost, who emphatically declared before the Agricultural and Mechanics' Association of Louisiana, "The innate faculty of our people to subdue the physical world, their energy and self-reliance . . . have made other nations say of us, that we alone could instill heroism in the common pursuits of life. With heroic determination, then, speed the plow; bear in mind that to go ahead without ever taking difficulties into account, and by that means to succeed, when others dare not undertake, is emphatically the AMERICAN SYSTEM."[3]

These claims resonated with the hollow ring of antebellum boosterism, for Louisiana's sugar industry underwent a profound transformation in the early nineteenth century. Following Etienne Boré's successful granulation of sugar in 1795, the nascent industry spread swiftly from its original core in New Orleans. By 1812—when Louisiana entered the Union as a slave state—sugar occupied a premier position among agriculturalists on the lower reaches of the Mississippi River. Exporting their sugars to the American market, a small coterie of planters brought science and innovation to cane farming. Boré himself overcame the environmental limits to successful sugar farming in Louisiana by experimenting with novel methods of cultivation and irrigation. To protect his crop from the risk of frost damage, he densely packed his canes, and to prevent waterlogging—particularly away from the levee crest—he additionally constructed a system of gates and sluices to drain the fields. The Creole planter Jean Noel D'Estrehan similarly experimented with new strains of sugarcane and novel harvesting techniques. Lauded as one of the most active and enterprising sugar planters of the colony, D'Estrehan replaced wood as the principal fuel beneath the sugar kettles, burning crushed sugarcanes as a cheaper alternative. He also foreshadowed later plantation

3. U.S. Patent Office, *Annual Report of the Commissioner of Patents for the Year 1848* (Washington, D.C., 1849), 275 (first quote); *Baton Rouge Gazette,* 2 December 1843 (second quote); *Franklin Planters' Banner,* 5 January 1854 (third quote); *De Bow's Review* 15 (December 1853): 648; ibid. 1 (February 1846): 166 (fourth quote); ibid. 4 (December 1847): 436 (fifth quote).

management by dividing his workers into three quarter-watches in which slaves worked six-hour shifts at the mill during harvest time. Thus "by a wise distribution of hours," colonial prefect Pierre Clement de Laussat uncritically remarked, the plantation lord "doubled the work of forty to fifty workers without overworking any of them." By introducing the order and regimentation of clock time, D'Estrehan established an innovative, modern, and apparently efficient labor system that later sugar masters emulated.[4]

These early experiments with labor and agricultural practices proved highly profitable. Louisiana's first territorial governor, William C. C. Claiborne, reflected on the bond between slave labor and commercial sugar when he reported to President Jefferson that the "facility with which the sugar Planters amass wealth is almost incredible." Claiborne added, "It is not uncommon with 20 working hands to make from 10 to 14 thousand Dollars and there are several planters whose field Negroes do not exceed forty who make more than 20,000 Dollars each year." The *Louisiana Gazette* concurred with Claiborne's rosy predictions, announcing that on an 800-acre estate with 60 hands, planters could expect to produce 250,000 pounds of sugar and 160 hogsheads of molasses, valued at over $22,000. After an initial, though substantial, investment of $84,000, typical running costs seldom exceeded $3,000 annually, ensuring that a planter could realize a complete return on their expenditure well before the fifth year of production. With the local press publicizing such impressive profit margins, Anglo-Americans flocked to the sugar belt to cash in on Louisiana's apparent riches and for the chance to double or triple their fortunes in the process.[5]

The maturation of the plantation complex in the New Orleans district gave way to increased production along the Mississippi River and on higher lands north of

---

4. Charles Gayarré, *History of Louisiana*, 4th ed., 4 vols. (1854; reprint, New Orleans, 1903), 3:349–50; Georges-Henri-Victor Collot, *A Journey in North America* (1826; reprint, Florence, 1924), 169, 173; James Pitot, *Observations on the Colony of Louisiana*, trans. Henry C. Pitot, ed. Robert D. Bush (Baton Rouge, La., 1979); Pierre Clément de Laussat, *Memoirs of My Life,* trans. Sister Agnes-Josephina Pastwa, ed. Robert D. Bush (Baton Rouge, La., 1978), 60–61 (quote).

5. William C. C. Claiborne to Thomas Jefferson, 10 July 1806, in *Official Letter Books of W. C. C. Claiborne, 1801–1816,* 6 vols., ed. Dunbar Rowland (Jackson, Miss., 1917), 3:363 (first quote); *Louisiana Gazette*, 19 September 1806, quoted in Thomas N. Ingersoll, *Mammon and Manon in Early New Orleans: The First Slave Society in the Deep South, 1718–1819* (Knoxville, Tenn., 1999), 275. On settlement, see Timothy Flint, *Recollections of the Last Ten Years in the Valley of the Mississippi,* ed. George R. Brooks (1826; reprint, Carbondale, Ill., 1968), 216; William Darby, *The Emigrant's Guide to the Western and Southwestern States and Territories* (New York, 1818), 59, 75; Levin Wailes Letter, LSU.

Baton Rouge. By 1805, a coterie of sugar planters had pioneered a small, dynamic industry. Snaking along the alluvial corridors, the sugar industry monopolized most of the rich river bottomlands and, as Thomas Ashe observed, transformed the "frigid character of North America [with] the drapery of the West Indies." In a similar vein, Henry Brackenridge declared that the plantation belt between New Orleans and Baton Rouge proved "what may be done by the art and industry of man." It afforded no less than "one of the strongest arguments in favor of civilization." This self-styled "civilization" promised untold riches for settlers to the region. If only a fourth of the sugar lands along the Mississippi River were cultivated, an observer calculated, the annual Louisiana crop would constitute 25,000 hogsheads of sugar. By 1818, Louisiana sugar output surpassed this figure; by the close of the antebellum era, it had increased fifteen-fold. Little wonder then that Anglo-Americans "were swarming in from the northern states," Laussat observed, "invading Louisiana as the holy tribes invaded the land of Canaan."[6]

Within the plantation belt, Anglo- and African Americans added their respective traditions to an ethnically diverse culture. English-speaking bondsmen and women from the Atlantic coast rubbed shoulders with Afro-Haitians and older communities of Bambara, Mandingo, and Wolof peoples. This diffuse cultural mix made for a vibrant slave community where overlapping traditions flourished and ethnically enriched the world beyond the planter's house. Louisiana's colonial heritage of French and Spanish rule further contributed to the cultural development of the region. French—and to a lesser extent, Spanish—continued to be widely spoken languages in southern portions of the state, Roman law prevailed, and the Catholic Church remained the region's principal religious denomination. The Louisiana Purchase of 1803 and the flood of Anglo-American settlers to the new territory did not eradicate its colonial past or its Franco-Iberian culture. Yet American settlement introduced Protestantism, Anglocentric political traditions, greater trade possibilities with the U.S., and a model of slaveholding derived largely from the British. This model upheld rigid definitions of race, slaveholding hegemony, and a thoroughly articulated, racialized defense of black slavery. In 1806, the passage of a rigorous slave code clamped down on potentially rebellious West Indians and

6. Thomas Ashe, *Travels in America* (London, 1809), 296 (first quote); Henry Marie Brackenridge, *Views of Louisiana* (1814; reprint, Chicago, 1962), 173 (second quote). On the Anglo-American influx, see de Laussat, *Memoirs of My Life*, quoted in Sitterson, *Sugar Country*, 23 (third quote); W. W. Pugh, "Bayou Lafourche from 1835 to 1840—Its Inhabitants, Customs, and Pursuits," *Louisiana Planter and Sugar Manufacturer* 1 (October 1888): 167.

ceded power away from the state to the planters. This transformation in legal precedent effectively entrenched the Anglo tradition of planter hegemony and stripped
the bondsperson of Iberian and Francophone legal protection. Roman law had long
established the de jure right of slaves to purchase themselves (*coartación*) and had
protected manumission through court action. Slaves in the Spanish empire also
possessed the right to challenge their owners' excessive cruelties, though under the
1806 Black Code these privileges collapsed before the dominating influence of the
slaveholders. Grimly, the Anglo-American pattern of ownership and control cast its
imprint on Louisiana, where institutionalized bondage strengthened its iron grip
over slaves. Equipped with the legal instruments of exploitation and an organized
supply of enslaved workers, Anglo-Americans planters pushed ahead with their
compelling combination of sugar and slavery.[7]

Beyond all other factors, the rapid expansion of Louisiana's sugar industry
hinged on the growing domestic demand for sugar, which was increasingly part of

7. On the origins of slavery in Louisiana, see James Thomas McGowan, "Creation of a Slave Society: Louisiana Plantations in the Eighteenth Century" (Ph.D. diss., University of Rochester, 1976). On
African ethnicity, see Hall, *Africans in Colonial Louisiana;* Thomas Ingersoll, "The Slave Trade and the
Ethnic Diversity of Louisiana's Slave Community," *LH* (spring 1996): 133–61; Peter Caron, "'Of a Nation Which Others Do Not Understand': Bambara Slaves and African Ethnicity in Colonial Louisiana,
1718–60," in *Routes to Slavery,* ed. Eltis and Richardson (London, 1997), 98–121; Michael A. Gomez,
*Exchanging Our Country Marks: The Transformation of African Identities in the Colonial and Antebellum
South* (Chapel Hill, N.C., 1998), 23–24, 43–55, 151. On the Native American population and ethnic in-
teraction, see Daniel H. Usner Jr., *Indians, Settlers, and Slaves in a Frontier Exchange Economy: The Lower
Mississippi Valley before 1783* (Chapel Hill, N.C., 1992), and Usner Jr., "Indian-Black Relations in Colo-
nial and Antebellum Louisiana," in *Slave Cultures and the Cultures of Slavery,* ed. Palmié (Knoxville,
Tenn., 1995), 145–61; Jennifer M. Spear, "Colonial Intimacies: Legislating Sex in French Louisiana,"
William and Mary Quarterly 60 (January 2003): 75–98. On the transformation from Franco-Iberian
slavery to the Anglo-American model, see Paul F. Lachance, "The Politics of Fear: French Louisianans
and the Slave Trade, 1786–1809," *Plantation Society* 1 (June 1979): 184; Judith Kelleher Schafer, *Slavery, the Civil Law, and the Supreme Court of Louisiana* (Baton Rouge, La., 1994), 5–6; Kimberly S. Hanger,
"Greedy French Masters and Color-Conscious, Legal-Minded Spaniards in Colonial Louisiana," in
Slavery in the Caribbean Francophone World, ed. Kadish (Athens, Ga., 2000), 106–21. On the strength-
ening legal precedents in antebellum Louisiana, see Taylor, *Negro Slavery in Louisiana,* 155–57. On
the ethnic conflict between Creole/French Louisianans and American settlers, see Sarah Russell,
"Ethnicity, Commerce, and Community on Lower Louisiana's Plantation Frontier, 1803–1828,"
LH 40 (fall 1999): 389–405; Russell, "Cultural Conflicts and Common Interests: The Making of the
Sugar Planter Class in Louisiana, 1795–1853" (Ph.D. diss., University of Maryland, 2000); Joseph G.
Tregle Jr., "Creoles and Americans," in *Creole New Orleans,* ed. Hirsch and Logsdon (Baton Rouge, La.,
1992), 131–85.

the standard laborer's diet. The growth of personal income, combined with declining real and relative costs of sugar, triggered a demand boom for sucrose between the 1820s and the eve of the Civil War, as sugar consumption ballooned from 161 million pounds in 1837 to almost 900 million pounds in 1854. Per capita consumption increased sharply from thirteen pounds of sugar per annum in 1831 to thirty pounds by midcentury.[8]

Faced with a burgeoning market for sugar and assisted by federal tariffs, Louisiana planters intensified cultivation and enlarged production to exploit this market opportunity. Within the space of a few years, the sugar industry surged ahead as the number of estates more than doubled, from 308 in 1827 to 691 in 1830. Production increased sharply in the 1840s, when Whig tariff support and lean cotton prices stimulated new sugar concerns from the Gulf Coast to central Louisiana. The core zone along the Mississippi "coast" (between New Orleans and Baton Rouge) continued to dominate production, but four other areas emerged during the 1830s and 1840s. The first encompassed properties along Bayou Lafourche; the second centered on the western Attakapas district (along Bayou Teche); the third included properties north of Baton Rouge (toward Pointe Coupee and Bayou Sara); and the fourth was comprised of mixed cotton- and sugar-producing farms in Rapides and Avoyelles Parishes. In these areas, particularly in the rich western sugar districts, ambitious planters like Weeks and Richardson "converted waste lands into verdant fields and reaped . . . stores of gold and silver from the glebe they turned up." As one prominent planter observed, Anglo-Americans were "rushing to the sugar gold fields, each with his own idea of working them to their best advantage." As their slaves cleared land, drained swamps, and erected plantation complexes, the sugar masters oversaw a flourishing trade in which both the scale and scope of production advanced briskly. Less

8. On sugar and dietary change, see Sidney W. Mintz, *Sweetness and Power: The Place of Sugar in Modern History* (New York, 1985); Wendy A. Woloson, *Refined Tastes: Sugar, Confectionary, and Consumers in Nineteenth-Century America* (Baltimore, 2002). On U.S. sugar consumption, see *Farmer's Cabinet and American Herd Book* 2 (October 1837): 78; *Journal of Agriculture* 1 (December 1845): 281; *Hunt's Merchants' Magazine* 27 (December 1852): 679–81; ibid. 39 (November 1858): 550; U.S. Patent Office, *Annual Report of the Commissioner of Patents for the Year 1858* (Washington, D.C., 1859), 233. On rising wages, see Robert A. Margo, "Wages and Prices during the Antebellum Period," in *American Economic Growth and Standards of Living before the Civil War*, ed. Gallman and Wallis (Chicago, 1992), 173–210. On sugar prices, consult Arthur Harrison Cole, *Wholesale Commodity Prices in the United States, 1700–1861* (Cambridge, Mass., 1938), 192–357; Noël Deerr, *The History of Sugar*, 2 vols. (London, 1950), 2:524–33.

than sixty years after Boré's celebrated granulation of Louisiana sugar, the cane industry had reached its geographic limits, with 1,536 estates annually producing 250,000 hogsheads of raw sugar by midcentury. In the following decade, planters consolidated their holdings, smaller producers left the industry, and the total number of sugar estates declined to 1,308. Despite conolidation, however, the industry posted quantitative gains. In 1852, Louisiana planters sold over 320,000 hogsheads; in 1853, they produced a quarter of the world's exportable sugar. Enthusiastically fanning regional pride, Representative Miles Taylor announced that such progress "is without parallel in the United States, or indeed in the world in any branch of industry." But Taylor's encomia proved premature. Annual yields crashed in 1855 and 1856, when drought, frost, and hurricane damage brought the industry to its knees. Production recovered in the final antebellum years, however, and as the nation descended into Civil War, planters celebrated the bumper 1861 harvest of 460,000 hogsheads of sugar. It was the last crop made entirely by slave labor and the crowning moment of the antebellum sugar masters.[9]

Despite their impressive record, nineteenth-century sugar planters faced a series of overlapping ecological and environmental problems. Louisiana's swampy landscape restricted the volume of available land for cane cultivation, and the state's sporadically icy climate restricted the agricultural calendar. Most planters partially circumvented these ecological obstacles by windrowing their canes and by introducing new plant varieties. To windrow, slaves cut the canes, laid them in furrows, and covered them with leaves until they were ready for grinding or planting in the New Year. Not all planters agreed on the efficacy of windrowing, but most believed that it was the only effective means to store their crop in cold weather and evade the risks incumbent in grinding frost-damaged cane. The introduction of ribbon cane in 1817, however, transformed the nascent sugar industry and enabled planters to extend the agricultural belt. Early planters, like Boré and D'Estrehan,

9. *De Bow's Review* 1 (January 1846): 55–56; Sitterson, *Sugar Country,* 28; P. A. Champomier, *Statement of the Sugar Crop, 1845–1846,* 35; Champomier, *Statement of the Sugar Crop, 1849–1850,* 51; Champomier, *Statement of the Sugar Crop, 1859–1860,* 39; Champomier, *Statement of the Sugar Crop, 1861–1862,* vi; *Franklin Planters' Banner,* 16 March 1848 (first quote); F. D. Richardson, "The Teche Country Fifty Years Ago," *Southern Bivouac* 4 (March 1886): 593; *De Bow's Review* 22 (April 1857): 435 (second quote). On northern expansion and increased yields, see *De Bow's Review* 2 (December 1846): 442; *Alexandria Democrat,* reprinted in *Franklin Planters' Banner,* 13 January 1848; *New Orleans National,* reprinted in *Franklin Planters' Banner,* 30 September 1847; *De Bow's Review* 3 (May 1847): 414; *American Agriculturist* 9 (November 1850): 351.

TABLE 1. SUGAR PRODUCTION IN LOUISIANA
(HOGSHEADS)

Year	Number	Year	Number
1832	70,000	1847	240,000
1833	75,000	1848	220,000
1834	100,000	1849	247,923
1835	30,000	1850	211,201
1836	70,000	1851	236,547
1837	65,000	1852	321,934
1838	70,000	1853	449,324
1839	115,000	1854	346,635
1840	87,000	1855	231,427
1841	90,000	1856	73,976
1842	140,000	1857	279,697
1843	100,000	1858	362,296
1844	200,000	1859	221,840
1845	186,000	1860	228,758
1846	140,000	1861	459,410

had cultivated Creole and Otaheite cane. The latter was an improvement on earlier cane varieties, but its vulnerability to frost damage ensured that sugar farming remained confined to the southerly portions of the state. Ribbon cane had several advantages over earlier varieties. It was larger, James McCoy observed in the *Southern Agriculturist,* "and makes from three to four hogsheads per acre, grows further north, and matures at least one month earlier." Importantly, the cane proved significantly more frost- and cold-resistant, enabling sugar cultivation to spread into central portions of the state.[10]

If ribbon cane partially resolved the geographical limits to cane farming in Louisiana, it also compelled technological improvement, since its tougher bark was difficult to crush with animal-powered mills. The introduction of steam-powered technology in 1822, however, boosted the mills' productive capacity and allowed planters to expand cultivation, confident that their machinery would grind

10. On frost risks, see Evans, *Sugar Planter's Manual,* 82; Norman J. King, *Manual of Cane-Growing* (New York, 1965), 180–85. On ribbon cane, see Noël Deerr, *Cane Sugar* (London, 1921), 49; Lewis Cecil Gray, *History of Agriculture in the Southern United States to 1860,* 2 vols. (Washington, D.C., 1933), 1:740; *Southern Agriculturist* 1 (April 1828): 179 (quote).

the crop before the first hard freeze struck. Until the 1830s, the newer technology proved prohibitively expensive, though when occasion and opportunity allowed, planters gradually replaced their slow, inefficient, and oxen-driven mills with steam-powered equipment. This process accelerated once the price of a new steam-powered mill began to decline (from approximately $12,000 in 1822 to about $6,000 a decade later). As late as 1847, Cincinnati foundries continued to charge on average $7,000 for their mills, even as the New Orleans–based Leeds and Co. marketed their cheaper mills at just over $3,000. While still expensive, Leeds's machinery remained within the grasp of moderately wealthy individuals, especially when paid for over two or three years. The growing affordability of steam power proved critical for its dissemination. In 1828, the sugar economy remained overwhelmingly horse-powered: just 82 of 308 estates utilized steam to drive their mills. Thirteen years later, steam powered the mills of 361 of 668 Louisiana sugar estates, and by 1850 steam engines were operating in over 900 plantations and grinding most commercially-produced cane. By the end of the antebellum era, almost 80 percent of all sugar plantations possessed steam engines and mills, which crushed and ground ribbon cane with increasing efficiency.[11]

Steam power profoundly shaped the sugar industry, but its economic success rested primarily on the mass importation of African American bondspeople to Louisiana. With relatively accessible credit, federal tariff protection, and a compliant state, planters tapped the domestic slave trade for young, strong men on whom the sugar production rested. The number of slaves cultivating sugar steadily increased: 21,000 in 1827, 36,000 in 1830, 50,600 in 1841, and 65,000 in 1844. By midcentury, the slave population had almost quadrupled in the space of twenty years, as 125,000 men and women toiled in the oppressive heat of Louisiana's sugar bowl. In conjunction with the expansion of the aggregate slave population, plantations featured ever-larger slave crews. In 1830, 52 slaves labored on the average sugar plantation; just over a decade later, most slave crews consisted of 76 bondspeople. By the early 1850s, most plantation quarters across the sugar belt housed

---

11. On the transformation from horse to steam power, see *De Bow's Review* 1 (January 1846): 55; U.S. Patent Office, *Annual Report of the Commissioner of Patents for the Year 1848*, 294; *Franklin Planters' Banner*, 29 July 1847; Edward J. Forstall, *Agricultural Productions of Louisiana* (New Orleans, 1845), 4; J. A. Leon, *On Sugar Cultivation in Louisiana, Cuba, Etc., and the British Possessions*, 26; P. A. Champomier, *Statement of the Sugar Crop, 1850–51*, 43; Champomier, *Statement of the Sugar Crop, 1860–1861*, 39.

85 slaves, and by the Civil War, most large sugar plantations listed as many as 110 enslaved African Americans on their inventories. Although Louisiana posted a small white majority in 1860, African Americans overwhelmingly dominated the sugar country. In Ascension Parish on the Mississippi River, slaves outnumbered whites two to one, while in the western St. Mary Parish, over 13,000 slaves outnumbered less than 3,500 whites. Like almost every sugar lord, William F. Weeks resided among a black majority, and whether on Grand Cote or visiting the family residence at Shadows-on-the-Teche, his life and identity were anchored firmly in slaveholding.[12]

Both steam power and slave labor underpinned agricultural expansion. Farm output multiplied from a mean of 108 hogsheads in 1830 to over 250 hogsheads in 1845 and 310 hogsheads in the bumper crop of 1853. Productivity per field hand also increased. The number of acres cultivated per hand rose from approximately 2 acres in 1802 to 3.5 in 1822. This figure climbed to 5 acres by the latter years of the antebellum era, and on the largest estates planters might expect their slaves to cultivate as many as 6 acres of sugarcane per hand. While man-land ratios rose, so did individual output, with estate managers measuring appreciable increases in productivity from the 1830s to the Civil War. In 1831, efficient sugar masters cultivated and manufactured approximately 2.5 hogsheads per slave or 4 per plantation worker. Sixteen years later, the St. James Parish sugar magnate Valcour Aime estimated that the average yield of sugar per hand in the late 1840s varied between 5 and 8 hogsheads. Edward Forstall concurred, noting that where the slaves' tasks were "made to harmonize, so as to insure rapidity and constant working," the sugar master might reasonably hope to produce 7 hogsheads of sugar per working hand—a figure matched only by the finest Cuban estates, and 2 to 3 times the yield of West Indian plantations in 1808. The sugar lords ultimately produced more sugar per worker than their predecessors and Caribbean competitors principally because of their commitment to steam power and, as we shall see, regimented slave management.[13]

12. *De Bow's Review* 1 (January 1846): 55–56; *Hunt's Merchants' Magazine* 30 (April 1854): 499; McDonald, *Economy and Material Culture of Slaves*, 3; Joseph Karl Menn, *The Large Slaveholders of Louisiana, 1860* (New Orleans, 1964), 23. On slave importation, see Michael Tadman, "The Demographic Cost of Sugar: Debates on Societies and Natural Increase in the Americas," *AHR* 105 (December 2000): 1570–73.

13. *De Bow's Review* 1 (January 1846): 55; *Hunt's Merchants' Magazine* 30 (May 1854): 499; P. A. Champomier, *Statement of the Sugar Crop, 1853–1854,* 42. On productivity, see Gray, *History of*

Despite these indices of economic progress, the continued success of the early sugar boom hinged on the availability of capital and credit to facilitate growth; it was also critically dependent on the federal protection of Louisiana sugar. In 1803, sugar planters enjoyed the benefits accrued from the U.S. sugar rate of 2.5¢ per pound on brown, white clayed, and powdered sugars. These revenue-raising tariffs had the additional advantage of allowing Louisiana planters and merchants to increase the price of domestic sugar by at least the duty charged on imported goods. Under such protection, Louisiana planters enjoyed assured profit margins, which rose during the War of 1812 and remained at a buoyant 3¢ per pound with the passage of the protectionist 1816 tariff. In the wake of the prosperous wartime years, Louisianans badgered their congressmen to protect their agricultural interests and retain American independence in supplying sucrose. Asking whether there is "any manufacture more important to the nation" than sugar, Thomas Bolling Robertson announced that with adequate tariff support, the state could supply all the nation's sugar needs. But to do so, Louisiana planters required protection against cheap foreign competition and assistance against obstacles they faced—most notably, their dependence on other states for manufactured goods. Despite vigorous debate and marked opposition, the 3¢ per pound duty remained unchanged until 1832 and provided the base for sustained growth in the 1820s, during which sugar revenues tripled from $2 to $6 million. Mincing few words on the importance of the domestic sugar tariff, Henry Clay gloomily predicted that if protectionism failed, it "would be almost as fatal . . . as if Congress were to order the dykes to be razed from Pointe Coupee to the Balize." Strong words, to be

---

Agriculture, 2 : 750–51; J. S. Johnston, *Letter of Mr. Johnston of Louisiana to the Secretary of the Treasury . . . Relative to the Culture of the Sugar Cane* (Washington, D.C., 1831), 8; *De Bow's Review* 4 (November 1847): 385–86; Forstall, *Agricultural Productions of Louisiana,* 6 (quote). On Magnolia Plantation, slaves produced six hogsheads per hand, which was considered "very good." J. Carlyle Sitterson, "Magnolia Plantation, 1852–1862: A Decade of a Louisiana Sugar Estate," *Mississippi Valley Historical Review* 25 (September 1938): 204; Michael Craton, *Searching for the Invisible Man: Slaves and Plantation Life in Jamaica* (Cambridge, Mass., 1978), 143. For contemporary descriptions on the relative technical condition of the Louisiana and Caribbean sugar industries, see *Franklin Planters' Banner,* 5 January 1854; *De Bow's Review* 15 (December 1853): 648. On technical innovation in the Caribbean, see Manuel Moreno Fraginals, *The Sugar Mill: The Socioeconomic Complex of Sugar in Cuba, 1760–1860* (New York, 1976), 82–102; Stuart B. Schwartz, *Sugar Plantations in the Formation of Brazilian Society: Bahia, 1550–1835* (Cambridge, 1985), 125–31; Richard S. Dunn, *Sugar and Slaves: The Rise of the Planter Class in the English West Indies* (New York, 1972), 191–95.

sure, but as the subsequent history of Louisiana sugar indicates, Clay was basically right.[14]

Planters who had gained as much as 8 to 10¢ per pound in the 1820s found their profit margins slashed after the tariffs of 1832, 1833, and 1841 reduced sugar protection. Widely condemned for gouging the American public merely "to render the labor of twenty thousand slaves more profitable to their owners," the sugar interest weakened before repeated waves of congressional criticism and proved too feeble to stem the tide of antitariff sentiment. The 1832 tariff lowered the duty on imported sugar to 2.5¢ per pound, and the following year Congress reduced the sugar levy still further in a sliding scale that terminated in a flat duty rate of 20 percent. Reduced protectionism and increased Cuban competition during the latter half of the 1830s dealt a hammer blow to Louisiana's agricultural interests, forcing politician and planter alike to recalculate the potential effect of the final 20 percent duty. Revived protectionism under the 1842 tariff restored the 2.5¢ per pound rate and briefly injected enthusiasm and optimism into the industry. James K. Polk's election as president in 1844 heralded the reversal of Whiggish tariff measures and the introduction of a 30 percent ad valorem rate. Senators Alexander Barrow and Henry Johnson fought to stem the challenge to sugar, but the Walker tariff remained and set the tone for protectionism during the remaining antebellum years.

Federal protection was a lifeline for an industry that could not openly compete with its rivals in the Caribbean. The sugar duty guaranteed reasonably good profits of 6 to 12 percent and offset the notoriously heavy costs of cane sugar production. In Louisiana, these costs proved exceptionally high. Indeed, escalating competition and rising overheads ensured that the sugar trade remained solely the pursuit of the wealthiest slaveholders. "It is true," Moses Liddell observed, "that some very large fortunes have been realized at sugar planting but with an immense exertion and capital to commence with or a strong mind and over laborious perseverance." Sternly though unsuccessfully counseling his son against sugar farming, Liddell catalogued the multiple expenses incurred in establishing a productive

14. For a discussion of tariffs, see Joseph G. Tregle Jr., "Louisiana and the Tariff, 1816–1846," *LHQ* 25 (January 1942): 24–148 (first quote, 30); Philip D. Shea, "The Spatial Impact of Governmental Decisions on the Production and Distribution of Sugar Cane, 1751–1972" (Ph.D. diss., Michigan State University, 1974), chap. 2; John Sacher, "'A Perfect War': Politics and Parties in Louisiana, 1824–1861" (Ph.D. diss., Louisiana State University, 1999), 79; Forstall, *Agricultural Productions of Louisiana,* 7; Henry Clay to Anonymous, 16 February 1831, in *The Private Correspondence of Henry Clay,* ed. Calvin Colton (New York, 1856), 294 (second quote).

cane estate. "If you go at sugar," he warned, "it will take you three years before you can procure seed or plant cane to make a full crop—you have an extensive building—you must have a steam engine and mill[,] $4500 expenses putting it up and keeping it in order, risk of crops, and continuous unforeseen . . . expenses that will eat up the profits." Belaboring his point, he flatly concluded, "I am rather sick of sugar growing, there is such a succession of labor to perform the whole season round and so much anxiety prevails . . . every year improvements are to be made, repairs to be done, new fixtures to be added—there is never an end of these things as it is with cotton." Capital investment in land and labor added to the planter's substantial debt and entailed the accumulation of massive credit and loans. As one successful sugar master observed:

> Buying a plantation . . . is essentially a gambling operation. The capital invested in a sugar plantation of the size of mine ought not to be less than $150,000. The purchaser pays down what he can, and usually gives security for the payment of the balance in six annual installments, with interest (10 per cent. per annum) from the date of the purchase. Success in sugar as well as cotton planting is dependent on so many circumstances, that it is as much trusting to luck as betting on a throw of a dice. If his first crop proves a bad one, he must borrow money of the Jews in New Orleans to pay his first note; they will sell him this on the best terms they can, and often at not less than twenty-five per cent. per annum. If three or four bad crops follow one another, he is ruined.[15]

Despite the risks of sugar farming, Louisiana's cane interests were valued at $60 million by 1844, even as spiraling investment in steam-powered grinding facilities, vast slave crews, and costly real estate amplified the industry's collective debt. Capital costs continued to escalate in later antebellum years. Planters who wished to puncture the hundred-hogshead ceiling had to double their investments from

15. On profits, see Johnston, *Letter of Mr. Johnston of Louisiana*, 8; *De Bow's Review* 8 (January 1850): 36; David O. Whitten, "Antebellum Sugar and Rice Plantations, Louisiana and South Carolina: A Profitability Study" (Ph.D. diss., Tulane University, 1970), 81–96; Whitten, "Tariff and Profit in the Antebellum Sugar Industry," 226–33. On the escalating costs of sugar farming, see Moses Liddell to John R. Liddell, 28 July 1845, Moses and St. John Richardson Liddell Family Papers, LSU (first quote); Olmsted, *Seaboard Slave States*, 2:318–19 (second quote); Russell, "Cultural Conflicts and Common Interests," 182–89.

$40,000 to $75,000 and purchase land and slaves, frequently delaying payment to their creditors and accruing interest at 6 to 10 percent per annum. Ambitious and wealthy planters who sought to join the elite sugar lords needed to secure loans of between a third- and a half-million dollars. These financial requirements left the productive thrust of the industry in the hands of planters who could acquire significant capital in the New Orleans banks or who could fuse their assets through capital-accumulating partnerships.[16]

In the late 1840s, spiraling land, labor, and capital costs, combined with reduced tariff protection, prompted a wave of mergers. Across the cane world, planters wisely consolidated their investments by creating cooperative partnerships that allowed them to utilize financial advantages in ways similar to those with greater individual assets. Mergers allowed smaller operators and urban residents to enjoy the social prestige of slaveholding without wildly risking their savings and future on the sugar economy. By spreading the risk of sugar investments, planters could additionally cushion themselves from a potentially bankrupting run of bad crops. While cooperative estates were often no larger than single-unit enterprises, they enabled smaller planters to strengthen their position in the plantation economy and moderately sized operators to become masters of giant (albeit shared) estates.[17]

Louisiana's comparatively advanced financial sector underpinned the extensive credit and debt chains that linked the rural planter through the factor-broker to urban banks and lenders. Within the nation's financial structure, commercial banks in New Orleans enjoyed a privileged position. Louisiana ranked third in banking capital (after New York and Massachusetts) in 1840, while in 1859 state bankers still commanded fourth place (behind Pennsylvania as well). Building upon the city's original banks, the Louisiana legislature chartered the state's first property banks in the 1820s. Specifically designed to finance development in the rural plantation belts, these property banks enabled planters to mortgage their own property as collateral for specie reserves and thus partially fulfilled the growing demand for rural credit. During the early 1830s, British lenders extended their lucrative credit lines across the Atlantic, and the availability of European finance provided a welcome jolt to southwestern expansion in the United States. Property banks were joined by a host of other financial houses, boosting capitalization from $9 to $46

---

16. Forstall, *Agricultural Productions of Louisiana*, 4–7; Gray, *History of Agriculture*. 2:743.

17. See Richard Follett and Rick Halpern's ongoing "Documenting the Louisiana Sugar Economy, 1844–1917," http://www.utoronto.ca/csus/sugar (accessed 11 April 2004).

million between 1831 and 1837. The banking crisis that dominated Martin Van Buren's presidency drastically squeezed the availability of credit, and the contraction of banking in the 1840s replaced the easy credit mentality with one of restraint. The banking crisis affected every sector of Louisiana's economy, but even during those relatively credit-lean years, planters nonetheless gained financing from foreign sources, other states, and private unchartered banks; in addition, they utilized their own capital reserves for expansion. The slave stock alone represented tens of millions of dollars—capital that planters mortgaged for loans and debt repayment or utilized to spread the enormous costs and risks of cane farming. These fiscal devices ensured that the state's sugar interests remained comparatively buoyant during the financial crisis and economic depression of the 1840s. By 1853, the demand for credit proved so strong that the Louisiana legislature passed a free banking law to ease bank incorporation. This liberalization policy bore immediate fruit; the volume of bank loans in New Orleans increased from $18.6 million in 1850 to $31 million in 1856. Having "been tested and not found wanting," *De Bow's Review* trumpeted, the New Orleans banking system underpinned a decade of commercial growth among the cotton and sugar fields of the region.[18]

Progress in the nineteenth-century sugar country fused colossal investment in land, labor, and machinery. To ensure maximum production, the labor lords financed a tripartite division of capital in which investment in the primary factors of production dwarfed those of the cotton South. Funneling their assets into a capitalized labor system that transformed the productive capacity of the Louisiana swamps, the sugar masters sponsored a technological revolution that bore profitable fruit in higher yields and enhanced sugar quality. In contrast to the cotton South—where, some historians argue, slaves "crowded physical capital out of the portfolios of southern capitalists" and farms remained technologically backward—capital-rich sugar estates towered above their neighbors in almost

18. On antebellum banks and regional economic growth, see Howard Bodenhorn, *A History of Banking in Antebellum America: Financial Development in an Era of Nation-Building* (Cambridge, 2000); Larry Schweikart, *Banking in the American South from the Age of Jackson to Reconstruction* (Baton Rouge, La., 1987). On Louisiana banking, see George D. Green, *Finance and Economic Development in the Old South: Louisiana Banking, 1804–1861* (Stanford, 1972); Merl E. Reed, "Boom or Bust—Louisiana's Economy during the 1830s," *LH* 4 (winter 1963): 46–49; Stephen A. Caldwell, *A Banking History of Louisiana* (Baton Rouge, La., 1935), 30–31, 127; Sacher, " 'A Perfect War,' " 133; *Hunt's Merchant's Magazine* 42 (February 1860): 156–57; *De Bow's Review* 25 (November 1858): 559 (quote). On slaves as credit instruments, see Richard Holcombe Kilbourne Jr., *Debt, Investment, Slaves: Credit Relations in East Feliciana Parish, Louisiana, 1825–1885* (Tuscaloosa, Ala., 1995), 5.

every index of agricultural investment. Unlike the cotton complex in the South or the wheat culture of the Midwest, where farmers invested $1.60 and $1.46 respectively on farm implements for each cultivated acre, the late antebellum sugar masters spent approximately $20 on machinery per improved acre. Even in the relatively advanced rice industry, where steam-powered threshers and advancing technology necessitated considerable investment, planters in Georgetown County, South Carolina, invested approximately $10 per acre on implements and machinery. Moreover, Louisiana sugar masters led the nation in technological investment per farm, with almost $19 million disbursed in agricultural capital. Only New York and Pennsylvania surpassed Louisiana in the total value of farm implements, but these states had 195,000 and 156,000 farms respectively, while Louisiana had just 17,000 farms, of which less than 10 percent produced sugar. By the eve of the Civil War, then, Louisiana's sugar industry was the most heavily capitalized and investment-rich agricultural region in the country.[19]

Within the sugar country, the value of farm implements per parish frequently outstripped those of free-labor states. West Baton Rouge Parish featured more mechanical capital than all of Minnesota in 1860, while Ascension Parish's 232 estates almost matched the combined total of 5,600 farms in Oregon. Although not every cent was invested in the sugar industry, the northern sugar parishes of Rapides, Avoyelles, and Pointe Coupee amassed nearly $3.5 million in farm implements—a figure just a few hundred thousand less than the entire state of Massachusetts. Louisiana's capital in farm implements surpassed the state's investment in manufacturing ($7.1 million) and outstripped manufacturing investment in several states of the Old Northwest. But despite its primacy in steam engines, Louisiana—like its fellow southern states—was appallingly prepared for sectional conflict and entered the Civil War with plenty of sugar mills but no manufacturing base. In the case of Louisiana's sugar country, slavery and the capital-rich plantation complexes drained available resources for industry and checked the development of the state's infrastructure.[20]

19. On capitalized land and labor systems, see Dusinberre, *Them Dark Days*, 6–7; Ralph Anderson and Robert E. Gallman, "Slaves as Fixed Capital: Slave Labor and Southern Economic Development," *JAH* 64 (June 1977): 25–46. For contrast with the cotton belt, see Roger Ransom and Richard Sutch, "Capitalists without Capital: The Burden of Slavery and the Impact of Emancipation," *AH* 62 (summer 1988): 133–60, esp. 138–39; Wright, *Political Economy of the Cotton South*, 52.

20. Data on state implement investment drawn from U.S. Bureau of the Census, *Eighth Census of the United States* (Washington, D.C., 1864–1866).

Writing in the early 1850s, the editors of *Le Pioneer de L'Assomption* stated an obvious truism for those on the margins of the industry when they concluded that "it is necessary to have great capital or immense credit" to succeed in the antebellum sugar trade. Few in the western Attakapas needed such counsel. A few sugar magnates like William F. Weeks emerged, but the main focus of the industry lay with moderately sized farmers who lacked the vast capital reserves of the Mississippi Coast planters. A moderate sugar plantation, however, was large by anyone else's standards. Whether in the western St. Mary Parish or along the Mississippi in Ascension Parish, even the smallest sugar producers dwarfed the average cotton farm, which possessed 130 acres and was valued at $4,378. In both these sugar-producing regions, farm acreage and value expanded from 1850 to 1860 as planters seized the rich levee crests, drained the backswamp for cane farming, and acquired their smaller neighbors' property. In 1850, the average sugar planter in Ascension Parish cultivated 460 acres. A decade later, fourteen fewer sugar planters recorded that 29,149 acres—approximately 800 acres per estate—lay under crop. This pattern was repeated further west. In St. Mary Parish, cane farmers almost doubled the number of improved acres per farm (from 230 to 413 acres) and increased production by 70 percent. Rapid growth also turned on investment in machinery. In Ascension Parish, the average plantation included $10,884 worth of machinery at midcentury; a decade later, that figure had risen to almost $19,000. Capital investment kept pace with the increase in acreage. In both 1850 and 1860, for instance, Ascension Parish sugar lords possessed over $22 in implements per cultivated acre. In the less capital-rich Attakapas district, sugar planters increased machine investment from $4,034 to $7,612 and boasted $18 worth of machinery per improved acre. What changed, however, was the type of machinery used. During the 1850s, Ascension Parish farmers invested in expensive vacuum pans and evaporation techniques that produced a higher grade of sugar. Bagasse burners, which allowed planters to replace wood with dried cane husks as their primary fuel, dotted the landscape; draining machines reclaimed lower portions of their land; and improved steam engines were shipped to riverside plantations from foundries in the North. To the west, planters remained technologically at least five to ten years behind their neighbors in Ascension Parish. They replaced their horse-drawn mills with steam-powered facilities, but they continued to produce sugar in open kettles. Others did not wait for their neighbors. Some western planters— like William Weeks—built themselves modern agro-industrial sugarhouses with copper granulating pans and a bagasse furnace for recycling crushed canes. Weeks embodied the ambition of many within his class, and his capital disbursement kept pace with acreage expansion as the industry developed during the late ante-bellum era.[21]

By 1860, the productive thrust of the Louisiana sugar industry lay in the hands of William Weeks, John Hampden Randolph, and approximately five hundred elite sugar masters who controlled over two-thirds of the slaves and available acreage in Louisiana's cane world. Although these planters represented less than 13 percent of the slaveholders in the sugar parishes, they produced three-quarters of the region's cane yield. In Ascension Parish alone, elite sugar planters possessed on average 191 slaves, 1,205 improved acres, and over $25,000 worth of farm equipment. The St. Mary Parish lords trailed their neighbors in every index, but even on the western frontier, elite sugar masters held 105 slaves, 620 acres of improved farmland, and $12,500 of machinery.[22]

The planters who sought to tap the growing demand for white sugar invested in evaporation technology and clarification facilities that produced "large and brilliant crystals . . . [of] any size required by the caprice of the customer." Developed by Charles Derosne and Jean-François Cail, though subsequently adapted for plantation use by Norbert Rillieux, vacuum processing transformed the manufacturing of sugar in the second half of the nineteenth century. Born the son of a white plantation owner and his slave mistress, Rillieux trained in Paris as a mechanical engineer before returning to New Orleans in 1833. Keenly aware that the open-kettle method of sugar production was both costly in terms of timber for the furnaces and hazardous for laborers on the mill floor, Rillieux sought to develop

21. Napoleonville *Le Pioneer de l'Assomption*, 26 October 1851 (quote); Wright, *Political Economy of the Cotton South*, 48. U.S. Bureau of the Census, *Seventh Census of the United States (1850)*, manuscript agricultural schedules, Ascension Parish and St. Mary Parish, Louisiana, and U.S. Bureau of the Census, *Eighth Census of the United States (1860)*, manuscript agricultural schedules, Ascension and St. Mary Parish, Louisiana. Sample: Ascension Parish, 1850 ($n = 50$); Ascension Parish, 1860 ($n = 36$); St. Mary Parish, 1850 ($n = 174$); St. Mary Parish, 1860 ($n = 152$). Also see Work Contract, W. F. Weeks and David Edwards, 1 April 1858, and Patent License, 10 April 1855, both in David Weeks and Family Papers, LSU.

22. "Elite" is defined as owning a minimum of fifty slaves. On land and wealth concentration, see Albert W. Niemi Jr., "Inequality in the Distribution of Slave Wealth: The Cotton South and Other Southern Agricultural Regions," *JEH* 37 (September 1977): 747–53; Roger W. Shugg, *Origins of Class Struggle: A Social History of White Farmers and Laborers during Slavery and After, 1840–1875* (Baton Rouge, La., 1939), esp. 76–156. It is untrue that the planters entirely eradicated smallholders from the sugar bowl, but it is accurate that the sugar planters increasingly dominated the most costly and productive land in the southernmost Louisiana parishes. Harry L. Coles, "Some Notes on Slaveownership and Landownership in Louisiana, 1850–1860," *JSH* 9 (August 1943): 381–94; Menn, *Large Slaveholders of Louisiana*, 121–24, 380–89; Rodrigue, *Reconstruction in the Cane Fields*, 20–21.

an alternative method of processing that utilized the heat generated by the exhaust of a steam engine rather than the direct heat of a furnace. In 1843, he patented his multiple-effect vacuum pan. This apparatus had many advantages for the largest cane lords who sought to manufacture higher-grade sugars. First, the use of a steam vacuum minimized the risk of scorching or discoloring the sugar. Second, vacuum pans maintained a lower average temperature than open kettles. By evaporating the sugar in a sealed unit, the quality and quantity of the final product surpassed the caliber of sugar made by all previous methods. For planters like Maunsell White, it produced sugar of a superior quality, worth fully two cents a pound higher than the best previous sugars. Vacuum-produced sugar, White concluded, was, in short, a "fancy article."[23]

For the very richest planters, like Judah Benjamin and Maunsell White, the demand for improved sugar and a whiter product necessitated the shift toward costly vacuum processing. Acquiring Rillieux's multiple-effect evaporator for use on Bellechasse Plantation—an investment of over $30,000—Benjamin formed a partnership with Theodore Packwood for the production of sugar using the new technology. Benjamin calculated that for the largest sugar masters, the Rillieux apparatus would generate a profit of $14,531 every season over the open-kettle method of production. Not only would planters save considerably on timber and produce higher quality sugar, he enthusiastically declared, but the vacuum system would produce 25 percent more sugar. Despite its expense, over sixty-five prominent sugar cultivators invested in the Rillieux apparatus and similar vacuum evaporators, producing an ever-increasing volume of crystalline and snowy sugar that Princeton chemist R. S. McCulloh praised as "equal to those of the best double-refined sugar of our northern refineries." Less affluent planters longed for the latest machinery and envisioned an economic future in which slaves would operate cost-effective steam pans and annual revenues would soar as saleable sugar was wrung from even the sourest cane juices.[24]

23. James D. B. De Bow, *The Industrial Resources, Etc., of the Southern and Western Estates*, 3 vols. (New Orleans, 1853), 2 : 206 (first quote); Maunsell White to Dunlop, Moncure & Co., 14 March 1845, Maunsell White Papers, UNC (second quote). On technological improvements in sugar production, see Heitmann, *Modernization of the Louisiana Sugar Industry*, 8–48; Deerr, *History of Sugar*, 2 : 561–69.

24. *De Bow's Review* 5 (February 1848): 292–93; De Bow, *Industrial Resources*, 2 : 206; U.S. Senate, *Report of the Secretary of the Treasury: Investigations in Relation to Cane Sugar*, 29th Cong., 2nd sess., Senate Doc. No. 209 (Washington, D.C., 1847), 121 (quote); Schmitz, "Economic Analysis of Antebellum Sugar Plantations in Louisiana," 39. Also see Daniel Dudley Avery to George Marsh, 8 December 1859, Avery Family Papers, UNC.

In addition to vacuum processing, the development of cost-effective bagasse furnaces proved to be another key technological advance for the sugar industry, and one which resolved the looming energy crisis. By the mid-1850s, twenty years of agro-industrial sugar production had diminished timber supplies throughout southern Louisiana. Albeit a side effect, the shift from horse to steam power had a profound ecological impact, as planters stripped the cypress swamps to supply early steam mills. Hardly fuel efficient, these rudimentary steam engines consumed three to four cords of wood for every one hogshead of sugar produced, and over an extended period, these fuel requirements placed a significant burden on regional wood supplies. Planters supplemented their timber resources with lumber cut beyond the plantation belt, but increasingly they required an alternative energy supply. Some experimented with coal, but high prices and uncertain supply made it uneconomical for long-term use. Louisiana's slave masters accordingly turned to burning bagasse—dried cane husks—as a readily available and almost cost-free alternative. The first commercially available bagasse burners emerged in the late 1840s, when James H. Dakin, a Baton Rouge sugar planter, invented a machine for desiccating bagasse and converting it into an inexpensive fuel. The applicability of Dakin's design was immediately evident. As *De Bow's Review* explained, "Wood is daily becoming more scarce, and, in many cases on plantations fronting the Mississippi river . . . not a cord is to be obtained." Faced with the potential catastrophe of running short of wood during the grinding season, local inventors and rural engineers—like A. J. Chapman, Evan Skelly, Samuel Gillman, and Moses Thompson—improved upon Dakin's design, and by 1860 a range of bagasse furnaces and drying machines supplied the industry's need for an alternative fuel source. Equipped with a bagasse burner (valued between $1,500 and $2,500), planters could rid themselves of costly wood purchases, assure absolute stability in their fuel supply, save on overtime payments to slave woodcutters who chopped timber on their own time, and redirect labor resources toward potentially more lucrative tasks. Local publications lauded the new technology, noting that Skelly's furnace burned so effectively as to keep the sugar kettles at full blast during grinding season. In a similar vein, *De Bow's Review* reported that green bagasse was readily burned in Thompson's wet and dry bagasse burner "without the aid of wood or blowers, furnishing sufficient steam for running the engine and other purposes." Never one to miss an opportunity within a competitive market where patent holders and local engineers vied for the planters' custom, William F. Weeks purchased the patent right to use Moses Thompson's bagasse burner in 1855 for $1,200.[25]

Despite a few planters' forays into expensive capital investment, almost 90 percent of planters continued to produce sugar in open kettles. Having brought steam power to the region, the sugar masters were ambivalent about the expensive vacuum pans and improved granulating facilities and were ultimately unwilling to gamble their future on the promise of tomorrow's technology. Moreover, while many planters were confident that African Americans could operate the new mills and toil at the metered cadence of the steam age, they shared some reservations about the social and racial implications of slaves managing complex vacuum equipment. Planter Theodore Packwood, for instance, installed a Rillieux apparatus as early as 1844 and reported that his slaves swiftly learned how to use it. Yet within a few years, other planters—notably Edward Forstall and Valcour Aime—publicly expressed their view that slaves alone could not be relied on and that the employment of white sugar-makers was imperative in utilizing vacuum equipment. Although most late antebellum sugar masters increasingly relied on their bonded crews to manage the steam engines and kettle trains, lingering racial concerns and planter conservatism mitigated against the widespread use of the newest and most costly technology. Furthermore, local institutions to disseminate the latest scientific findings failed because of the relative isolation of the region's cane lords and the lack of community-wide support. But above all, most individuals were not willing to shoulder the enormous costs that were required to advance beyond steam-powered mills and open-kettle production. As William P. Bradburn of the *Southern Sentinel* emphasized, the risk of misfortune ultimately defined antebellum sugar farming. "In our countryside," he observed, "the people seem run mad upon the culture of staple products. . . . They turn the farmers' life into that of a gambler and speculator. They are dependent upon chance, and an

25. On coal, see *De Bow's Review* 13 (December 1852): 624–26. On bagasse, see *Franklin Planters' Banner*, 6 December 1849; *De Bow's Review* 8 (April 1850): 401–2 (first quote). On the presence of locally manufactured bagasse furnaces, see *Plaquemine Southern Sentinel*, 5 December 1857; *Houma Ceres*, 24 January 1857; *Plaquemine Gazette and Sentinel*, 20 November 1858; ibid., 5 December 1857; *Baton Rouge Weekly Comet*, 30 March 1856; *De Bow's Review* 18 (January 1855): 60 (second quote); Patent License, 10 April 1855, David Weeks and Family Papers, LSU; G. M. Butler to Daniel Turnbull, 5 November 1858, J. P. Bowman Papers, LSU; Samuel H. Gilman to Charles L. Mathews, 11 February 1856, and J. Cook to Charles L. Mathews, December, n.d., 1856, Charles L. Mathews and Family Papers, LSU.

evil turn of the cards—a bad season, a fall in prices, or some such usual calamity."[26]

Bradburn's point was a sound one, but in their quest for higher production, planters nonetheless poured capital into land and slaves and upgraded their engines and farm equipment, even as they decried the staggering costs involved. As Edward Butler observed, the sugar masters' "apparent determination to always be in debt" in anticipation of a favorable crop proved infectious. Unlike many of their slaveholding brethren, he complained, Louisiana planters rarely appropriated a portion of their income for their families; rather, they invested it all in buying slaves and extending the fronts of their plantations. Butler's critique of the industry remained broadly valid even in 1860, as the penchant for agro-industrial sugar production mounted in the later antebellum years. While most planters drew a line at the vast expense of vacuum-processed sugar, they nevertheless occasioned sweeping changes to Louisiana's landscape. In March 1831, when Joseph Lyman settled in St. Mary Parish, the land appeared as "one immense flat surface, intersected by bayous running in every direction and bearing on its surface, almost every vegetable." Describing the physical attributes of the Teche country, Lyman noted that the "soil is entirely alluvial and very productive [producing] sugar cane, corn, sweet potatoes, melons, and most articles in the gardening line." He explained that after plowing a shallow furrow, farmers cultivated sugarcane largely as they would corn, and that during the grinding season they utilized primitive sugar mills that appeared closer in design to a cider mill than to an industrial or processing plant. These simple mills, Lyman related, were powered by the slow pace of a horse and operated by a slave who fed the cane shoots into the revolving cylinders. Having extracted only a small portion of the potential cane juice, slaves then began the process of evaporating the sugar-rich liquid in four large iron kettles. Once the cane juice reached striking point and began to granulate, which occurred every hour, slaves drained the semi-molten syrup before finally packing the (hopefully dry) sugar into wooden hogsheads.[27]

26. On the limits to innovation, see Heitmann, *Modernization of the Louisiana Sugar Industry,* chap. 2, esp. 38–39; *De Bow's Review* 2 (November 1846): 344; ibid. 4 (December 1847): 425; *Plaquemine Southern Sentinel,* 22 June 1850.

27. E. G. W. Butler to Thomas Butler, 7 February 1830, Thomas Butler and Family Papers, LSU (first quote); Joseph Lyman Letter, UNC (second quote). Also see *Southern Agriculturist* 4 (May 1831): 225–32; ibid. 5 (June 1832): 281–85.

Lyman's account of sugar making reflected an industry that remained bound to the timeworn traditions of the seventeenth and eighteenth centuries. Work advanced at the pace of a plodding mule or oxen, and the bondsperson required few technical skills. Thirty years later, however, visitors to the cane country remarked on an industry that combined science, technology, and industrialization:

> The apparatus used upon the better class of plantations is very admirable, and improvements are yearly being made, which indicate high scientific acquirements, and much mechanical ingenuity on the part of the inventors. The whole process of sugar manufacturing . . . has been within a few years greatly improved, principally by reason of the experiments and discoveries of the French chemists, whose labors have been directed by the purpose to lessen the cost of beet-sugar. Apparatus for various processes in the manufacture, which they have invented or recommended, has been improved, and brought into practical operation on a large scale, the owners of which are among the most intelligent, enterprising, and wealthy men of business in the United States.

While these manufacturing facilities clearly impressed travelers to the cane world, most estates in the late antebellum era combined elements of agriculture and industry, in which open-kettle production and a mechanized production schedule defined the pace of work in the sugarhouse and cane fields. Skilled slaves managed the multiple operations of the steam mill and, in many cases, manufactured the sugar, controlled the fire beneath the kettles, and watched over the entire production process. Locked within an agro-industrial order that advanced at the relentless speed of the steam engine, African Americans now found themselves enslaved to a business regime that advanced at a regimented and methodic pace.[28]

Driven by the lure of the market, the plantation elite singlemindedly pursued their economic aspirations and established hundreds of well-developed, capital-intensive estates. They exploited the growing demand for sucrose, profited from technical gains in productivity, utilized sophisticated financial instruments, and thrived behind the lofty walls of federal tariff protection. Yet for all their modernity,

28. Olmsted, *Seaboard Slave States*, 2:328–29.

progress in estate organization, crop yields, and productivity could not mask Louisiana's poor infrastructure, its moribund agricultural societies, and the dearth of public-spiritedness in the cane world. Private initiative flourished on Louisiana's sugar estates, but in essence the planters lorded over an extractive industry, and their economic vision remained short-sighted and exceptionally individualistic. Although the state legislature attempted to provide a favorable environment for internal improvements, long-term regional development languished as the sugar masters failed to provide matching funds for transportation links or to support agricultural societies.[29]

In the early 1800s, transportation in the sugar country remained slow and time-consuming at best. Driftwood-clogged bayous and low water frequently made access to smaller rivers and canals almost impossible. To alleviate transportation difficulties and bring developing economic regions, such as the Attakapas, within reach of the New Orleans market, private individuals and the state government sponsored a sporadic program of canal construction and river-clearance projects. Yet these projects frequently failed to meet expectations. Plans for the Barataria and Lafourche Canal, for instance, called for it to span southern Louisiana and become a permanent and reliable link to Terrebonne Parish. Bedeviled by managerial incompetence, corruption, inadequate support, and the high rate of $3.75 for shipping a hogshead of sugar to New Orleans, the canal failed to meet expectations, and in 1859 Governor Robert Wickliffe removed its last state support. Smaller state-sponsored projects proved considerably more successful in securing the backing of planters. Snag clearance and the closure of Bayou Plaquemine in 1858, for instance, followed significant lobbying from local planters and residents. St. Mary Parish planters similarly appealed for state assistance in draining the Grand Marais swamp and in constructing canals and dredging rivers through their

---

29. For recent interpretations that highlight capitalistic features in the South, see Tom Downey, "Riparian Rights in Antebellum South Carolina: William Gregg and the Origins of the 'Industrial Mind,'" *JSH* 65 (February 1999): 78–108; Frederick E. Siegel, *The Roots of Southern Distinctiveness: Tobacco and Society in Danville, Virginia, 1780–1865* (Chapel Hill, N.C., 1987), 105–19; Wilma A. Dunaway, *The First American Frontier: Transition to Capitalism in Southern Appalachia, 1700–1860* (Chapel Hill, N.C., 1996), 195–223. Contrast these works with Douglas R. Egerton, "Markets without a Market Revolution: Southern Planters and Capitalism," *Journal of the Early Republic* 16 (summer 1996): 207–8; Gavin Wright, *Old South, New South: Revolutions in the Southern Economy since the Civil War* (New York, 1986), 18.

district. The Grand Marais project and other internal improvements proved especially attractive when they directly affected the sugar masters' self-interest and the economic well-being of their plantations. While the cane lords recognized the potential in larger construction projects for personal monetary gain, they ultimately lacked the resolve to complete Louisiana's transportation revolution.[30]

In the early 1850s, the sugar masters alighted on railroad construction as a means to expand their market access and improve transportation speed. As they had done with canal and river improvements, state, municipal, and parish governments attempted to support railroad companies by offering land grants, tax exemptions, and over $7 million in aid. Several railways departed from New Orleans, but from the perspective of planters in the western sugar country, the most important line was the New Orleans, Opelousas, and Great Western Railroad (NOOGWR). Supported by Senator John Moore and Francis Dubose Richardson, among others, the projected new railroad "nerve[d] the arm of everyone," the *Louisiana Spectator* trumpeted, "who cultivates the soil in that healthy, rich, and beautiful section." Buoyed by the thought of increasing land values and rapid access to the New Orleans sugar market, planters along the route pledged almost $760,000 in private subscriptions during the first year of the railroad's incorporation. Evidently impressed by the company's commitment to "develop large agricultural districts" and the railroad's projected route through 933 plantations, the cane lords pledged funds for their collective future, though as costs escalated interest faltered and funding petered out. After the first few miles of track were opened in December 1853, the difficulties inherent in laying rails through silty and geologically unstable swamps made for slow progress and high costs. Hamstrung by the lack of sufficient funding, escalating costs, and a perhaps overly grandiose scheme, the NOOGWR failed to attract adequate support. By 1857, private investors furnished only 20 percent of the construction costs. Of the three-quarters of a

---

30. On transportation and internal improvements, see Donald J. Millet, "The Saga of Water Transportation into Southwest Louisiana to 1900," *LH* 15 (fall 1974): 339–56; Thomas A. Becnel, *The Barrow Family and the Barataria and Lafourche Canal: The Transportation Revolution in Louisiana, 1829–1925* (Baton Rouge, La., 1989), 41–65; *New Roads Pointe Coupee Democrat*, 27 February 1858; "Annual Report of the Board of Swamp Land Commissions to the Legislature of the State of Louisiana, January 1860," in *Documents of the First Session of the Fifth Legislature, State of Louisiana* (Baton Rouge, La., 1860), 97. On the issue of snag clearance, see Paul F. Paskoff, "Hazard Removal on the Western Rivers as a Problem of Public Policy," *LH* 40 (summer 1999): 261–82.

million dollars initially promised by rural planters residing along the railroad, just $250,000 was actually paid.[31]

Despite boosterism from the local press, the slaveholding barons rarely cooperated for their mutual good. To be sure, the sugar masters maintained slave patrols and contributed to the upkeep of the levees—anything less would have been suicidal on the Mississippi floodplain—but they proved unwilling converts to communal progress. Because of their relative isolation and prosperity, antebellum sugar lords were indifferent to the nascent Agriculturists and Mechanics Association; they proved similarly unresponsive to the University of Louisiana, despite the fact that it offered engineering courses specifically tailored to the local economy. In both cases, institutions committed to statewide commercial progress failed due to want of public support. By contrast, the legislature established a military academy, under the supervision of William Tecumseh Sherman, for the training of cadets. Honor and militarism ultimately proved more compelling than the university's courses in chemical engineering! Private initiative, however, seldom languished. Pouring capital into land, labor, and machinery, the plantation elite exploited the market for sugar and reallocated their resources to maximize their harvest yields. Planters could modernize their equipment and embrace the market economy, but their identity as slaveholders ultimately defined progress in the sugar world. Speaking a lingua franca of rationality and modernization, Louisiana's slaveholders remained wedded to individualistic notions of progress. Their social ethic and economic ideology exalted their authority and emphasized personal liberty. Above all, their collective self-identity remained anchored to the plantation and their role as slaveholders and labor lords. Capital expenditure on a plantation mansion, more slaves, or the latest machinery heightened the planters' sense of mastery over land, labor, and sugar; public investment, on the other hand, did little to exalt their power. By contrast, bricks and morar, steam engines, and above

---

31. On railroad building, see Merl E. Reed, "Government Investment and Economic Growth: Louisiana's Antebellum Railroads," *JSH* 28 (May 1962): 198; Reed, *New Orleans and the Railroads: The Struggle for Commercial Empire, 1830–1860* (Baton Rouge, La., 1966); "A Century of Progress in Louisiana, 1852–1952," *Southern Pacific Bulletin* (October 1952): 1–55; Walter Prichard, ed., "A Forgotten Louisiana Engineer: G. W. R. Bayley and his 'History of the Railroads of Louisiana,'" *LHQ* 30 (October 1947): 1065–85. For a contemporary description, see *Franklin Planters' Banner,* 7 March (quote), 2 May, and 29 August 1850, 6 February 1851; *Report of the President and Directors of the New Orleans, Opelousas, and Great Western Railroad Company to the Stockholders, at their First Annual Meeting, 24th January, 1853* (New Orleans, 1853), 10; *Thibodaux Minerva,* 3 November 1855.

all slaves symbolized the lord's sway and reflected his apparent modernity and his mastery of the sugar country.[32]

Localized investment on one's own estate carried ideological value, but it also proved relatively prudent, given the volatility of the sugar economy. Capital investment in land, labor, or machinery could always be liquidized and the collateral moved into cotton farming if protective tariffs dissipated, or else transferred to the newly emerging sugar lands of east Texas. Given the vagaries of Louisiana's climate and the insecurity of sugar farming, it is not entirely surprising that the regional elite chose private investment over the massive public spending required to build railroads, maintain rivers, and shore up the colossal power of the Mississippi. As Moses Liddell well understood, the never-ending costs of sugar production drained the planters' coffers and constrained the economic and public vision of all save the largest sugar masters. The result of the planters' self-absorbed commercialism was to create a region pockmarked by hundreds of highly developed plantations, but lacking the completed infrastructure of an advanced economy.

New Yorker Frederick Law Olmsted accurately pointed to the contradictory elements of private economic progress and public stagnation when he observed in his travels through central Louisiana that "there was certainly progress and improvement at the South . . . but it was much more limited, and less calculated upon than at the North." Olmsted's observations were harsh, though his central argument that there was "no *atmosphere* of progress and improvement" proved tellingly accurate. As he saw it, "There was a constant electric current of progress" in the North, but in the South, "every second man was a non-conductor and broke the chain." Characteristically, Olmsted put his finger on the central problem. "Individuals at the South," he wrote, "were enterprising, but they could only move themselves." Herein lay Louisiana's greatest flaw. Sugar planters invested thousands of dollars in New Orleans slave pens and in new steam-powered mills, yet their vision of economic growth remained bound to the estate and their own profit

32. On slave patrols and planter involvement, see Moody, *Slavery on Louisiana Sugar Plantations,* 24–27; Sally E. Hadden, *Slave Patrols: Law and Violence in Virginia and the Carolinas* (Cambridge, Mass., 2001), 71–104. On the state university and planter diffidence to collective policy, see Heitmann, *Modernization of the Louisiana Sugar Industry,* 40–48; Walter L. Fleming, *Louisiana State University, 1860–1986* (Baton Rouge, La., 1936), 3–20. On the slaveholders' identity, see (among others) Daniel Dupre, "Ambivalent Capitalists on the Cotton Frontier: Settlement and Development in the Tennessee Valley of Alabama," *JSH* 56 (May 1990): 215–40, esp. 237–39; Bowman, *Masters and Lords;* Morris, *Becoming Southern.*

margins. As sugar lord Maunsell White declared, "I am one of those who believe in protecting myself independent of any aid from the government by using skill, industry, and economy." Few in the sugar interest would have agreed to eradicate tariff protection, but White's headstrong independence and blunt individualism struck a chord. Wealth accumulation and social status were gained, planters believed, from perseverance, tough-minded land and slave management, an occasional stroke of luck, and, above all, self-reliance and private initiative.[33]

Yet added to this was the lure of the market and the burgeoning demand for sugar. Increased sugar consumption stimulated domestic production, but the nation's predilection for sweetened food and drinks swiftly overwhelmed the capacity of Louisiana's sugar producers to satiate the republic's appetite. In the acid test of supply and demand, the sugar masters seldom could supply more than half the domestic demand, and planters faced stiff international competition by the close of the antebellum era. When the crops failed in 1855 and 1856, merchants simply filled the void with Caribbean sugar. The mass importation of foreign sucrose during those particular years underscored Louisiana's weak position in the domestic marketplace and the relative insignificance of the state's sugar output to national consumption. Ill-placed to dictate prices—as they perhaps could have done if they had been able to produce enough sugar to dominate the U.S. sugar market—Louisiana planters nonetheless remained buoyant about an industry that reported three decades of relatively sustained growth.[34]

Further difficulties were on the horizon for Louisiana's cane lords, as slavery and the sugar tariff increasingly fueled northern indignation. Reeling from public criticism, Louisiana's planter elite defended their industry as a strategic necessity that saved the consumer thousands of dollars. Portraying a grim future in which Louisiana's fields lay bare and Cuban sugar magnates fleeced the American public, the Baton Rouge pressmen asserted that any reduction in the sugar duty would not trigger a countervailing reduction in price, "for the Spaniard will extort from us all he can get; and what now goes into the United States Treasury will be added to the profits of the West India planter." The *Weekly Comet* pressed local planters to consider the implications of tariff withdrawal, charging that if the quantity of Cuban

33. Olmsted, *Seaboard Slave States*, 2:274–75 (first quote); Maunsell White to Hamilton Smith, 17 September 1849, Maunsell White Papers, UNC (second quote).

34. On sugar planters as price-takers wielding little power over the market, see Schmitz, "Economies of Scale and Farm Size in the Antebellum Sugar Sector," 961–62; J. Carlyle Sitterson, "Financing and Marketing the Sugar Crop of the Old South," *JSH* 10 (May 1944): 188–99.

sugar doubled or tripled in the U.S. market, the price would fall below the cost of producing Louisiana sugar. Faced with this potentially parlous scenario, the *Comet* bluntly asked planters to consult their account books and consider a dark future in which sugar, cotton, and slaves would all diminish in value. These were not idle concerns. By the eve of secession, Cuban and Puerto Rican imports surpassed the entirety of Louisiana's increasingly marginal crop. Scotching any discussion over the acquisition of Cuba and lambasting the keenly competitive Texas sugar country as a "naked land, exposed to the unmitigated fury of north-westers," the sugar press urged planters to seek price stability in a world of volatility. Shadowboxing with their Caribbean and domestic competitors, the sugar masters declared, "[We] have more efficient machinery, cheaper transportation, better labor[,] more abundant supplies of food, and greater skill and enterprise than the planters in the islands and the South American States." Ever stalwart in defense of his native industry, Henry Hyams of the *West Baton Rouge Sugar Planter* announced emphatically that with or without the tariff on foreign sugar, Louisiana's cane interests "will continue to be remunerative"; a like-minded contemporary avowed, "Louisiana must remain the great sugar region; her climate and her soil are the best, and her geographical position unrivaled." Hyams's zealous defensiveness was born of unease. Crowded by Caribbean competition and protected solely by crumbling tariff walls, Louisiana's planters experienced the cut and thrust of global competition in a market where political volatility, price instability, commercial rivalry, and icy economic and meteorological winds buffeted the industry.[35]

Those planters who weathered the storms saw steam power, racial slavery, and aggressive business growth as key to the antebellum sugar industry. Residing on rich sugar land in Pointe Coupee Parish, William Hamilton expressed the quintessential

35. *Baton Rouge Weekly Advocate*, 18 January 1857 (first quote); *Baton Rouge Weekly Comet*, 20 July 1854 (second quote); *De Bow's Review* 4 (December 1847): 434 (third quote); *Plaquemine Southern Sentinel*, 15 April 1854; *Port Allen Sugar Planter*, 18 December 1858 (fourth quote). Louisiana was not alone in raising sugarcane in the American South. Planters in Georgia produced a small volume of horse-milled sugar in the 1830s, and cultivators along the Gulf Coast experimented with cane farming in the 1840s. Only Texas began to make inroads into the national market, but even its bumper 1852 crop represented less than 3.5 percent of Louisiana's output. See Sitterson, *Sugar Country*, 30–35; *Monthly Journal of Agriculture* 1 (March 1846): 462–63; P. A. Champomier, *Statement of the Sugar Crop, 1852–1853*, 44. For Georgia's sugar industry, see E. Merton Coulter, *Thomas Spalding of Sapelo* (Baton Rouge, La., 1940), 111–27. On sugar in Alabama, Mississippi, and Florida, see *Scientific American* 4 (December 1848): 93; *American Farmer* 8 (January 1832): 352; B. L. C. Wailes, *Report on the Agriculture and Geology of Mississippi* (Jackson, Miss., 1854), 190.

values of the antebellum sugar master. "I am a lover of order and system," Hamilton affirmed, "to have a certain way of doing everything and a regular time for doing everything." Like Hamilton, the sugar masters valued discipline, diligence, and—as banker Hughes Lavergne observed—"industry and good management." Finding little incongruity between slave labor and the pressures of a capitalist economy, the sugar masters modernized their immense agricultural enterprises while inscribing their brand of economic, social, and racial mastery onto the landscape. Whether in the sugarhouse, the plantation quarters, or the domestic slave trade, their world view was defined by agro-industrial slaveholding—in which economic power, calculating rationality, authority, and social prestige informed their every action.[36]

36. William B. Hamilton to William S. Hamilton, 27 September 1858, William S. Hamilton Papers, LSU (first quote); Hughes Lavergne to Alexander Gordon, 6 March 1829, de la Vergne Family Papers (Hughes Lavergne Letterbooks), 1829–1842, TUL (second quote).

two

"HEATHEN PART O' DE COUNTRY"

*T*HE MEMORIES OF slavery haunted Ceceil George long into old age. "I come up in hard times—slavery times," the former bondswoman recalled of her childhood in the Louisiana sugar country. Delving into the recesses of her memory, George bitterly remarked, "Everybody worked, young, an ole', if yo' could carry two or three sugar cane yo' worked. Sunday, Monday, it all de same . . . it like a heathen part o' de country." Louisiana's fearsome reputation was well deserved. Throughout southern portions of the state, cane farming and sugar production fused on industrial-scale plantations. Facing acute ecological risks and a volatile Gulf climate, planters held speed and human strength at an absolute premium as they strove to fashion muscular work crews who would speedily harvest the crop before the first killing frosts descended in November or December. To increase output and plantation productivity, Louisiana's sugar lords demographically shaped their slave crews to meet their annual labor requirements. In the process, they forged a punishing labor regime that reaped appalling human costs.[1]

Observing the order and discipline of late antebellum sugar estates, visitors to south Louisiana astutely gauged the onset of agro-industrial capitalism in the giant brick-and-mortar production facilities that rose throughout the region. One New Yorker remarked that the imposing sugarhouse on one plantation seemed akin to a New England factory buzzing with energy, noting that the "factory-looking sugar houses with their towering chimneys" dominated the landscape. Within the mill, the industrialized pace of the sugarhouse during the annual grinding season thrilled and horrified visitors. *Harper's New Monthly Magazine* vividly portrayed the

---

1. Interview with Ceceil George, 15 February 1940, WPA Ex-Slave Narrative Collection, LSU.

expectant energy of the grinding season, albeit in a description filled with crudely racial overtones:

> Everything is hurry and bustle. . . . The teams, the negroes, the vegetation, the very air, in fact, that has been for months dragging out a quiescent existence . . . now start as if touched by fire. The negro becomes supple, the mules throw up their heads and paw the earth with impatience, the sluggish air frolics in swift currents . . . while the once silent sugar house is open, windows and doors. The carrier shed is full of children and women, the tall chimneys are belching out smoke, and the huge engine as if waking from a benumbing nap, has stretched out its long arms, given one long-drawn respiration, and is alive.

Marching to the methodical beat of pistons and flywheels, slaves and masters alike faced an arduous schedule in which labor continued at a frenzied pace until the grinding season drew to a close. Solomon Northup, a former slave in the cane fields, graphically described the sugar masters' agro-industrial order, recalling that his fellow bondsmen supplied the mill house with freshly cut canes at double-quick speed. Enmeshed within a mechanical system of production, Northup and his fellow slaves labored at the metered cadence of the industrial age and toiled feverishly within the system that brutally exploited them.[2]

Like the writer Cora Montgomery, antebellum planters racialized the logic underpinning slave labor in the cane world. "This whole region is so noxious to white constitutions," Montgomery noted, that without slavery "we should have to resign altogether the production of sugar and rice, until we have reared in starving poverty a Paria class of whites miserable enough to undertake it." Most planters agreed, viewing sugar work firmly through the prism of racial slavery. In the aftermath of the Civil War, planters introduced Italian and Chinese laborers as a "pariah" class, but—like their slaveholding predecessors—they could never quite divorce themselves from their racialized constructions or gendered notions of sugar work. Be it in the eighteenth-century Caribbean or nineteenth-century Louisiana, cane planters ascribed distinct economic, sexual, and social roles to both male and

2. A. Oakey Hall, *The Manhattaner in New Orleans* (New York, 1851), 121; Robert Everest, *A Journey through the United States and Part of Canada* (London, 1855), 107; Charles Lanman, *Adventures in the Wilds of the United States and British American Provinces*, 2 vols. (Philadelphia, 1856), 2:209 (first quote); *Harper's New Monthly Magazine* 7 (November 1853): 761 (second quote); Solomon Northup, *Twelve Years a Slave*, ed. Sue Eakin and Joseph Logsdon (1853; reprint, Baton Rouge, La., 1968), 161.

female workers. Internalizing their gendered constructions of slavery and giving shape to their social and ideological beliefs, planters sought to convert these gender prescriptions into empirical definitions of task and labor. In the eighteenth-century Antilles, for instance, plantation organization was distinctly sexist; women overwhelmingly populated the field gangs and men frequently served as skilled or specialist laborers.[3]

By the early nineteenth century, the social and gender construction of women's work diminished as planters reconstituted their notion of field work to favor males. In all probability, this transformation in the gendered definition of work derived not only from a heightened awareness of biological difference, but also from changing perceptions of womanhood and planters' efforts to project their own patriarchal conceptions of sexuality and appropriate gender spheres onto their slaves. Nineteenth-century medical thought, which viewed women as naturally frail, further engraved a gender line in the workplace. Women, contemporaries believed, were fit for childbearing and nurturing a family but less equipped for the rigors of work. Above all, nineteenth-century physicians believed that the ovaries dictated a woman's life, shaping her character, destining her for motherhood, and defining her place and role in society. Although these conventions and expectations reinforced patriarchy in the white world, changing perceptions of the woman's social role and her physical attributes altered the slaveholders' bias in favor of robust male workers.[4]

3. Cora Montgomery, *The Queen of Islands and the King of Rivers* (New York, 1850), 35 (quote). On the fusion of "social convention" with physical and sexual definitions of work, see Bernard Moitt, "Women, Work, and Resistance in the French Caribbean during Slavery, 1700–1848," in *Engendering History*, ed. Shepherd, Brereton, and Bailey (Kingston, 1995), 160–62; Gwendolyn Midlo Hall, "African Women in French and Spanish Louisiana: Origins, Roles, Family, Work, Treatment," in *The Devil's Lane*, ed. Clinton and Gillespie (New York, 1997), 257–58; David P. Geggus, "Slave and Free Colored Women in Saint Domingue," in *More Than Chattel*, ed. Gaspar and Hine (Bloomington, Ind., 1996), 261–62.

4. On "domesticity" and changing expectations attached to white and black womanhood, see esp. Elizabeth Fox-Genovese, *Within the Plantation Household: Black and White Women of the Old South* (Chapel Hill, N.C., 1988); Marli F. Weiner, *Mistresses and Slaves: Plantation Women in South Carolina, 1830–1860* (Urbana, Ill., 1998), 53–71, 89–112. On biological and gender determinism, see Carroll Smith-Rosenberg and Charles Rosenberg, "The Female Animal: Medical and Biological Views of Woman and Her Role in Nineteenth-Century America," *JAH* 60 (September 1973): 334–56; Carroll Smith-Rosenberg, *Disorderly Conduct: Visions of Gender in Victorian America* (New York, 1985), 182–87. On gender as a historical category, see Linda K. Kerber, "Separate Spheres, Female Worlds, Woman's Place: The Rhetoric of Women's History," in *Towards an Intellectual History of Women* (Chapel Hill, N.C., 1997), 159–99; Joan Wallach Scott, "Gender: A Useful Category of Historical Analysis," in

Planter-historian Bryan Edwards expressed these social priorities and gender essentialism when he observed, "Though it is impossible to conduct the business, either of a house or plantation[,] without a number of females, the nature of the slave-service in the West Indies (being chiefly field labour) requires, for the immediate interest of the planter, a greater number of males." In Cuba, slaveholding logic similarly impelled the acquisition and maintenance of all-male crews, particularly for the mill and boiler house. Planters incorrectly assumed—in their misogynistic parlance—that "as low-yield animals" it made no sense to purchase women. Louisiana's sugar lords concurred. Like their tropical brethren, they shared an ardent commitment to physical might and retained a persistent sexual imbalance in the slave population.[5]

Establishing a gendered definition of labor and ascribing value to the ideal sugar worker, slaveholders in Louisiana's cane world sought to maximize both physical might and childbirth on their estates by coldly but shrewdly purchasing young men and women from the New Orleans slave pens. This intrusive policy of demographic engineering reflected the planter's overwhelming desire to shape and define every aspect of the slaves' lives and simultaneously profit from the fruits of every slave's labors. "Nowhere," historian James Oakes writes, "did the force of capitalism appear more strongly . . . than in the masters' efforts to extend their rationalizing impulse beyond the workplace and into the private life of the slaves." Tragically, this rationalizing impulse led the sugar masters toward slave breeding and the abject exploitation of enslaved men and women in the cane world.[6]

By acquiring young men for the sugar fields and teenage women with their reproductive lives ahead of them, planters demographically engineered their slave forces through gender-selective and age-discriminatory purchasing. Perhaps more so than their less gender-focused neighbors in Mississippi, Louisiana's slaveholders entered the New Orleans slave market with a set of overlapping prejudices. They gauged strength from the slaves' physique, measured age through careful

Gender and the Politics of History (New York, 1989), 28–50. On patriarchy as the institutionalization of male dominance over women, see Gerda Lerner, *The Creation of Patriarchy* (New York, 1986), 212–29.

5. Bryan Edwards, *The History, Civil and Commercial, of the British Colonies in the West Indies,* 5 vols. (London, 1801), 2:36, quoted in Barbara Bush, *Slave Women in Caribbean Society, 1650–1838* (Bloomington, Ind., 1990), 36 (first quote); Moreno Fraginals, *Sugar Mill,* 142; Franklin W. Knight, *Slave Society in Cuba during the Nineteenth Century* (Madison, Wis., 1970), 72–74 (second quote); Michael Tadman, *Speculators and Slaves: Masters, Traders, and Slaves in the Old South* (Madison, Wis., 1989), 23.

6. Oakes, *Slavery and Freedom,* 144.

probing questions, assessed personality traits by scars, and estimated the bonds-woman's reproductive potential and market value by scanning her body. The sugar lords were masters of physiognomy: they understood the barbarous nature of sugar work and the almost constant need for fresh bodies. In the sugar world, this requirement translated into a lucrative trade for a narrow band of workers who would farm the fields and bear the next generation of cane hands. Men accordingly represented as much as 85 percent of all slaves sold to sugar planters, and those individuals probably stood a full inch taller than most African American slaves. Although profitable, the social impact of such age- and gender-selective purchasing proved devastating in the long run. Women were relatively scarce, and south Louisiana's birthrates proved among the lowest in the Old South. In sum, the region mirrored the slave gulags of the Caribbean, where nineteenth-century sugar planters demonstrated a similar preference for males as agricultural workers.[7]

The skewed gender imbalance of sugar production was reflected in the plantation demographics of Nicolas Reggio, who acquired part ownership of Habitation Pointe aux Chenès, south of New Orleans. With slavery property valued at over $34,000 in 1824, Reggio's estate included twice as many adult males as females—an imbalance that sharpened over time. Less then a decade later, his plantation numbered over fifty male slaves and twenty-one women. Reggio's preference for

7. On assessing bodies, see Johnson, *Soul by Soul;* Ariela J. Gross, *Double Character: Slavery and Mastery in the Antebellum Southern Courtroom* (Princeton, N.J., 2000), 122–52; Jonathan D. Prude, "To Look upon the 'Lower Sort': Runaway Ads and the Appearance of Unfree Laborers," *JAH* 78 (June 1991): 140–43. The slave's body was assessed, Prude argues, according to "universal social characteristics," which were in turn shaped by a transatlantic discourse on slaves' physiognomy. On the latter, see Marcus Wood, *Blind Memory: The Visual Representation of Slavery in England and America, 1780–1865* (Manchester, 2000). On the specifics of the New Orleans market, see Herman Freudenberger and Jonathan B. Pritchett, "The Domestic Slave Trade: New Evidence," *JIH* 21 (winter 1991): 452; Pritchett and Freudenberger, "A Peculiar Sample: The Selection of Slaves for the New Orleans Market," *JEH* 52 (March 1992): 115–16; Richard H. Steckel, "Slave Height Profiles from Coastwise Manifests," *Explorations in Economic History* 16 (October 1979): 368–69. On the imagery of the "slave gulag," see Orlando Patterson, *Rituals of Blood: Consequences of Slavery in Two American Centuries* (Washington, D.C., 1998), 28. On the sexual imbalance of Caribbean slaves, see Barry W. Higman, *Slave Populations of the British Caribbean, 1807–1834* (Baltimore, 1984), 281; Schwartz, *Sugar Plantations,* 348; David P. Geggus, "Sugar and Coffee Cultivation in Saint Domingue and the Shaping of the Slave Labor Force," in *Cultivation and Culture,* ed. Berlin and Morgan (Charlottesville, Va., 1993), 73–98. For parallels to Cuban slave trading, see Manuel Moreno Fraginals, Herbert S. Klein, and Stanley L. Engerman, The Level and Structure of Slave Prices on Cuban Plantations in the Mid-Nineteenth Century: Some Comparative Perspectives," *AHR* 88 (December 1983): 1203; Knight, *Slave Society in Cuba,* 79.

male labor paralleled a broader demographic pattern in the Louisiana and Cuban sugar country, where young men constituted approximately 60 percent of all sugar workers. On Eugenie Dardenne's estate along Bayou Plaquemine, working-aged men outnumbered women by almost three to one. The owners of Wilton Plantation followed their neighbors' predilection for male workers; men represented over two-thirds of their adult slave force. Fifty men and thirty-three women were prime sugar workers, aged eighteen to thirty-nine, who formed the principal work gangs on the plantation. With a good supply of thirty-nine slaves under the age of eighteen, Wilton Plantation possessed both youthful muscle power for its immediate needs and a labor force for the future.[8]

Fully cognizant of the symbiotic relationship between long-term agricultural growth and the expansion of slave crews, the planter Andrew Durnford maintained not only a gender-imbalanced workforce but a distinctly youthful population on his plantation. Despite being a free man of color, Durnford defied antebellum racial constructions by demonstrating shrewd business acumen and a calculating regard for the productivity of his slaves. Following the advice of fellow slaveholder Theodore Packwood, who "advised me very much to get people, and says that [I] cannot do as my neighbors to make 3 and 400 hogsheads without augmenting my force," Durnford increased his slave crews to include forty-six men and thirty-one females by 1860. On Oaklands Plantation, Samuel McCutchon similarly favored adult males, though, like Durnford, he retained thirty-one boys under fifteen years of age to satiate his long-term labor requirements. At Variety Plantation in Iberville Parish, Joseph Kleinpeter similarly maintained a good supply of fifteen slaves between ten and nineteen years to furnish his labor needs for the immediate future.[9]

To supply the Louisiana slave market with prime individuals, a complex trading network emerged, in which males constituted 70 percent of all slaves imported

8. Jonathan B. Pritchett, "The Interregional Trade and the Selection of Slaves for the New Orleans Market," *JIH* 28 (summer 1997): 57–85; Division of Estate of Charlotte Constance and Hélène Jorda, 11 February 1824, and Certificat et Serment des Appréciateurs, Banque de l'Union de la Louisiane, 28 September 1832, Reggio Family Papers, LSU; Eugenie Dardenne Document, LSU; Cash Book 1854–1862 (vol. 6), Bruce, Seddon, and Wilkins Plantation Records, LSU.

9. Andrew Durnford to John McDonogh, 18 February 1834, John McDonogh Papers, LSM; Whitten, *Andrew Durnford* 33 (quote); David O. Whitten, "Slave Buying in Virginia as Revealed by the Letters of a Louisiana Negro Sugar Planter," *LH* 11 (summer 1970): 231–44; Whitten, "A Black Entrepreneur in Antebellum Louisiana," *Business History Review* 45 (summer 1971): 207; Inventory and Valuation of Slaves, Stock, and Farming Utensils of Oaklands Plantation, 1859, Samuel D. McCutchon Papers, LSU; Slave List, 1856, Joseph Kleinpeter and Family Papers, LSU.

to New Orleans. In explaining this regional peculiarity, scholar Michael Tadman contends that the labor-intensive work of the nearby sugar plantations directly shaped the Louisiana slave trade by placing a premium on strength and brawn. Yet as slave-trade historians observe, the traffic was not only gender specific, but age specific as well. Ninety-three percent of slaves imported to New Orleans were prime slaves, aged between eleven and thirty. In addition, physiologically informed analyses of sexual fertility, adolescent growth patterns, and slave profitability further defined slave trading and purchasing in the sugar country. As Frederick Law Olmsted observed in his turn through the sugar country, planters shared "an almost universal passion . . . for increasing their negro stock." They drew slaves from the New Orleans slave pens not only to add to their slave rosters, but also demographically to shape their crews to maximize future reproductive potential.[10]

The experience of slave traders Isaac Franklin and John Armfield illustrates that in supplying the New Orleans market—and by extension, the sugar parishes—traders combined profit impulses with a discerning eye for gender and youth. Upon opening their business venture in May 1828, Franklin and Armfield advertised that they would surpass their competitors and pay at the highest rate for "likely young negroes" aged between twelve and twenty-five. Proferring cash payments for physically fit males and females, Franklin and Armfield focused their slave buying on slaves who would sell rapidly in the New Orleans market and, importantly, who would survive "seasoning" and the rigors of Louisiana's unhealthy climate. Procuring slaves in the upper South, Franklin and Armfield's network of agents shipped over a thousand slaves a year from the Chesapeake to their main entrepôt in New Orleans. Of the 3,600 bondspeople dispatched to New Orleans from 1828 to 1836, almost 2,000 were males. Of these men and young boys, historian Donald Sweig concludes, a staggering 84 percent were single and an overwhelming number were less than twenty-five years of age. Specializing in slaves aged thirteen to twenty-eight, Franklin and Armfield effectively supplied the sugar masters' needs for young, strong, and fertile workers; almost 80 percent of all single males and females belonged to this age group. Viewing Franklin and Armfield's operations from their northern Virginia headquarters, Ethan Andrews remarked upon the gender and age of bondspeople purchased for the southwest trade in the mid-1830s. He noted that approximately fifty to sixty men and thirty

10. On the peculiarities of the Louisiana market, see Tadman, *Speculators and Slaves,* 68; Tadman, "Demographic Cost of Sugar," 1534–75; Freudenberger and Pritchett, "Domestic Slave Trade," 453; Olmsted, *Seaboard Slave States,* 2:308 (quote).

TABLE 2. SLAVES IMPORTED BY FRANKLIN AND
ARMFIELD TO NEW ORLEANS, 1828–36

Age	Single females	Single males	Family females	Family males
0–10	63	83	250	219
11–12	73	92	14	24
13–16	329	243	29	26
17–20	367	401	77	12
21–24	87	433	49	4
25–28	54	238	55	8
29–30	9	51	20	3
31+	33	129	74	32
Total	1,015	1,670	568	328

Source: Sweig 1990

to forty women filled the slave pen, most of whom were between eighteen and thirty years old. In accordance with Louisiana law, which forbade the sale of young children separately from their mothers, Franklin and Armfield also purchased and traded a small number of slaves in family units. These family groups usually comprised women aged seventeen to twenty-eight, with between two and three children under the age of ten. Young children and babies represented the overwhelming majority of slaves purchased in family units, but the total number of adult slaves traded as families remained relatively slight compared with the purchasing and transshipment of single slaves.[11]

Franklin and Armfield's experience proved representative of interregional slave trading. Austin Woolfolk, a professional slave trader in Maryland, similarly tailored his slave purchases to meet the burgeoning demand from New Orleans. Purchasing

11. Wendell Holmes Stephenson, *Isaac Franklin: Slave Trader and Planter of the Old South* (Baton Rouge, La., 1938), 25; Frederic Bancroft, *Slave Trading in the Old South* (1931; reprint, Columbia, S.C., 1996), 30, 59; Donald M. Sweig, "Reassessing the Human Dimension of the Interstate Slave Trade," *Prologue* 12 (spring 1980): 9–11; Robert H. Gudmestad, "A Troublesome Commerce: The Interstate Slave Trade, 1808–1840" (Ph.D. diss., Louisiana State University, 1999), 46–48, 53–56. Jonathan B. Pritchett and Richard M. Chamberlain, "Selection in the Market for Slaves: New Orleans, 1830–1860," *Quarterly Journal of Economics* 108 (May 1993): 469–70, calculate that 69.8 percent of all imported slaves to New Orleans were aged between fifteen and thirty-five. On the protections afforded mother and child in the Louisiana *Code Noir*, see Schafer, *Slavery, the Civil Law, and the Supreme Court of Louisiana*, 165–68.

surplus slaves in Baltimore and along Maryland's Eastern Shore, Woolfolk vigorously trawled the local market to acquire youthful individuals for the southwest trade. During the 1820s, Woolfolk dominated the Chesapeake–New Orleans traffic, and, like Franklin and Armfield, he purposely focused on teenagers—a demographic in which males outnumbered females by eight to five. Planter Farish Carter bought his slaves in Charleston, but his penchant for youth marked every sugar master from the 1820s to the Civil War. Purchasing the "likeliest lot of Negroes I ever saw together," Carter returned to Bayou Salé with eighteen men and four women. Of those with recorded ages, the average slave in Carter's lot was twenty-two years old, but significantly, three of the women—Fanny, Hannah, and Daphne—were between seventeen and eighteen. By the early 1850s, New Orleans traders still retained their signature gender predilections and penchant for "likely" muscular bodies. Strolling past a clothing store, Frederick Law Olmsted's attention was caught by a line of twenty-two identically clad young men, standing at ease like a rank of soldiers. Dressed in suits of blue cloth, wearing black hats, and holding a pair of shoes in their hands, all but one of these slaves (who was probably a driver) were between eighteen and twenty-two years old; the oldest of them was no more than twenty-five. "They were silent and sober," Olmsted noted, "and, perhaps, were gratified by the admiration their fine manly figures and uniform dress obtained from the passers by." The slaves' masculine figures clearly impressed the visiting journalist, but slave traders in the Crescent City had been buying, selling, and marketing men like these for decades. Whether one draws upon illustrative portraits such as Olmsted's or the colder records of Franklin and Armfield, young single men—and to a lesser extent, teenage women—overwhelmingly dominated the slave traders' portfolios. The evidence from Franklin and Armfield's partnership indicates that slave trading between 1828 and 1836 focused on two broadly disparate cohorts: young men aged seventeen to twenty-four and teenage girls aged thirteen to twenty. In both cases, age, gender, physical maturity, and reproductive fitness received marked attention, as the slaveholders orchestrated the internal slave trade— just as they had shaped the Atlantic trade—with frightening efficiency and with a rational eye for profit and expansion.[12]

Louisiana's unique agricultural calendar imposed an inflexible framework of

---

12. William Calderhead, "The Role of the Professional Slave Trader in a Slave Economy: Austin Woolfolk, A Case Study," *Civil War History* 23 (September 1977): 198; Ralph B. Flanders, "An Experiment in Louisiana Sugar, 1829–1833," *North Carolina Historical Review* 9 (April 1932): 155–56 (first quote); Olmsted, *Seaboard Slave States*, 2:232 (second quote).

gender-defined labor patterns. Planters firmly maintained that young men were best suited to sugar and that to convert woodland into cultivable acreage required the strongest and hardiest labor crews. Bondswomen—like Ceceil George—also worked in the fields and mill house; but for the sugar masters, they were also a valuable source of future workers. A discriminating search for brawn and fertility accordingly defined slave trading in south Louisiana, as slaveholders incorporated expectations of sexual and physical maturity into their calculations. As historian Walter Johnson has brilliantly shown, the slave market held "transformative possibilities" and proffered a route to economic, political, social, and racial mastery. Within the confines of the auction room, white Louisianans gauged one another by their acumen in the slave pens and projected visions of their own mastery onto those who trudged out before them. Among the planters visiting the auction rooms, the sugar masters envisioned themselves as savvy businessmen who maximized plantation output by acquiring another dozen men. Simultaneously, they must also have seen themselves as the breeders of slaves. Typically, planters eagerly read the slaves' body on the rostrum; they stripped the bondsman's back to judge his character, mentally made note of a slave woman's hips, listened for hacking coughs and wheezing chests, and attributed behavior and comportment to racial categories. In the cane parishes, these coarsely imposed constructions and categorizations of African American labor fused with long-standing gender discrimination and bias to create or imagine the ideal body for sugar. If muscle and sinew defined the contours of the black male—and indeed, black masculinity—then reproductive potential tragically defined the black female. Occidental notions of the ideal female fetishized the bondswoman's body, imputing fertility to the greased hourglass figure that stood on the auction block. The slaveholders' license to prod the muscles, knead the joints, and eye the scars of their prospective purchases made their power all the more real, but social conventions, Victorian mores, and their own cloying sense of themselves as paternalists mitigated against open, crass statements of the bondswoman's sexual worth. Planters and traders nonetheless shared a discourse of racial categorization and human differentiation. "Likely," "choice," "of consequence," and "sound" served as convenient labels through which planters and traders could mutually converse without recourse to nomenclature that coarsely articulated their communal standards.[13]

13. On slave trading, see Johnson, *Soul by Soul*, 78 (quote), 149. On planters' epithets for physical health, see Sharla M. Fett, *Working Cures: Healing, Health, and Power on Southern Slave Plantations*

In the bodies they fingered both before and after purchase, slaveholders commodified their human cargoes, differentiating them by shape, size, skill, and gender and ascribing monetary and nonmarket values to racial phenotype and sexual appearance. Planters told themselves and others that they were teaching "good discipline and management" to their inferiors, but their paternalistic charade did not disguise the commercial benefits accrued from slave trading. Youth, brawn, and reproductive potential characterized the sugar masters' approach to the slave market, out of which they sought to shape their crews demographically, maximize plantation output, and increase their chattel population. As such, the overwhelming majority of all women imported by Franklin and Armfield were aged between 13 and 20, while the mass of single males were 17 to 24. In 1850, the age specificity of slave trading continued to favor youth, with the peak in sales occurring for women aged 15 to 25 and for men aged 20 to 29. Such age- and gender-specific purchasing closely mirrored human development, in which the growth rate followed a gradual process before rising sharply during adolescence. Among female slaves, the adolescent growth spurt occurred between 12.5 and 13.5 years; in boys, growth peaked between 14.5 and 15.5 years. Slaves matured slightly earlier (and grew two to three inches taller) than contemporary Europeans, reaching adult height at approximately 17 years for females and 19 years for males. These figures partially explain the pattern of slave trading to Louisiana. Franklin and Armfield shipped the largest number of slaves in the two age categories directly after the peak in both male and female adolescent growth rates. In the case of women under the age of eighteen, this policy proved coldly profitable, since their projected earnings would at least temporarily exceed those of physically immature teenage boys. Indeed, as historians Robert Fogel and Stanley Engerman have concluded, America's slaveholders behaved "with as much shrewdness as . . . any northern capitalist." To be sure, Franklin and Armfield provided slaveholders with prime-aged laborers who had just reached or were reaching their full height, but, unlike a northern capitalist, they were also supplying the sugar masters with

---

(Chapel Hill, N.C., 2002), 21. For a fascinating parallel with the "hand test," in which post–World War II Florida sugar companies assessed prospective sugar workers by feeling men's hands to see if they were calloused enough to indicate a history of hard labor and conducted physical examinations of their bodies to test their aptness for sugar work, see Cindy Hahamovitch, "No Man's Land: Jamaican Guest Workers in Post–World War II America" (work in progress).

sexually maturing and fertile women with their reproductive lives far ahead of them.[14]

Demographic historians estimate that most slave girls experienced menarche at fifteen, though they remained effectively sterile until they reached their seventeenth to eighteenth birthdays. Since weight and body fat largely regulated the onset of menarche, chronic childhood diseases and the slaves' low-calorie and low-protein diet ensured that slave girls would not reach the average weight for menarche (approximately one hundred pounds) until their mid-teenage years. Although it is difficult to state with specificity the average age at which African American bondswomen could bear children the manuscript record from the sugar parishes indicates that many slave women experienced their first pregnancy between seventeen and eighteen years of age. Franklin and Armfield's fixation on young women ensured a ready supply of physically maturing women who in all probability had passed through menarche. Of the thousand single females dispatched to New Orleans by Franklin and Armfield, almost 70 percent came from two age groups—thirteen to sixteen and seventeen to twenty—suggesting that they supplied a discerning market with women who had their reproductive lives unmistakably before them. Significantly, interest in slave females swiftly declined with age; Franklin and Armfield purchased only 141 single women aged twenty-one to twenty-eight. By contrast, their obsession with youth—put frankly, teenage girls—ensured that the sugar masters purchased mature young adults who were both physically and reproductively fit.[15]

14. Donald D. Avery to Sarah Craig Marsh, 16 May 1846, Avery Family Papers, UNC (first quote); Sweig, "Reassessing the Human Dimensions of the Slave Trade," 9–11; Judith Kelleher Schafer, "New Orleans Slavery in 1850 as Seen in Advertisements," *JSH* 47 (February 1981): 35–36; Fogel and Engerman, *Time on the Cross,* 73 (second quote). On growth rates, see J. M. Tanner, *Foetus into Man: Physical Growth from Conception to Maturity* (Cambridge, Mass., 1990), 6; Robert A. Margo and Richard H. Steckel, "The Heights of American Slaves: New Evidence on Slave Nutrition and Health," *Social Science History* 6 (fall 1982): 519. Robert Fogel suggests that adolescent slaves experienced very rapid growth after age sixteen, partly due to their improved diet after they entered the primary adult work gangs. See Fogel, *Without Consent or Contract,* 142–44, and Richard H. Steckel, "A Peculiar Population: The Nutrition, Health, and Mortality of American Slaves from Childhood to Maturity," *JEH* 46 (September 1986): 721–26.

15. On menarche, see James Trussell and Richard Steckel, "The Age of Slaves at Menarche and Their First Birth," *JIH* 8 (winter 1978): 477–505; Rose E. Frisch, "Demographic Implications of Female Fecundity," *Social Biology* 22 (spring 1975): 17–22. Frisch estimates that "adolescent sterility after menarche is approximately about 3.5 years." Since slave women probably experienced a

Historical evidence for slave breeding in the Old South remains patchy, but planters throughout the slave states certainly cajoled, prodded, and at times coerced their bondswomen to reproduce. Whether using subtle encouragement, crass bullying, or open rape, the slaveholders' manipulation of African American fertility suggests that they valued "breeding wenches" and sought to appropriate their bodies for reproductive potential. Mincing few words about the planters' sexual politics, historian Catherine Clinton concludes that the business of slave reproduction was ultimately "too vital" to be left to the slaves themselves. This dreadful practice reached its apex in the Louisiana sugar country. Franklin and Armfield did not conduct blatant stock farming or slave breeding, but they sent to the New Orleans slave market—and by extension, to the sugar fields of Louisiana—young female slaves who would soon bear children and continue to do so for many years to come.[16]

---

similarly long period of adolescent sterility, we can conclude that most slave women were capable of bearing their first child at approximately eighteen years of age. For a parallel study of women with a low-protein diet, see R. J. W. Burrell, M. J. R. Healy and J. M. Tanner, "Age at Menarche in South African Bantu Schoolgirls Living in the Transkei Reserve," *Human Biology* 33 (September 1961): 250–61. On the ideal weight and height for menarche, see Rose E. Frisch, *Female Fertility and the Body Fat Connection* (Chicago, 2002), chaps. 3, 6. On slaves' first births, see Malone, *Sweet Chariot*, 176.

16. The key works on the slave-breeding thesis are Richard Sutch, "The Breeding of Slaves for Sale and the Westward Expansion of Slavery, 1850–1860," in *Race and Slavery in the Western Hemisphere*, ed. Engerman and Genovese (Princeton, N.J., 1975), 173–210; Herbert G. Gutman and Richard Sutch, "Victorians All? The Sexual Mores and Conduct of Slaves and Their Masters," in *Reckoning with Slavery*, ed. David et al. (New York, 1976), 134–64; Robert Fogel and Stanley L. Engerman, "The Slave Breeding Thesis," in *Without Consent or Contract: The Rise and Fall of American Slavery (Technical Papers)*, vol. 2, *Conditions of Slave Life and the Transition to Freedom*, ed. Fogel and Engerman (New York, 1992), 455–72. Most recently, Christopher Morris has argued that slaveholders sought to breed slaves or at the very least to create conditions that would favor births. See Morris, *Becoming Southern*, 70–78; Morris, "Articulation of Two Worlds," 992–94. By contrast, David Whitten argues against slave breeding in his "Medical Care of Slaves: Louisiana Sugar Region and South Carolina Rice District," *Southern Studies* 16 (summer 1977): 162. From a more slave-centered perspective, see Deborah Gray White, *Ar'n't I a Woman? Female Slaves in the Plantation South* (New York, 1985), 98–106; Catherine Clinton, "'Southern Dishonor': Flesh, Blood, Race, and Bondage," in *In Joy and Sorrow*, ed. Bleser (New York, 1991), 54 (quote); Nell Irvin Painter, "Soul Murder and Slavery: Toward a Fully Loaded Cost Accounting," in *U.S. History as Women's History*, ed. Kerber, Kessler-Harris, and Kish-Sklar (Chapel Hill, N.C., 1995), 134–36. For sporadic examples of slave breeding, see Larry Eugene Rivers, *Slavery in Florida: Terriorial Days to Emancipation* (Gainesville, Fla., 2000), 152; Brenda E. Stevenson, *Life in Black and White: Family and Community in the Slave South* (New York, 1996), 181; Wilma King, *Stolen Childhood: Slave Youth in Nineteenth-Century America* (Bloomington, Ind., 1995), 109–10.

The female slave was not alone in being judged on the auction block. While youthful masculinity proved particularly appealing for the sugar masters, their focus on physically impressive slave men also included a rather crude—but nonetheless essentially valid—calculation about male virility. Modern physiologists contend that taller men are socially considered more attractive and have greater reproductive success than men of average height. Planters would not have been able to quantify the positive effects of stature on reproductive ability, but, just as they characterized the slave woman as a "breeding wench," their nomenclature for males as "buck" and "mose" in all probability carried sexual connotations. Indeed, the etymology of "buck" in the American South derived from its Revolutionary meaning of a "dashing, young, virile man," which by the nineteenth-century denoted a "self-proclaimed fascinator of women." "Mose" likewise carried his sexuality on his shirtsleeves, while the tough "B'hoy," or boy, evolved into an urban dandy and a gritty lover who thrilled minstrel-going audiences on the eve of the Civil War.[17]

If such overtly sexualized characters walked the antebellum stage, it is probable that slaveholders projected sexual and erotic connotations onto those before them on the rostrum block. Greased and polished for sale, these enslaved men and women symbolized the vibrant, virile sexuality that planters sought to acquire from Franklin and Armfield's auctions. In a society in which racial constructions of blackness veered between hostility, fear, and patronizing notions of childlikeness, slaves' lithe bodies and sleek physiques signified an animalistic promiscuousness that drew upon and contributed to planters' cultural marketplace of misogynistic language and signs. Perhaps Valsin Mermillion shared these cultural values, for his former slaves observed that he "prided himself in having only handsome slaves" and that if he heard of a particularly "fine physique," he would pursue that individual at any price. Whether male or female, handsome or not, the sexualized slave body exoticized difference and visibly reflected the slaveholder's projection of his or her productive and reproductive desires. In the so-called

17. On height and reproduction, see Daniel Nettle, "Women's Height, Reproductive Success, and the Evolution of Sexual Dimorphism in Modern Humans," *Proceedings of the Royal Society of London,* ser. B, 269 (2002): 1919–23; B. Pawlowski, R. I. M. Dunbar, and A. Lipowicz, "Tall Men Have More Reproductive Success," *Nature* 403 (2000): 156. On racial epithets, see David R. Roediger, *The Wages of Whiteness: Race and the Making of the American Working Class,* rev. ed. (London, 1999), 99–100; Shane White and Graham White, *Stylin': African American Expressive Culture from Its Beginnings to the Zoot Suit* (Ithaca, N.Y., 1998), 116–21.

fancy-girl markets these values took human shape, as, planters paid up to $5,000 to satiate their desires for young concubines who would enhance their own sense of power and prestige. While social pretension motivated some sugar masters, others coveted their young purchases and led them back to the quarters for their own sexual pleasure. Slightly older women and younger men consequently received scant attention in the slave market, as traders and their eager clients set prices according to the reproductive potential and physical fitness of their chattel.[18]

Notarial records of slave prices in New Orleans quantify planters' perceived and real preference for specific age and sex profiles. Importantly, the records emphasize the relative value of gender, adolescence, and physical maturity, which is reflected in the fact that slave prices for young men were 15 to 20 percent higher than those for comparably aged females. Former slave Octavia George provided a stark comment on these cold price differentials, noting that the "big fine healthy slaves were worth more than those that were not quite so good." In both 1856 and 1859, age shaped price, and bondsmen aged twenty-one to thirty commanded a higher price than slaves between sixteen to twenty years old. Valued at $1,419 in 1856 and $1,639 three years later, these prime-aged men had evidently reached full strength, and they consequently cost more than their adolescent brothers. By contrast, teenage females who were physically mature received a higher valuation than women in their twenties. These conclusions suggest that planters calculated or estimated the relative earning power of both sexes over their life cycle and particularly valued teenage women who could toil as productive and reproductive capital. Yet these slave prices included a still more elusive estimation—the value

18. On misogynistic and racial slander in the circum-Caribbean, see Diana Paton, "Enslaved Women's Bodies and Gendered Languages of Insult in Jamaica during Late Slavery and Apprenticeship" (paper presented at the Fourth Avignon Conference on Slavery and Forced Labor, Women in Slavery, in Honor of Suzanne Miers, October 2002); Kirsten Fischer, "'False, Feigned, and Scandalous Words': Sexual Slander and Racial Ideology among Whites in Colonial North Carolina," in *The Devil's Lane,* ed. Clinton and Gillespie (New York, 1997), 139–53. On the evolving gendered and "animalized" image of the black female slave, see Karl Jacoby, "Slaves By Nature? Domestic Animals and Human Slaves," *Slavery and Abolition* 15 (April 1994): 89–99. The term "cultural marketplace" symbolizes the planters' ideological view of the appropriation of the slave woman's body. See Young, *Domesticating Slavery,* 118; Colette Guillaumin, *Racism, Sexism, Power, and Ideology* (London, 1995), 176–207. On Mermillion, the fancy-girl market, and sexual abuse, see Interview with Mrs. Webb, 17 August 1937, WPA Ex-Slave Narrative Collection, LSU (quote); Genovese, *Roll, Jordan, Roll,* 416, Stevenson, *Life in Black and White,* 180. On the treatment of older and infertile women, see White, *Ar'n't I a Woman?* 101–2.

attached to the slave woman's fertility. In the southwest, slaves' ability to bear children received especially high appraisal. Young women aged twenty received a premium evaluation of $170 in Louisiana, compared with $80 in the upper South. Although the exact value attached to potential childbearing capacity remains problematic to quantify, slaveholders calculated that at least 5 or 6 percent of their profit would accrue from their slaves' issue and that raising slaves and staple crops were far from inimical. Like their brethren elsewhere in the slave states, sugar planters shared a powerful economic motivation to encourage fertility among their slaves, and they paid the highest sums to procure young women at the height of their reproductive potential. Repellant though their business was, slave traders such as Franklin and Armfield, Austin Woolfolk, and their competitors materially advanced the potential for child rearing in the sugar parishes by meeting the peculiar age, gender, and labor requirements of the sugar country.[19]

The signature marks of discriminatory pricing in the New Orleans slave trade visibly confirmed Karl Marx's conclusions that the "price paid for a slave is nothing but the anticipated and capitalized surplus-value or profit to be wrung out of the slave." In the cane fields of Louisiana, the slaves' surplus value could be measured in the volume of sugar processed and the human capital borne by the pregnant bondswoman. Sugar lord William J. Minor proved representative of his class in his fixation on youth. Listing almost 200 slaves on his Waterloo estate in 1847, Minor maintained a slight male majority, with 76 men, 65 women, and 50 children. Thirteen years later, the census enumerator recorded that almost 220 slaves resided at Waterloo. In that year, William Minor counted 28 young women aged between 17 and 28 on his plantation, but, significantly, he maintained an especially large number of females in the age cohort 17 to 20. The men, by contrast, were a little older, with the majority of his prime sugar hands being between 21 and 32 years old. As a calculating businessman who surely valued the twin attributes of physical

19. Interview with Octavia George, WPA Slave Narrative Project, U.S. Work Projects Administration (USWPA), Manuscript Division, Library of Congress. See "Born in Slavery: Slave Narratives from the Federal Writers' Project, 1936–1938," http://memory.loc.gov/ammem/snhtml/snhome. html (quote; accessed 15 April 2004). Prices from Richard Tansey, "Bernard Kendig and the New Orleans Slave Trade," *LH* 23 (spring 1982): 167. The differential in male-female prices (approximately 20 percent) mirrors that of most slave societies. See Barry W. Higman, *Slave Population and Economy in Jamaica, 1807–1834* (1976; reprint, Kingston, 1995), 192; Moreno Fraginals, Klein, and Engerman, "Level and Structure of Slave Prices on Cuban Plantations," 1210, 1216; Fogel and Engerman, *Time on the Cross,* 75, 81.

strength and fertility, Minor, like his fellow sugar planters, purchased and maintained slave crews who were in their physical prime and at the peak of their reproductive capacity. Minor's predilections were hardly unique. In the late 1820s and 1830s, as the plantation economy accelerated, planter Valcour Aime purchased twenty-six women for his St. James Parish estate. Aime's Creole background did not distinguish him from the gender and age concerns of his Anglo counterparts; on average, his young purchases were less than twenty-three years old. Whether in the 1820s or towards the end of the slave era, the planters' partiality for youth, brawn, and reproductive promise molded the slave trade and markedly skewed the demography of the sugar country.

In Ascension Parish, for instance, the planters' attachment to youthful labor must have been noticeable to the census enumerator, for almost one-third of the slave population aged eleven to thirty in 1850 comprised of young men or women in their late teenage years. Above all, the sugar lords retained a persistent interest in slaves aged sixteen to twenty. Throughout the parish, plenty of young, maturing males resided on the sugar plantations, and whether aged sixteen or twenty-eight, significant numbers of them worked in the prime gangs. The number of females, however, declined quite sharply with age, suggesting that planters deliberately retained bondswomen within a relatively narrow and youthful age group. On Henry Doyal and John Preston's properties, for example, young women between the ages of sixteen and twenty-five defined the slave population. Like their neighbors, both Doyal and Preston maintained large numbers of prime-age slaves but kept a dwindling population of women in their late twenties.[20]

Natural increase cannot explain the distribution of the slave population in Ascension Parish, with its large group of teenage women, superfluity of men in their twenties, and the small number of women in their late twenties relative to those who had celebrated their eighteenth birthday. These imbalances and skewed sex ratios, however, were no mistake. Unlike their cotton-growing neighbors, who tended to balance sex ratios on their plantations, the sugar masters chose to

20. Karl Marx quoted in Ransom and Sutch, "Capitalists without Capital," 133. The margin of difference between the cost of a worker's wage and food and the value of his or her output is the surplus value of each worker. See Karl Marx, *Capital: A Critique of Political Economy*, vol. 3, *The Process of Capitalist Production as a Whole* (1894; reprint, New York, 1936), 342–53. Minor Diary No. 2, 1847–1848, 20 March 1847, William J. Minor and Family Papers, LSU; U.S. Bureau of the Census, *Seventh Census of the United States (1850)*, 6th Ward, Ascension Parish, Louisiana, and *Eighth Census of the United States (1860)*, 6th Ward, Ascension Parish, Louisiana; Record of the Purchases, Births, Deaths, and Assignments of the Slaves of the Plantation of Valcour Aime, 1821–1850, Valcour Aime Slave Records, LSM.

maintain the peculiar demographics of their estates by supplementing their chattel population with new blood purchased in the New Orleans slave trade. Whether shackled together on the deck of a Mississippi steamboat or shuffling along levee paths, the coffles that returned to the sugar country featured two main groups. Large numbers of young men climbed aboard the steam packets that would convey them to the sugar country, but young women in their late teens stood among them, sold to the planters for the fruits of their labors. Many of these men and women would die before their thirtieth birthdays, forcing the planters to return to New Orleans every year for thousands of newly imported hands. This constant influx of new workers ensured that plantations maintained their peculiar sex ratios and age profiles to the close of the slave era.

While concerns over age, gender, physique, and procreation informed planters' selection of slaves, they often had to balance their preferences against the escalating prices of the 1850s. The cost of a prime field hand rose from almost $700 in 1850 to just over $1,000 in 1856 and to $1,400 on the eve of the Civil War, forcing planters to focus less on gender and youth in favor of resolving immediate labor requirements and long-term demographic stability. Some planters responded to the rapid price rises by calling for the revival of the Atlantic slave trade and, in some cases, the transshipment of contraband African slaves to Louisiana via Cuba. Various possible apprenticeship schemes were floated, but the call for an additional hundred thousand enslaved laborers for Louisiana's planting interests ultimately fell on deaf ears. The sugar masters could not wait for these assorted plans to materialize, and so they returned to the slave markets and reassessed their former preferences. As the price for young men rose, women increasingly proved relatively better value for money, and experience swiftly demonstrated that sexist assumptions about the relative underperformance of women in the cane world were false. Women in their thirties accordingly increased in value by over one-third in the late antebellum years, although they still cost only 70 percent of the average price of equivalent males in 1859. Older individuals were also recategorized as attractive workers for the labor-hungry sugar lords.[21]

On a slave-purchasing trip to New Orleans in 1859 during the seasonal peak in trading, Alexander Franklin Pugh rejected the former planter bias in favor of men

21. Laurence J. Kotlikoff, "The Structure of Slave Prices in New Orleans, 1804 to 1862," *Economic Inquiry* 17 (October 1979): 506; James Paisley Hendrix Jr., "The Efforts to Reopen the African Slave Trade in Louisiana," *LH* 10 (spring 1969): 97–123; Barton J. Bernstein, "Southern Politics and Attempts to Reopen the African Slave Trade," *Journal of Negro History* 51 (January 1966): 16–35; *De Bow's Review* 25 (November 1858): 491–506. On contraband slave traffic, see John S. Kendall, "New Orleans'

by buying eight women and six men. For these slaves, Pugh paid between $1,325 and $1,400 for the women and between $1,600 and $1,700 for the men. Clearly looking for a skilled hand, Pugh expended an additional $2,500 for a blacksmith. Complaining extensively, Pugh wrestled with the high slave prices but nonetheless purchased additional hands, noting, "It seems we must have them at any price . . . for [I] fear they will go still higher when we require them, and [I] cannot do without them." Facing escalating costs and a scarcity of "good-ones," Pugh forsook his initial desire for men in favor of more readily available women. Only five days after he jotted in his diary that "we must have them at any price," Pugh gave up his idea of purchasing six more men when he spotted a woman and her child "offered low at $1,400." With the fear of rising slave costs and his own agricultural requirements clearly in mind, Pugh responded by meeting his labor demands prior to a further price surge in the value of prime field hands. A year later, he was back in the slave market, grumbling that prices had risen but nonetheless purchasing a dozen costly female slaves and rounding out his crews with several more men. Escalating slave costs placed an additional premium on long-term demographic stability and further encouraged natural increase over slave purchasing. As well as introducing pre- and postnatal care on the plantations to boost childbirth rates, planters also shifted the demographic balance of their slaves in favor of a more equitable mix between the sexes. Like several leading sugar magnates, Robert Ruffin Barrow of Terrebonne Parish responded to the spiraling increase in slave prices by focusing on long-term growth. In July 1857, he purchased thirty-four slaves for Residence Plantation, which comprised eight sexually balanced nuclear families. Since children and young adults constituted over 50 percent of his purchase, Barrow doubtless was pleased by his prospects and the promise of a productive return on his investment. Operating as calculating transactors in a highly developed market in human beings, Pugh and Barrow both pursued an astute economic course in light of rising slave prices by increasing their number of bondswomen and consequently strengthening the potential for natural increase among their workforce.[22]

'Peculiar Institution,'" *LHQ* 23 (July 1940): 18–25. On changing Cuban views over gender, see Moreno Fraginals, *Sugar Mill*, 142.

22. Diary for 1859–1860 (vol. 2), 5, 10 February 1859, 13–15 March 1860, 8 April 1860, Alexander Franklin Pugh Papers, LSU (all quotes); Residence Journal of R. R. Barrow, 10 July 1857, Robert Ruffin Barrow Papers, UNC.

Not all individuals, however, proved so willing to forego established gender conventions, especially when they sought to expand into the margins of the sugar belt. Even in the late 1850s, William B. Hamilton required bondsmen for clearing land and opening a sugar plantation near Red River Landing. Searching for "likely" males to meet his labor demands, Hamilton received timely counsel from his brother at Bayou Sara, a small settlement that enjoyed preferred connections with larger towns on the Mississippi. With an eager eye for seasoned or acclimated slaves, Douglas Hamilton urged his kinsman to venture beyond New Orleans and examine the Charleston slave market, as the "negroes of South Carolina generally stand our swamps remarkably well, living in a climate at home very much like ours." Cautioning his sibling to purchase "choice negroes" from Florida instead of expensive Virginia slaves, Hamilton advised against the purchase of "Northern negroes on account of sickness." Seasoned individuals, moreover, often appeared healthier; to a planter, clear eyes and greased, supple skin hinted at a bondsman's robust immune system and his suitability for the sugar country. As one Kentucky trader observed about the acclimation and seasoning of slaves to Louisiana bondage, the "longer it's put off, the harder it goes with him." "Seasoning," of course, conveniently reduced a slave's suffering and death to a single euphemism, but, like their predilection for pithy epithets such as "sound" and "choice," the slaveholders astutely gauged the bondsman's life expectancy by the state of his seasoning. Hamilton evidently agreed and counseled his brother to acquire slaves who would neither fall sick upon their arrival in Louisiana nor face the slow and hazardous acclimation process, in which new arrivals were often spared from the heat of the midday sun. Disappointed that "there are no negroes of consequence to be had in this Parish," Hamilton advised family members to purchase an intact slave community before splitting it up and selling it off. "The only plan to get hold of negroes now, without paying very high prices," he concluded, "is to buy a whole gang in some of the more eastern states," pick the strongest and those of choice, and dispose of the remainder. In fact, Hamilton proposed purchasing a gang of eighty "choice negroes" from Jefferson County, Florida. Given the scarcity and cost of slaves in New Orleans, purchasing an entire gang, selecting the youngest males, and then breaking up families—selling off the women, the elderly, and the children—made raw economic sense. For Hamilton, agricultural demands overwhelmed considerations of gender and, like many sugar masters, he plumbed for brawn over childbearing capacity. Until the late antebellum period, planters seldom concerned themselves with balanced sex ratios; when they bought women, they unblushingly gauged their reproductive potential, but rarely did they acquire sufficient numbers to create a balanced society and a base for sustained demographic growth. The relative dearth of women in the sugar country and the planters' overwhelming commitment to male labor triggered an acute demographic decrease, in which comparatively few children and low fertility rates combined to replicate the dwindling populations of the West Indies. Despite this parlous trend, planters did not entirely ignore the question of demographic increase. They tried to acquire comparatively expensive young women at the beginning of their reproductive lives, and—as much of the slaves' testimony suggests—they made shocking and invasive attempts to shape the demography and fertility of their crews.[23]

All slave women faced the double bind of labor and gender. On the one hand, they were plantation laborers; on the other hand, they served as reproductive capital. Planters viewed their slaves' productive and reproductive functions not as polarized, opposing forces, but as dualistic aspects of slave labor. The continuum between bondswomen's sexual and productive value was complicated still more by the planters' racial politics, which further reduced the status of the slave women. The slaveholders' designation of the bondswoman's body as a vehicle for sexual reproduction encapsulated the relations of subordination and patriarchal hierarchy that planters strove to impose upon their female slaves, either by rape, insult, or directly orchestrated sexual behavior. Former slave Ellen Betts recalled that some overseers sought to impress their authority over the prostate slaves by "pull[ing] up their dress and whup[ping] on their bottoms." Such crass displays carried overt sexual overtones and broadly reflected the racist misogyny of plantation slavery. Yet despite such episodes of sexualized punishments, the slaveholders' discourse masked a panoply of complex and often contradictory emotions toward their

23. William S. Hamilton to William B. Hamilton, 29 November 1858, and Douglas M. Hamilton to William S. Hamilton, 20 and 24 December 1858, 24 January 1859, William S. Hamilton Papers, LSU (first, second, fourth quotes). On the process of acclimation and seasoning, see Fett, *Working Cures,* 18–22; Olmsted, *Seaboard Slave States,* 2:303 (third quote); Donald D. Avery to Sarah Craig Marsh, 7 May 1846, Avery Family Papers, UNC. On the dearth of slave women in Louisiana, see Tadman, *Speculators and Slaves,* 68; Tadman, "Demographic Cost of Sugar," 1554; Fogel, *Without Consent or Contract,* 123–26. On parallels to Caribbean demography, also see Higman, *Slave Population and Economy in Jamaica,* 375; Richard B. Sheridan, *Doctors and Slaves: A Medical and Demographic History of Slavery in the British West Indies, 1680–1834* (Cambridge, 1985), 225–28; Richard S. Dunn, "Sugar Production and Slave Women in Jamaica," in *Cultivation and Culture,* ed. Berlin and Morgan (Charlottesville, Va., 1993), 49–72.

slaves. Even as planters objectified and commodified their bondspeople, they be-lieved them to be part of an extended family over which they and God benevo-lently ruled. Both aspects of the slaveholders' ideology emerged in the cane world. More often than not, however, cold materialism muzzled compassion.[24]

Former slave women were fully cognizant of the roles their owners ascribed to them, and they were equally sensitive to the physiological burden their masters heaped upon them. Elizabeth Ross Hite bitterly gave vent to the psychological cost of such treatment when she snapped, "All de master wanted was fo' dem wimmen to hav children." Julia Woodrich similarly recalled that her mother had fifteen children, each one fathered by a different man. "You see," Woodrich explained, "ever' time she was sold she had to take another man." Depressingly, her frequent pregnancies classified her as a "good breeder." Evidently, Woodrich's sibling re-ceived similar attention, as Woodrich noted, "I 'member how my massa usete come an' get my sister, make her take a bath an' comb her hair an' take her down in the quarter all night—den have de nerve to come aroun' de next day an' ax her how she feel." Sporting a straw hat and red shoes, Master Guitlot extracted his lord's due over this unfortunate bondswoman. The compulsion to "take another man" (and all that it implied) haunted Henrietta Butler, who remembered with loathing that "my dam ol' Missus was mean as hell . . . she made me have a baby by one of dem mens on de plantation. De ole devil!" Butler considered the

24. Botkin, ed., *Lay My Burden Down*, 125 (quote). On the duality of slave women's labor, see Deborah Gray White, "Female Slaves: Sex Roles and Status in the Antebellum Plantation South," *Journal of Family History* 8 (fall 1983): 248–61. For comparison, see Digna Castañeda, "The Female Slave in Cuba during the First Half of the Nineteenth Century," in *Engendering History*, ed. Shepherd, Brereton, and Bailey (Kingston, 1995), 142; Claire Robertson, "Africa into the Americas? Slavery and Women, the Family, and the Gender Division of Labor," in *More Than Chattel*, ed. Gaspar and Hine (Bloomington, Ind., 1996), 20–28. On rape, miscegenation, and forced sex, see Joshua D. Rothman, *Notorious in the Neighborhood: Sex and Families across the Color Line in Virginia, 1787–1861* (Chapel Hill, N.C., 2003), esp. 133–63; Thelma Jennings, "'Us Colored Women Had to Go through a Plenty': Sexual Exploitation of African-American Slave Women," *Journal of Women's History* 1 (winter 1990): 45–74; Darlene Clark Hine, "Rape and the Inner Lives of Black Women: Thoughts on the Culture of Dissemblance," *Signs* 14 (summer 1989): 912–20; also see Martha Hodes, *White Women, Black Men: Illicit Sex in the Nineteenth-Century South* (New Haven, Conn., 1997), esp. 39–67. Finally, on the planters' vision of the master-slave relationship, see Eugene D. Genovese, "'Our Fam-ily, White and Black': Family and Household in the Southern Slaveholders' World View," in *In Joy and Sorrow*, ed. Bleser (New York, 1991), 69–87. Planter paternalism is examined in chapter 5 of this study.

extent of sexual coercion, noting that "they made my Ma have babies all de time; she was sellin' the boys and keepin' the girls."⎡Frances Doby's mother plainly faced a similar experiment in wanton demographic management⎦Recalling that Master DeGruy possessed slaves for "de breedin'" and others for fieldwork, Frances added that "ma was de kind dey keeps for makin' children" and that after each birth, she worked in the house and preserved her energy for the next cycle of childbirth.[25]

The slaveholders' predilection for women like Frances's mother soon bore fruit, as teenage pregnancy characterized life in the slave quarters. As J. B. Roudanez chillingly observed, "The young masters were criminally intimate with the negro girls; it was their custom." Fourteen-year-old girls were forced into sex, and the "practice of copulation was so frequent . . . that a chaste one at seventeen years was almost unknown." While some slaves were taken as concubines, most of them were field laborers who toiled among the growing canes late into pregnancy. On Walter Brashear's sugar estate at Tiger Island, for example, female slaves bore children from an early age. Unmarried girls delivered their first infants at 16.5 years of age, while their married counterparts had their first babies well before their eighteenth birthday. By contrast, in the cotton-producing Felicianas, both married and single women bore their first children an average of three years later than on the more reproduction-focused sugar plantations. Since young women were in relatively short supply on sugar estates, as they matured they undoubtedly faced substantial sexual pressure both to marry and have early pregnancies. This pressure must have been intense, particularly for inexperienced girls, as the male majority quite possibly compelled some of them to have sex. Historian Orlando Patterson maintains that in Jamaica, where male cane cutters similarly outnumbered women, most slaves followed stable sexual norms. However, a considerable number of slave women, he argues, must have had several partners and led a more permissive sexual life. While young men undoubtedly competed for the young women's affections, planters exploited this situation, engineering early pregnancies either by compulsion or by manipulating the demographic balance to place communal pressure on early sex. The impact of this policy was appalling. As Patterson observes, families were little more than "nuclear *reproductive* units

25. Interviews with Elizabeth Ross Hite, date unknown (first quote), Julia Woodrich, 13 May 1940 (second quote), Henrietta Butler, 28 May 1940 (third quote), and Frances Doby, 6 December 1938 (fourth quote), WPA Ex-Slave Narrative Collection, LSU.

facilitated by the slave owner who wanted his slaves to breed as many children as possible."[26]

In the Louisiana cane world, reproduction among slaves mirrored Patterson's depressing conclusions. Unlike their counterparts in cotton- and tobacco-producing states, slaves in the sugar district became pregnant relatively early. Throughout the slave states, bondswomen normally delivered their first child between the ages of twenty and twenty-three, although most were physiologically capable of bearing a child two to three years earlier. That they did not suggests that some slave women delayed childbearing and foiled their masters' designs of maximizing their reproductive potential. In the sugar country, however, women proved less successful at thwarting their owners. Sexual pressure from the predominantly male community might have made abstinence difficult, and the lack of common abortifacients like cotton roots also surely complicated the lives of those who sought to delay pregnancy. But above all, slave women faced a predatory master class who attempted to manipulate their sexual lives to optimize reproduction. Overseer S. B. Raby exposed the coldly rational ethos of demographic management when he observed, "Rachel had a 'fine boy' last Sunday. Our crop of negroes will I think make up any deficiencies there may be in the cane crop." Clearly, Raby took an overly keen interest in this part of his responsibilities, for planter J. P. Bowman icily asked him to explain the birth of three mixed-race children within the past two years. "Free intercourse," the slavemaster announced, was strictly prohibited. "I allow no man that privilege with my people," he added. Raby might have been eager to make up in humans what he failed to provide in sugar, but his behavior testified to a regime committed to intrusive management and profitable outputs.[27]

If slave women refused forced sexual contact—and many plainly did—the planters had a further weapon in their arsenal: they attempted to control birth

---

26. "Testimony of a New Orleans Free Man of Color before the American Freedmen's Inquiry Commission," in *Freedom: A Documentary History of Emancipation, 1861–1867,* ser. 1, vol. 3, *The Wartime Genesis of Free Labor: The Lower South,* ed. Berlin et al. (Cambridge, 1990), 522 (first quote). On Brashear, see Malone, *Sweet Chariot,* 177; Patterson, *Rituals of Blood,* 30, 36 (second quote). For a parallel study on the slavemasters' promotion of fertility, see Wilma A. Dunaway, *The African-American Family in Slavery and Emancipation* (Cambridge, 2003), 123–25.

27. S. B. Raby to James P. Bowman, 8 May 1855, and J. P. Bowman to S. B. Raby, 19 January 1856, Turnbull-Bowman-Lyons Family Papers, LSU (quotes); Trussell and Steckel, "Age of Slaves at Menarche and Their First Birth," 504.

intervals and the extent to which African American women breast-fed their infants. Lactation can serve as a natural contraceptive; while women nurse their babies, hormonal activity suppresses ovulation. Women who feed their infants milk-replacement formula experience the return of ovulation and menses within four to ten weeks of birth. In societies where late weaning and frequent nursing prevails, women can undergo lactational amenorrhea for up to three to four years. More realistically, intense suckling can delay the resumption of ovarian activity for one to two years.[28]

European landed elites had traditionally manipulated lactational amenorrhea in order to establish large families within a relatively limited timespan. Upon birth, infants were farmed out to local wet nurses while their mothers swiftly began to ovulate, and the cycle of pregnancy and childbirth began again. This pattern of demographic management ultimately maximized the reproductive capacity of the female. Although difficult to quantify with exact precision, some sugar planters seem to have promoted wet nursing during the harvest season, when work constrained and delimited extracurricular activities. Rebecca Fletcher hinted at the presence of wet nursing when she noted that slave women had no respite during the working day to feed their infants, who were cared for by "ole women" instead. Older women, however, were not alone in nursing the young; as Elizabeth Hines recalled, "My sister suckled me on her breast." Hines's mother died when Elizabeth was three years old, suggesting that master S. G. Laycock had other plans for a woman that her daughter described as "fine in children." Aptly concluding her mother's life story with due reference to her fertility, Hines added, "They buys women like that you know." Laycock's breast-feeding policy ensured that Elizabeth's mother soon would be ovulating again, like other women who either left their children with wet nurses or who had little time for suckling their young. Ceceil George emphasized the inadequate time accorded field hands for maternal care when she observed that most women resumed their field duties a mere nine days after giving birth, either leaving their infants with plantation nurses or rushing back to their quarters to provide occasional suckling. On Trinity Plantation, where all Pierre Landreaux wanted "was fo' dem wimmen to hav children," Elizabeth Ross Hite noted that two nurses "tak care of de children" at

28. The literature on lactation is extensive, but see R. V. Short, "Lactation—The Central Control of Reproduction," in *Breast-Feeding and the Mother,* Ciba Foundation Symposium 45 (Amsterdam, 1976), 73–86; A. S. McNeilly, "Breastfeeding and Fertility," in *Biomedical and Demographic Determinants of Reproduction,* ed. Gray, Leridon, and Spira (Oxford, 1993), 391–412.

the plantation hospital. Ruefully observing that "dere mudders did not hav time to tak care of dem," Hite quite possibly watched the raw impact of demographic engineering. Equally aware of the planters' attempts at population management, Frances Doby observed that older women who had passed their physical and reproductive prime cared for and fed the infants and children, so that when their mothers returned from a day's labors, "all dey got to do is push dem in de bed." Separated from the duties of child-rearing, these women returned to their field work, and with the onset of menses, they soon fell pregnant again.[29]

While not all plantation owners and their wives utilized wet nurses for their slaves, most white southerners shared a basic understanding of lactational amenorrhea and appreciated the reproductive implications of early weaning. The agricultural reformer and slave-management adviser Thomas Affleck, for instance, recommended that slave women breast-feed their infants for nine months. Most bondswomen ignored such recommendations and either breast-fed for a longer period or adopted other methods for delaying childbirth. In the Caribbean, estate managers recognized the predilection of slave women for retaining the African tradition of prolonged lactation, and on occasion they provided incentives to West Indian bondswomen to curtail breast-feeding and boost childbirth potential. In Louisiana, the clearest indication of demographic and lactational management lies in birth-spacing data and in the age of women at their first and last births. Research on the Tiger Island slave community suggests that on average slave women on this Bayou Teche estate carried seven babies to term during a reproductive career that began in their mid-teens and drew to a close by their thirty-fifth birthday. While the number of miscarriages, abortions, stillbirths, and non-recorded infant deaths defies quantification, most women at Tiger Island probably delivered a child every twenty-four months. This suggests that female slaves in the sugar

29. Interview with Rebecca Fletcher, 2 July 1940–24 September 1940, WPA Ex-Slave Narrative Collection, LSU (first quote); Interview with Elizabeth Hines, WPA Slave Narrative Project, Arkansas Narratives, vol. 2, pt. 3, Federal Writers' Project, U.S. Work Projects Administration (USWPA), Manuscript Division, Library of Congress. "Born in Slavery: Slave Narratives from the Federal Writers' Project, 1936–1938," www.memory.loc.gov/ammem/snhtml/snhome.html (second quote; accessed 15 April 2004); Interviews with Ceceil George, 15 February 1940, Elizabeth Ross Hite, date unknown (third quote), and Frances Doby, 6 December 1938 (fourth quote), WPA Ex-Slave Narrative Collection, LSU; Botkin, ed., *Lay My Burden Down*, 126. On breast-feeding and birth-spacing, see Valerie Fildes, "Historical Changes in Patterns of Breast Feeding," in *Natural Human Fertility,* ed. Diggory, Potts, and Teper (Basingstoke, 1988), 119–29; Shyam Thapa, Roger V. Short, and Malcolm Potts, "Breast Feeding, Birth Spacing, and Their Effects on Child Survival," *Nature* 335 (October 1988): 679–82.

fields experienced a birth interval that was nine to ten months shorter than for bondswomen in the cotton country. Birth intervals on other sugar estates further indicate that women delivered infants every twenty-five months and that a period of sixteen months elapsed between delivery and successful conception of the next baby. This phase was significantly foreshortened among mothers who lost their child in labor or soon after delivery. Without the contraceptive effect of breast-feeding, these women soon became pregnant again, delivering another infant approximately one year after the death of their offspring. Betsey, like many of the slave women working on William T. Palfrey's St. Mary and St. Martin Parish estates, experienced a busy reproductive life. It began in 1843 with the birth of her son Paul and ended in October 1861 when she delivered Nancy, an unfortunate child who died before the passage of the Thirteenth Amendment. During her eighteen-year reproductive career Betsey was pregnant ten times, though only five of her offspring survived beyond 1865. In most cases, two years usually elapsed between her childbirths, but in cases of the premature death of her offspring, Betsey ceased lactating and delivered another infant within the year.[30]

The relatively short birth intervals for slaves in Louisiana suggest that planters followed Affleck's strictures and sought to maximize the reproductive potential of their bondswomen by limiting suckling—or else that the high rates of infant death and the destructive effect of the sugar industry impaired breast-feeding. Given the sixteen-month birth-to-conception interval, it appears highly probable that slaves recommenced ovulation significantly earlier than their counterparts in the Caribbean. Research conducted into cotton-producing plantations reveals that slave women in the cotton region experienced birth intervals of thirty-four

30. Data on Tiger Island from Malone, *Sweet Chariot*, 176–77; Register of Births among W. J. Palfrey's Negroes, Commenced August 1843, Palfrey Family Papers, LSU; "A Memorandum of the Births of Negro Children," Record Book, 1817–1852, Joseph Kleinpeter and Family Papers, LSU. On weaning and breast-feeding, see Sally G. McMillen, "Mother's Sacred Duty: Breast-Feeding Patterns among Middle- and Upper-Class Women in the Antebellum South," *JSH* 52 (August 1985): 333–56. Thomas Affleck, "On the Hygiene of Cotton Plantations and the Management of Negro Slaves," *Southern Medical Reports* 2 (1851): 435, quoted in Richard H. Steckel, "A Dreadful Childhood: The Excess Mortality of American Slaves," *Social Science History* 10 (winter 1986): 449. On contemporary Caribbean practices, see Herbert S. Klein and Stanley L. Engerman, "Fertility Differentials between Slaves in the United States and the British West Indies: A Note on Lactation Practices and Their Possible Implications," *William and Mary Quarterly* 35 (April 1978): 370–74; Sheridan, *Doctors and Slaves*, 228–30; Bernard Moitt, "Slave Women and Resistance in the French Caribbean," in *More Than Chattel*, ed. Gaspar and Hine (Bloomington, Ind., 1996), 252–53.

months; in other words, approximately two years elapsed between birth and the next successful conception. Slaves in the sugar country accordingly marched to a different reproductive beat than women in both North and South. On average, they were pregnant more often than women in other slaveholding societies or free women in rural, agricultural communities with little experience of birth control.[31]

Although some planters might have exercised compulsion to coerce slave breeding, most did not systematically adopt alternative lactational practices beyond maternal nursing nor were they especially generous in pre- or postnatal care. Slaveholders valued their enslaved offspring and attempted to maximize the reproductive potential of their predominantly male estates, but they had to weigh their immediate labor requirements against lower workloads during the final trimester or during postpartum recuperation. Most sugar masters favored labor over leave, keeping their pregnant women in the fields and offering little respite from the sugar order. During the latter stages of pregnancy, women were occasionally assigned lighter duties—such as hospital work or maintenance of the slave quarters—but during the main cultivation and harvesting seasons, they did not receive a substantial reduction in their workloads until the final trimester. Even if slave women appealed to their owners to have their load reduced by one-quarter or even one-half, they still faced a burdensome workload and inadequate diet. Predictably, overseer S. B. Raby carefully monitored his pregnant slaves and ordered them to clear ditches, scrape the canes, or toil alongside the "sucklers" in the sugarhouse. While grimly attentive masters like Raby and Bowman reduced workloads for expectant slaves, most overseers and estate managers provided scant prenatal care and pushed ahead with their production-focused agendas. Masters frequently gave slave mothers a short break after childbirth, though it often proved woefully inadequate. Upon visiting one particularly "energetic and humane" planter's estate, a correspondent observed that slave women received two hours to be with their children at noon and that they left work an hour before other field hands to attend to their babies.

31. On longer birth intervals among slave women, see Joan Ezner Gundersen, "The Double Bonds of Race and Sex: Black and White Women in a Colonial Virginia Parish," *JSH* 52 (August 1986): 361–62. For white women, see Mary P. Ryan, *Cradle of the Middle Class: The Family in Oneida County, New York, 1790–1865* (Cambridge, 1981), 155–56; Tamara K. Hareven and Maris A. Vinovskis, "Introduction," to *Family and Population in Nineteenth-Century America*, ed. Hareven and Vinovskis (Princeton, N.J., 1978), 3–21.

William Howard Russell similarly noted that mothers left their newborns with an elderly woman and traveled from the cane fields to the nursery at scheduled times to breast-feed.[32]

While important on these estates, postnatal care proved sporadic. Although some slaveholders adopted maternal leave, others provided slave women with a parsimonious half an hour three times a day to suckle their young. Undoubtedly, such a modest allotment triggered an early return to normal luteal function and ovulation. Other planters, however, proved slightly more generous. Mistress Tryphena Holder Fox, for instance, gave her slaves a month off after parturition, but she predictably laced the allocation of leave with the sugar masters' penchant for work. Sharing no ground with her chattel, Fox reinforced her authority, warning her bondswoman to "work in downright good earnest" as soon as her period of postpartum convalescence ended. Fox followed legal precedent in allowing a month of recovery after childbirth; as former slaves testified, however, such leave was frequently honored more in principle than in practice. Free black J. B. Roudanez, for instance, observed that some pregnant women worked until delivery and that others returned to their duties within a week of confinement. Although few slaves testified to women delivering in the fields, the reminiscences of Edward De Buiew confirm the sugar masters' erratic approach to pregnancy and childcare. "My ma died 'bout three hours after I was born," the former bondsman recounted. "Pa always said they made my ma work too hard." De Buiew elaborated, "He said ma was hoein'. She told the driver she was sick; he told her to just hoe-right-on. Soon, I was born, and my ma die[d] a few minutes after dey brung her to the house." While some women, like De Buiew's mother, probably died from excess work and inadequate prenatal care, slaveholders in the sugar country

---

32. For examples in Louisiana of labor being reduced during confinement, see William Dosite Postell, *The Health of Slaves on Southern Plantations* (Baton Rouge, La., 1951), 113–14; Sitterson, "Magnolia Plantation," 199; Frogmoor Plantation Diary, 7 and 27 May, 24 July 1857, Turnbull-Bowman-Lyons Family Papers, LSU; Olmsted, *Seaboard Slave States*, 2:314 (quote); Russell, *My Diary North and South*, 263. On the way some southern masters reduced the working load of "lusty women in the family way," see John Campbell, "Work, Pregnancy, and Infant Mortality among Southern Slaves," *JIH* 14 (spring 1984): 801–2; Leslie A. Schwalm, *A Hard Fight for We: Women's Transition from Slavery to Freedom in South Carolina* (Urbana, Ill., 1997), 19–20, 42–44. From a more profit-minded perspective, Michael Johnson contends that planters "found it easier to ignore the risks of hard work for pregnant slave women" than to overlook the "promise of a cash crop safely harvested." Michael P. Johnson, "Smothered Slave Infants: Were Slave Mothers at Fault?" *JSH* 47 (November 1981): 519.

more often tried to balance the demands of cane production against the slave-woman's health. Tragically, sugar often won out.[33]

Whether planters paid greater or lesser attention to the condition of their pregnant bondswomen, they surely valued their chattel's offspring. The brutal whipping of a recalcitrant slave would flay the pregnant bondswoman's back but keep the unborn baby relatively protected. As Rebecca Fletcher observed, "They wanted slaves to have babies bekase they wuz valuable," so to protect their investment while simultaneously inscribing their authority, slaveholders dug a hole in the ground and ordered the pregnant woman to lie face down over it before applying the lash. Carlyle Stewart also charged that planters protected the unborn fetus during whipping "'cause the children were worth money." Affixing value to their unborn human property and the bondswoman's capacity to incubate the next generation of sugar workers, planters held reproductive fitness and sexual coercion to be key aspects of successful plantation management.[34]

Despite the enormous imposition on their dignity, slave women visibly reclaimed their own bodies and challenged the slaveholder's definition of them as reproductive machines. Through gynecological resistance, black women defied white constructions of their sexuality and wove contraception through the dark fabric of slave oppositional culture. Whether swallowing abortifacients such as calomel and turpentine or chewing on natural contraceptives like cotton roots, slaves ingested homemade remedies that provoked miscarriage. The extent to which slaves exercised this form of reproductive resistance is almost impossible to quantify, though the short birth intervals in the sugar country suggest that women who aborted their infants soon became pregnant again. Those women who resisted the planters' untrammeled sexual exploitation were probably lionized by fellow bondswomen as gritty survivalists, though few of them were able to

33. The evidence from the manuscript record suggests a patchy record on leave, though it is probable that the month-long postparturition break accounts for the relatively low post-neonatal mortality rate of 6.6 percent. Steckel, "Dreadful Childhood," 447. On Tryphena Holder Fox, see Wilma King, "The Mistress and Her Maids: White and Black Women in a Louisiana Household, 1858–1868," in *Discovering the Women in Slavery,* ed. Morton (Athens, Ga., 1996), 89 (first quote); "Testimony of a New Orleans Free Man," in *Freedom,* ser. 1, vol. 3: *The Wartime Genesis of Free Labor: The Lower South,* ed. Berlin et al., 522 (second quote); Interview with Edward De Buiew, 10 June 1940, WPA Ex-Slave Narrative Collection, LSU (third quote).

34. Interviews with Rebecca Fletcher, 2 July 1940–24 September 1940 (first quote), and Carlyle Stewart, 3 May 1940 (second quote), WPA Ex-Slave Narrative Collection, LSU.

control their fertility over long periods. Planters occasionally complained that slave mothers committed infanticide by smothering their babies, although it appears more likely that estate managers misdefined "smothering" for crib death or Sudden Infant Death Syndrome. At Variety Plantation, for example, the slaves Rachel and Marie allegedly smothered their children. Rachel's case is especially noteworthy, for during her reproductive life of thirteen years she apparently smothered two of her children. The first, Miles, was born in late September 1838, during a particularly early sugar-making season, before being smothered on December 1. Joseph Kleinpeter recorded in April 1845 that Rachel had again smothered her six-week-old baby. Whether these infants actually died of crib death or not is unknown. The tantalizing question remains of whether Rachel orchestrated these deaths as a rebellious symbol to her master or whether she chose to liberate her children from the murderous sugar regime by killing her babies. Although some women undoubtedly aborted their babies, consumed folk contraceptives, and even killed their newborns, the majority in all probability redoubled the potent attachment that slave women evidently had for their children.[35]

The evaluation of unborn infants and the rude calculation of the bondswoman's reproductive potential tragically defined slave trading and slave breeding in the sugar country. If the planters did not quite run stock farms, they surely engineered the demography of the plantation belt in order to maximize reproductive activity. As Ellen Betts chillingly observed, they derived considerable pride from "them black slick children" on their estates. "Marse's" pride, however, was insatiable; to stock his estate with another generation of sleek sugar workers, he purchased young women in the prime of their sexual lives and perhaps encouraged or bullied them to bear children more frequently than they either sought or wanted. Whether planters strove to be self-sufficient in their labor supply or not, some of

35. On contraception, see Liese M. Perrin, "Resisting Reproduction: Reconsidering Slave Contraception in the Old South," *Journal of American Studies* 35 (2001): 255–74; Fett, *Working Cures*, 176–77; Loretta J. Ross, "African-American Women and Abortion: A Neglected History," *Journal of Health Care for the Poor and Underserved* 3 (fall 1992): 274–84; Barbara Bush, "Hard Labor: Women, Childbirth, and Resistance in British Caribbean Slave Societies," in *More Than Chattel*, ed. Gaspar and Hine (Bloomington, Ind., 1996), 193–217. For examples of smothering, see Memorandum of the Births of Negro Children, Record Book, 1817–1852, Joseph Kleinpeter and Family Papers, LSU; Johnson, "Smothered Slave Infants," 495–520. For an example of infanticide where the newborn child was left to die in "de ole toilet hole," see Interview with Mariette Smith, 4 June 1940, WPA Ex-Slave Narrative Collection, LSU. It is not clear, however, whether this act was conducted under slavery, though it hints at established practices.

them undoubtedly took pride in their power and authority over the slave infants. Planters could preen their egos and magnify their self-worth by possessing the tallest, strongest bondsmen in the parish or by parading their pretty teenage girls to dewy-eyed visitors. Once their guests' backs were turned, however, planters could return to more mundane tasks, even as they coveted their young and attractive females with sickening glee.[36]

Yet however much they practiced gender- and age-selective purchasing and labeled their slaves with sexualized epithets, the sugar masters' coercive efforts at slave breeding failed. To be sure, they created conditions to favor population growth, but their overwhelming commitment to male labor and the punishing sugar regime undercut their cold-blooded approach to plantation demography. It is a tragic though inescapable fact that agro-industrial sugar production impinged on the slave woman's life in multiple ways. Extreme work and undernutrition taxed the bondswoman throughout most of the year, while disease and death eroded the adult slave population still further. Despite importing thousands of slaves a year to the sugar region, Louisiana was unique among the slave states in having a natural decrease in its population. While the slave population grew swiftly in the rest of the South, in Louisiana's cane world the natural growth rate may have been as low as 6 or 7 percent—a figure four times smaller than the regional average. Louisiana's appalling demographic record derived from the specific labor requirements of sugar production, the dearth of women, the virulence of disease, and high infant mortality rates. Slave women in particular suffered under the slaveholder's regime. The heavy physical labor and inadequate nutrition (particularly a protein-deficient diet) of the cane world led to abnormally low levels of conception, ovarian dysfunction, depressed libido, miscarriages, and physiological harm to the slave woman's capacity to conceive and bear children. Malnourished and fatigued by their labors, slave women gave birth to underweight babies and produced inadequate quantities of nutritionally inferior milk. Small wonder that slave infants died so frequently, often after suffering from typical signs of malnutrition. At first appearance, these infants looked fat, slick, and healthy to their owners, but their lack of protein calories mde them highly susceptible to a series of lethal infections that preyed on the nutritionally deprived. Upon weaning, the combination of carbohydrate-rich foods and the high frequency of lactose intolerance among African Americans conspired to create a mineral-deficient diet that

36. Botkin, ed., *Lay My Burden Down*, 125.

proved nutritionally disastrous for slave children. Maladies and disease flourished, leaving a crippling legacy of infant mortality. On some plantations, over 55 percent of children died in the course of their earliest years.[37]

Ultimately, the demands of the regional sugar economy overrode every other consideration. Although planters sought demographically to engineer their slave populations, their efforts were hamstrung by their total commitment to staple agriculture and their overreliance on men. The complete failure of the planters' moribund attempts to coax population growth is revealed in some startling statistics on relative fertility across the slave states. Throughout the South, the ratio between slave children aged 0 to 9 years and slave women aged 15 to 49 was 1,320 children to every 1,000 women. In Louisiana's cane world, however, this figure slumped to 922 children per 1,000 women on the eve of the Civil War. Decades of selecting young women and employing other sordid tactics to beget children had failed. Those women who survived the brutalities of the sugar regime faced incessant labor, tropical disease, and frequently poor health. On Oaklands Plantation, for instance, sugar wove its familiar and fatal signature in the terms "sickly," "unhealthy," and "ruptured," which were scattered throughout the slave inventory on this estate and dozens like it. These inexact epithets masked a host of maladies, from agonizingly painful hip arthritis to spinal hernias, especially among older slaves. Slave women, moreover, quite possibly suffered from anemia and fell prey to a host of maladies and infections that flourished in the swampy sugar lands. Exhausted by a punishing labor regime, taxed by malaria, sickened by cholera, aggravated by intestinal worms, and struck down by summer heat, yellow fever, and diarrhea-inducing infections, the slave population was ill-prepared to meet the slaveholders' demands. The grueling rigors of the sugar industry placed an un-

37. For a detailed discussion of the slaves' work culture and fecundity, see Richard Follett, "Heat, Sex, and Sugar: Pregnancy and Childbearing in the Slave Quarters," *Journal of Family History* 28 (October 2003): 510–39. The literature on nutrition, workload, and ovarian function is extensive. See esp. G. Jasieńska and P. T. Ellison, "Physical Work Causes Suppression of Ovarian Function in Women," *Proceedings of the Royal Society of London*, ser. B, 265 (1998): 1847–51. On maternal nutrition, see N. Tafari, R. L. Naeye, and A. Gobezie, "Effects of Maternal Undernutrition and Heavy Physical Work during Pregnancy on Birth Weight," *British Journal of Obstetrics and Gynecology* 87 (1980): 222–26; Richard H. Steckel, "Birth Weights and Infant Mortality among American Slaves," *Explorations in Economic History* 23 (April 1986): 182; Steckel, "Dreadful Childhood," 447; Ashland Plantation Book, Appendix K, LSU. On infections precipitated by nutritional deficiency, see Kenneth F. Kiple and Virginia H. Kiple, "Slave Child Mortality: Some Nutritional Answers to a Perennial Puzzle," *Journal of Social History* 10 (1977): 288–99.

fathomable strain on the human body. Even if the planters replaced their dead with likely men and young, fertile women, they could not stem their dwindling slave populations.[38]

Despite these horrific conditions and limitations, slaves in the cane world reconstituted their communities through familial and kin networks that provided stability and emotional support, in addition to protecting their integrity from the ravages of the slaveholders' order. As in other parts of the South, enslaved men and women resisted the plantocracy's attempts at controlling their public and private lives and forged relatively adaptable family and domestic structures to cope with family separation, death, and inestimable suffering. Although the nuclear family unit certainly flourished in the slave quarters of the lower Mississippi Valley, the very large numbers of unmarried men, orphaned children, and widowed adults in the cane world also led to the emergence of more flexible familial structures, based more often on consanguine (extended) rather than conjugal (nuclear) ties. Extended networks of both blood and fictive kin ensured cultural stability, provided a vital arena for love and sustenance, and provided an established structure into which new arrivals could be integrated.[39]

family stability

On the Petite Anse estate, family structure adapted to the peculiarities of age and gender and shifted markedly as the demographic profile of the slave community became more stable. Although nuclear families of parents and children initially dominated, they were increasingly replaced by multiple-family households,

38. On Louisiana's fertility, see Tadman, *Speculators and Slaves*, 69; Inventory and Valuation of Slaves, Stock, and Farming Utensils of Oaklands Plantation, 1859, Samuel D. McCutchon Papers, LSU (quote). On adult mortality, see Jennifer Olsen Kelley and J. Lawrence Angel, "Life Stresses of Slavery," *American Journal of Physical Anthropology* 74 (1987): 199–211; Katherine Bankole, *Slavery and Medicine: Enslavement and Medical Practices in Antebellum Louisiana* (New York, 1998), 116–17.

39. On family formation in Louisiana, see Malone, *Sweet Chariot*, 7–8, 144–49. Key reference works on the slave family in the antebellum American South include: Herbert G. Gutman, *The Black Family in Slavery and Freedom, 1750–1925* (New York, 1976); Larry E. Hudson Jr., *To Have and To Hold; Slave Work and Family Life in Antebellum South Carolina* (Athens, Ga., 1997), 165–76; John W. Blassingame, *The Slave Community: Plantation Life in the Antebellum South*, rev. ed. (New York, 1979); Charles Joyner, *Down by the Riverside: A South Carolina Slave Community* (Urbana, Ill., 1984). Works that underscore the planters' ability to destroy family stability include Brenda E. Stevenson, "Distress and Discord in Virginia Slave Families, 1830–1860," in *In Joy and Sorrow*, ed. Bleser (New York, 1991), 103–24. Also see Stevenson's longer *Life in Black and White*. Cheryll Ann Cody underscores the power of forced separation, but she also highlights efforts among slave women to preserve a semblance of family life, despite sale and subdivision, in "Sale and Separation: Four Crises for Enslaved Women on the Ball Plantations, 1764–1864," in *Working toward Freedom*, ed. Hudson (Rochester, N.Y. 1994), 119–42.

which were comprised of a nuclear family plus an unmarried daughter and grandchild forming a second conjugal unit. These extended family structures and household forms enabled most slaves and newcomers to be swiftly integrated into broader familial and kinship networks. Life on Tiger Island, however, proved much less stable. Planter Walter Brashear's predilection for young men ensured that the mean age of all his slaves in 1842 was just nineteen—twenty-two for the males and, predictably, sixteen for the females. Tiger Island's slave community embodied the productive and reproductive designs of the sugar masters, but it was in no sense a perfect breeding ground for stable nuclear family life. Given that men outnumbered women by almost two to one on the estate, some individuals inevitably went without formalized spouses, while some women undoubtedly faced sexual pressure from a number of potential partners. Indeed, on occasion a woman might divorce her husband and change partners. Despite these overlapping and potentially explosive pressures the African American family and kinship structure proved resilient. Slaves on the Brashear plantation eventually integrated almost 50 percent of their population into nuclear families. A variety of other complex family structures emerged during the 1840s and 1850s, the aim of which was to incorporate young single men into broader kinship networks. As a result, the number of single men living alone dropped by almost half from 1842 to 1860. In spite of the warped demographics of Tiger Island, the African American community had proceeded toward a more stable and family-centered society. On this plantation and others like it, the reconstituion of family life cushioned slaves from the savage excesses of the sugar masters' order and underscored the extent to which African Americans struggled to wrest some stability and decency from the violence of plantation slavery.[40]

Unquestionably, life in the sugar country remained profoundly compromised by the slave regime and the masters' overwhelming power. The invasive power of sugar farming eroded communal stability and left an enduring legacy of low fertility, demographic imbalance, intrusive sexual politics, and human suffering. Yet the slaveholders' penchant for youth and brawn ultimately optimized labor productivity and provided a degree of flexibility during the harvest season. Planters who boasted a very large slave force with a high man-to-land ratio evidently possessed

40. Malone, *Sweet Chariot*, 152, 154, 165. The presence of "divorce" proceedings on Minor's estates suggests a flexibility within slave marriage that allowed women to swap partners with relative ease and one month's notice. See J. Carlyle Sitterson, "The William J. Minor Plantations: A Study in Antebellum Absentee Ownership," *JSH* 9 (February 1943), 69.

a singular advantage, but few save the very largest could afford to maintain adequate crews to grind the crop at double-quick speed. As Theodore Weld observed, planters required twice the number of workers during the harvest than they did in other parts of the year, but they were unwilling to purchase and augment the slave crews for this season alone. Weld indicated that by driving their slaves day and night throughout the grinding season, planters could harvest their crops with the slaves they had on hand. Weld's views, although surely propagating the antislavery cause, proved often accurate, as planters visibly pushed their crews to work at a feverish pace for long and exhausting hours. When compulsion and the lash failed, estate managers turned to a vibrant and active slave rental market to augment their work crews for the harvest or to conduct particularly dangerous labor. Those slaveholders who combined rational calculation of labor costs with the mantle of self-gratifying paternalism found slave rental to be a suitable vehicle to soothe their consciences against extreme work or to pamper their religious sensibilities and still enable the rolling season to advance seamlessly. While self-constructed benevolence inured itself to slave rental, profit maximizers undoubtedly realized that extra hands would increase workfloor speed and ensure that every cane was cut before the autumn frost. Hiring extra hands for the sugar rolling was, W. H. White bluntly observed, "money well invested." By adding hired workers for their crews, planters could also ensure that their principal workers remained alert for the entire rolling season. Rested men would be likely to exert themselves more efficiently than a tired workforce, and rented labor could be pressed or exhausted at little long-term cost to the planter. The introduction of hired labor, moreover, enabled planters and estate managers to harvest their crop without recourse to a back-breaking and potentially destabilizing work schedule. Fearing that "it will be impossible to drive the hands through a long sugar rolling," White urged the acquisition of temporary hired labor to evade potential conflict with his crews, whose resistance to hard driving would prove infinitely more costly than a few extra hands. Slave rental thus emerged as an antidote to the region's labor problem. Whether adopted to boost plantation productivity, to protect valuable human property, or even to maintain an even current of workplace stability, it resolved labor shortfalls and effectively meshed within the exploitative structures of agro-industrial sugar production.[41]

41. Theodore Weld, *Slavery As It Is: Testimony of a Thousand Witnesses* (New York, 1839), 38; W. H. White to John Moore, 25 August 1858, David Weeks and Family Papers, LSU (quotes). On the practice of slave hiring, see Miranda C. Twiss, "Slave Hiring in Louisiana, 1719–1861"

Solomon Northup, a hired slave who journeyed to St. Mary Parish to cut cane and work in William Turner's sugarhouse, remembered that almost all plantations required one or more additional slaves. Following a disappointing cotton harvest, he recounted, local Avoyelles planters dispatched 147 laborers to the Attakapas district, where "wages were high and laborers in great demand" on the south Louisiana sugar fields. Northup recalled that the traffic in rented hands proved so heavy that the squad thinned as it advanced the 140 miles southward under the vigilant eye of overseers and planters, who rented laborers to sugar masters along the way. With a cane knife thrust into his hands, Northup joined Turner's field hands before being promoted to driver status, where his primary responsibilities extended to work discipline and coercing slave and hired hands to toil ever faster. As New Orleans sugar factor Martin Gordon emphasized, slave rental circumvented short-term labor scarcities that occasioned plantation underperformance. Meditating on the largest problem Benjamin Tureaud faced on Brulé Plantation, Gordon concluded that Tureaud's apparent dearth of labor hampered efficient time and labor management. "It is a pity," he noted, "that you have not hands sufficient to carry on all operations at once. . . . However, you must attend to the most important part—viz., the making of the sugar."[42]

To supply their plantations with adequate labor, the sugar masters hired extra hands either from neighboring cotton planters (who by late October and November could afford to minimize their slave crews) or from small local slaveholders who wished to profit from their larger peers' labor shortfall by renting

---

(M.A. thesis, University of London, 2002). Andrew W. Foshee argues against a widespread slave rental market in "Slave Hiring in Rural Louisiana," *LH* 26 (winter 1985): 63–73. On the efficacy of slave hiring elsewhere, see Jonathan D. Martin, "Divided Mastery: Slave Hiring in the Colonial and Antebellum South" (Ph.D. diss., New York University, 2000); Randolph B. Campbell, "Slave Hiring in Texas," *AHR* 93 (February 1988): 107–14; Keith C. Barton, "'Good Cooks and Washers': Slave Hiring, Domestic Labor, and the Market in Bourbon County, Kentucky," *JAH* 84 (September 1997): 437–48. For an interesting alternative to hiring, planters or agents could steal slaves to work in the sugar country; one prominent example of this is the New Yorker Solomon Northup, enslaved in Louisiana. On theft of slaves and free people of color, see Schafer, *Slavery, the Civil Law, and the Supreme Court of Louisiana*, 90–96; Schafer, *Becoming Free, Remaining Free*, 115–28.

42. Northup, *Twelve Years a Slave*, 145 (first quote); Martin Gordon Jr. to Benjamin Tureaud, 3 November 1849, Benjamin Tureaud Family Papers, LSU (second quote).

out their slave crews. Not infrequently, the need for extra slave labor during the grinding season preoccupied the planters during October and November as they feverishly prepared land, labor, and machinery for the impending agricultural onslaught. In a rather desperate letter to Attakapas sugar magnate David Weeks, Frederick Conrad beseeched his brother-in-law to supply him temporarily with extra slaves. These additional hands, Conrad implored, "will give me a great lift in my troubles." Turning to his sister, Rachel O'Connor, a modest cotton planter in West Feliciana Parish, Weeks finally hired five young men to assist Conrad in his ailing sugar operations. Conrad's request was not unusual. Planters with connections in the cotton industry frequently moved their slaves from cotton to sugar cultivation during the November and December grinding season after completing all cotton picking on their more northerly estates. Dual interest planters like Dr. William Webb Wilkins of Wilton Plantation, owner of a sugar estate in St. James Parish and a cotton plantation over a hundred miles to the north, possessed the unique capacity to move laborers from one agricultural pursuit to another with comparative ease. In 1847, Wilkins transferred a small number of slaves from his estate in north Louisiana downstream to help during harvest time and to bolster his sugar crews. Wilkins also turned to the local slave rental market, hiring a number of hands from his neighbor Octave Colomb to further augment his labor force. Patrick Keary similarly shifted his slaves between diverse agricultural pursuits, relocating twenty slaves from cotton production on Ben Lomand Plantation to increase the sugar yield at Catalpa Grove during the 1852 rolling season. Keary's decision to expand the slave crews at Catalpa Grove proved successful, as his estate produced over five hundred hogsheads of sugar in 1852, a figure representing over 10 percent of all sugar produced in Avoyelles Parish that year. Local slave hiring proved equally advantageous for William T. Palfrey, a part-owner of Ricahoc Plantation on Bayou Teche. Facing labor shortages prior to the 1835 harvest, Palfrey hired widely in the Attakapas slave rental market to expand his meager crews. He also elected to maintain the traditional Sabbath observance by preserving his slaves' customary vacation on Sundays. Whether Palfrey was forced to pursue this course by a vibrant slave agency, whether he chose to cede Sundays to his slaves from honest religious conviction, or whether he had a cloying desire to appear benevolent to his friends, neighbors, and self remains unclear. As he well knew, such an indulgence during the grinding season could easily prove fatal. To maximize production and yet retain the Sunday break, Palfrey turned directly to the slave rental market, where he engaged the services of twelve slave boys.[43]

With short contracts and the sugarcane to cut, conditions for hired slaves proved rough, demanding, and extremely trying during the grinding season. On Andrew Crane's plantation in St. James Parish, circumstances proved so grim that the hired slaves protested to their owner, Euphiman Hebert, that Crane provided scant food for them. Astutely protecting his own slave investments—and, significantly, placing African American testimony above that of a fellow sugar master—Hebert complained to Crane that "they would go anywhere before they would [go] to you." Unwilling to tolerate Crane's miserly approach to slave management for a second year, Hebert agreed to hire the slaves "according to their wishings." In addition to the bewildering costs of sugar production, hired labor frequently required a hefty outlay of cash and capital merely to resolve or insure against immediate workfloor needs. During the 1850s, Louisiana's railroad con- struction projects paid unskilled laborers $26 to $30 and carpenters $50 to $60 per month. Supervisors received significantly more than plantation overseers at about $100 per month. Increasingly, these wages placed white labor firmly beyond the planter's means. But the limits to white hiring extended well beyond relative wage rates into the tormented ground of racialized work patterns. In Louisiana, this proved exceptionally important, as planters—albeit unsuccessfully—sought to remove all symbols of white indolence and to separate their slaves from the "excessively apathetic, sleepy" residents along the river. Eager to squash this object lesson in free labor and shelter their hands from dangerous examples of indolent whites, planters racialized sugar production and fused an iron bond of prejudice that linked sugar and slavery deeply within the white conscience. In contrast to Cuba, where planters imported free contract laborers to toil alongside slaves in the cane economy, Louisiana slaveholders actively engraved a firm racial hierarchy between definably black sugar work and white labor. Occasionally, labor-hungry planters in isolated portions of the state utilized Creole or Cajun

43. F. A. Conrad to David Weeks, 9 October 1833, David Weeks and Family Papers, LSU (quote); Allie Bayne Windham Webb, ed., *Mistress of Evergreen Plantation: Rachel O'Connor's Legacy of Letters, 1823–1845* (Albany, N.Y., 1983), 122; Avery O. Craven, *Rachel of Old Louisiana* (Baton Rouge, La., 1975), 59; Cash Book, 1847–1851 (vol. 2), 5 and 10 December 1847, Bruce, Seddon, and Wilkins Plantation Records, LSU; Patrick F. Keary to Juan y de Egana, 30 November 1852, Patrick F. Keary Letters, LSU; P. A. Champomier, *Statement of the Sugar Crop, 1852–1853*, 3; Journal, 1834–1839 (vol. 1), 10 January 1836, Palfrey Family Papers, LSU.

swampers during the rolling season, but these occurrences were generally viewed with disfavor.[44]

Only the Irish sank as low on the sliding scale of Anglo-American condescension. Savagely exploited by English landlords and the British state in their homeland, Irish immigrants faced treacherous employment as ditchers and delvers for Louisiana's sugar lords. On Shady Grove Plantation, Isaac Erwin employed five Irishmen to dig the sugarhouse pond, which stored a ready supply of steam-engine water for the impending grinding season. Working alongside Erwin's slaves, the Irish laborers remained onsite until the harvest began in earnest, whereupon the sugar master reverted to an all-slave work crew, which he augmented in late November with an extra twenty-two hands. Other Irishmen toiled on plantation canals, drained the backswamp, cleared bayous, toiled on the levee, and dug estate ditches at a rate of approximately $15 per acre. Some Irish laborers worked for local contractors, who hired out teams of immigrants to conduct laborious and hazardous plantation maintenance knee-deep in disease-infested waters. Agents unscrupulously charged elevated prices for this type of work, but, as John Burnside's overseer observed, "It was much cheaper to have Irish to do it, who cost nothing to the planter, if they died, than to use-up good field hands in such severe employment." With the exception of Andrew Durnford—an African American sugar

44. Euphiman Hebert to Andrew E. Crane, 6 October 1858, Andrew E. Crane Papers, LSU (first quote); Olmsted, *Seaboard Slave States*, 2:333 (second quote). On labor markets, see J. Carlyle Sitterson, "Hired Labor on Sugar Plantations of the Ante-Bellum South," *JSH* 14 (May 1948): 200. On experimentation with white antebellum laborers, see Shugg, *Origins of Class Struggle*, 100; Sitterson, *Sugar Country*, 65–66. On the use of other immigrant groups in the postbellum era, see Moon-Ho Jung, "'Coolies' and Cane: Race, Labor, and Sugar Production in Louisiana, 1852–1877" (Ph.D. diss., Cornell University, 2000); Richard Follett and Rick Halpern, "From Slavery to Freedom in Louisiana's Sugar Country: Changing Labor Systems and Workers' Power, 1861–1913," in *Sugar, Slavery, and Society*, ed. Moitt (Gainesville, Fla., 2004). In Cuba, where new technology was deemed too complex for slaves, sugar planters recruited waged (mainly male) Spanish and Chinese workers to labor alongside bondspeople. See Moreno Fraginals, *Sugar Mill*, 134–42; Moreno Fraginals, Klein, and Engerman, "Level and Structure of Slave Prices on Cuban Plantations,"1206 n. 15. On racialization and the utilization of skin color as a principle of "social classification" and a tool in establishing "racial hierarchies" in cane economies, see esp. Madhavi Kale, *Fragments of Empire: Capital, Slavery, and Indian Indentured Labor Migration in the British Caribbean* (Philadelphia, 1998), 174; Verena Martinez-Alier, *Marriage, Class, and Colour in Nineteenth-Century Cuba: A Study of Racial Attitudes and Sexual Values in a Slave Society* (Cambridge, 1974), 80–81. For a key overview of the role of contract labor in cane-producing societies, see Stanley L. Engerman, "Contract Labor, Sugar, and Technology in the Nineteenth Century," *JEH* 43 (September 1983): 635–59.

planter—who hired Irish, German, and Dutch laborers to cut cane at between $10 and $12 per month, few slaveholders extensively employed free immigrants for the sugar harvest. Ultimately, the risk of strike action or the fear of labor instability during the harvest put paid to the use of alternative workers. In due course, Durnford also found free labor too insecure and unreliable for sugar work. Slave rental, by contrast, was more guaranteed. Although bondspeople could join others in communal work slowdowns, they remained an infinitely safer option than free white workers. Moreover, by compartmentalizing sugar as a black and slave-only occupation, planters fortified the institution of slavery and relegated African Americans to seemingly perpetual service in the Louisiana cane fields.[45]

Plantation stability and successful productivity thus hinged on the existence of a deep slave rental market. The extent to which planters drew upon this fluctuating pool varied, though some planters clearly and substantially increased their slave force as harvest approached. This strategy enabled the largest planters to exploit economies of scale—albeit temporarily—and efficiently utilize the grinding potential of the mechanized sugar mill. By tapping the rental market and injecting extra labor into the grinding season, overseers could plant extensively with a modicum of confidence that their temporarily enlarged slave crews could cut and mill the canes within the brief harvest window. Like other sugar lords, Samuel Fagot capitalized on the availability of short-term labor, spending almost $850 on renting slaves during the 1854 rolling season. Edward Gay similarly hired extra slaves to cut cane in early December 1854, though to assure optimum speed he established a payment scheme in which each slave received $1 for a full day's labor. Children received half pay, and Gay paid an additional 50¢ to those who labored through the night. He also docked each man's wage for slow or inefficient work. Few, however, fell behind the required pace. During the rolling season, Gay hired

45. Isaac Erwin Diary, 28 October–23 November 1851, LSU; Sitterson, *Sugar Country,* 66 (quote); Whitten, *Andrew Durnford,* 64–66. Also see Diary (vol. 2), 1 May 1848, Kenner Family Papers, LSU; Plantation Record Book, 1849–1860 (vol. 36), Edward J. Gay and Family Papers, LSU; Plantation Journal, 1859–1872 (vol. 28), 24 February 1860, Uncle Sam Plantation Papers, LSU; Diary of E. E. McCollam of Ellendale Plantation, 1842–1846, 28 May 1845, Andrew McCollam Papers, UNC; Octave Colomb Plantation Journal, 17 March 1854, TUL. On the Irish in the sugar belt, see Earl F. Niehaus, *The Irish in New Orleans, 1800–1860* (Baton Rouge, La., 1956), 44–49; David T. Gleeson, *The Irish in the South, 1815–1877* (Chapel Hill, N.C., 2001), 33, 53.

slaves to work literally hundreds of additional hours on his plantation. Requiring extra labor during summer and fall, Maunsell White contracted with a fellow planter whose farming operations had fallen victim to spring flooding. Unable to save his crop, this farmer willingly agreed to rent eighteen or twenty slaves, provided that his slaves labored solely on high and dry ground and engaged in "perfectly healthy work." Anxious that they clear land, cut cane, and fence in a tract of land, White agreed to hire the slaves at the rate of $10 per month for the women and $12 per month for the men. His decision to augment his regular slave crews during the peak harvest period ensured optimal performance for a relatively modest price. Financially, the decision to hire slaves during the grinding season was a logical one, since the cost of renting extra labor was considerably lower than the annual cost of purchasing and maintaining a slave.[46]

The presence of individual slaves and whole gangs for hire gave the planters substantial flexibility; they could plant extra cane or delay the harvest, confident that come November hired hands would be available to assist in the grinding season. Renting slaves was not cheap, but it ultimately proved better value than purchasing a further, potentially underemployed hand. Furthermore, the rental market enabled planters to concentrate labor costs in a manner not usually associated with lifelong slave investments, and it gave sugar farmers some of the flexibility of temporary wage labor. Daily wage rates for rented slaves in the sugar belt usually averaged between 50¢ and $1, or between $28 and $56 per slave for the entire grinding season. By contrast, to purchase and maintain an adult male field hand required an average capital outlay of $2,398, or $80 per annum during the peak thirty years of his working life. Daily rental costs were more than twice as expensive as purchasing a bondsman outright, but for $80 per annum a planter could

---

46. Cashbook, 1845–1859 (vol. 45), 15 and 22 March 1854, Uncle Sam Plantation Papers, LSU; Plantation Record Book, 1849–1860 (vol. 36), Edward J. Gay and Family Papers, LSU; Maunsell White to N. C. Hall, 30 March 1849, Letter Book, 1845–1850 (vol. 1), Maunsell White Papers, UNC (quote). Masters who rented their slaves could sue hirers for negligence and liability in case of death or physical harm, depending on the health and condition of the slave upon hiring. See Schafer, *Slavery, the Civil Law, and the Supreme Court of Louisiana*, 35–36, 101–7. On slave rental law, see Thomas D. Morris, *Southern Slavery and the Law, 1619–1860* (Chapel Hill, N.C., 1996), 132–58; Jenny Bourne Wahl, *The Bondsman's Burden: An Economic Analysis of the Common Law of Southern Slavery* (Cambridge, 1998), 49–77. On slave insurance to protect property, see Eugene D. Genovese, "The Medical and Insurance Costs of Slaveholding in the Cotton Belt," *Journal of Negro History* 45 (July 1960): 141–55; Flanders, "Experiment in Louisiana Sugar," 155.

hire between two and three workers. To price-conscious farmers, rental clearly appealed as an adaptable labor system that enabled them to match their exact agricultural needs with a compatible workforce.[47]

Slave rental added significant elasticity to a labor regime packed with expensive and young slaves. Purchasing a bondsman or woman entailed a considerable long-term capital investment, and it was one that planters did not make lightly. In the slave auction, the sugar masters—like all slaveholders—calculated what benefit the new purchase might bring. An additional laborer would strengthen the planters' crews come harvest time, but would that individual be adequately engaged for the whole year? Once the sugar-making season commenced, one planter observed, "It must be pushed without cessation, night and day . . . [but] we cannot afford to keep a sufficient number of slaves to do the extra work at the time of sugar-making as we could not profitably employ them the rest of the year." These words masked a panoply of white fears over the redundant slave who might turn to intrigue, deception, and dangerous plotting if not kept busy with enough work. Locked within a fixed labor system, slaveholders could neither afford to underemploy their slaves nor to provide African Americans with the liberty to invert the plantation order. Fully cognizant of the racial and economic pressures that impelled the logic of chattel slavery, sugar planters drove their bonded workers forward and sealed the flaws within the plantation order by renting additional hands when necessary.[48]

The mode of capitalist sugar production made a direct impact on bondspeople and profoundly shaped life in the plantation quarters. Young men—many of whom were unusually tall—overwhelmingly populated the sugar fields and gave human shape to the planters' gendered and racial perception of the ideal sugar worker. Women were conspicuous by their relative absence, but those that entered Louisiana's slave communities often proved younger than their fellow bondsmen

47. On slave-owning costs, see Whitten, "Sugar Slavery," 423; Alfred Conrad and John Meyer, "The Economics of Slavery in the Antebellum South," *Journal of Political Economy* 66 (April 1958): 95–123; Shugg, *Origins of Class Struggle,* 88. For a parallel study where slaveholders measured comparative costs of purchasing a slave and hiring free workers or renting slaves, see Carville Earle, "To Enslave or Not to Enslave: Crop Seasonality, Labor Choice, and the Urgency of the Civil War," in *Geographical Inquiry and American Historical Problems* (Stanford, 1992), 226–57; Laird W. Bergad, "The Economic Viability of Sugar Production Based on Slave Labor in Cuba, 1859–1878," *Latin American Research Review* 24 (1989): 107.

48. Weld, *American Slavery As It Is,* 39.

and were pregnant more often than bondswomen elsewhere. Imported to the sugar fields to raise the planter's future labor force, these young women faced the cruelest imposition of all on their independence and dignity. For the planters, demographic management underscored their universal power and heartlessly imprinted their authority upon human relationships. Through exploitative management and savage demographic control, the planters wove sugar's peculiar signature through the New Orleans slave pens and shaped the human traffic to conform to their gender, racial, and labor assumptions.

Although the slave lords of Louisiana practiced crude population control and sexual management, the mores of antebellum society prevented them from detailing their actions in letters, diaries, or other written forms. The slaves' faintly audible words and their reproductive histories, however, highlight a dark and seamy side of antebellum life—one in which conscious and semiconscious steps were taken to promote female fertility from the earliest age possible and to maximize the sexual potential of young bondswomen. Spotting young females who would potentially liberate them from their reliance on the domestic slave trade and young men who would conduct back-breaking labor, planters turned their keen and perceptive eyes toward selective slave purchasing and grisly population management. And when their gangs proved incapable, the sugar masters either brandished the whip more forcefully, purchased additional hands, or trawled the slave rental market until their labor needs were filled. Flexible, adaptable, and savagely manipulative, the sugar lords ultimately failed to breed slaves in sufficient numbers; but in their efforts to do so, they tailored chattel bondage to the uneven and constantly shifting labor requirements of plantation capitalism and bound slave labor to the undeniably hellish demands of agro-industry in the cane world.

three

"AN INTELLIGENT EYE"

ᴀ̶NDREW MᴄCOLLAM ᴡᴀꜱ unimpressed by Brazil. In the wake of slave emancipation, he traveled from his native Louisiana to the rich cane estates near Rio de Janeiro. Examining the lands and cane operations with the intention of commercially speculating in proslave Brazil, McCollam's shrewd business eye quickly focused on the deficiencies of Brazilian land and slave management. Upon visiting Julian Rebeiro de Castro's plantation, the Louisiana sugar lord mused on the regimented plantation order of his own native state, noting that he could "do more work with the same number of hands than was being done" on his Brazilian competitor's lands. Those favoring the annexation of Cuba in the prewar years similarly believed that when directed with American energy, the rich lands east of Havana would yields vast fortunes within a few years. Self-assured in their plantation management, their greater skill in sugar making, their better machinery, and the value of forced labor, McCollam and his ilk accurately reflected the conceited but prevailing logic of Louisiana's slaveholders.[1]

Like McCollam, Louisiana planters would have agreed with John Hampden Randolph's assertion that the key to prosperity in the sugar bowl lay in perseverance and, above all, good management. Through prudent supervision and a sagacious division of labor, Robert Russell concluded, "Free labor cannot compete, in the manufacture of sugar, with better organized slave labor." These sentiments

---

1. Brazilian Diary of Andrew McCollam, 1866–1867 (vol. 1), 13 July 1866, Andrew McCollam Papers, UNC (first quote). Also see J. Carlyle Sitterson, "The McCollams: A Planter Family of the Old and New South," *JSH* 6 (August 1940): 360–61. On the pro-Cuba movement, see Olmsted, *Seaboard Slave States*, 2:311–12; Robert E. May, *The Southern Dream of a Caribbean Empire, 1854–1861* (Baton Rouge, La., 1973), 46–76.

proved valid, for when alternative workers entered the labor market, the sugar masters soon found them wanting. Recalling the story of one planter who dispensed with slave labor in favor of Irish and German immigrants, Sir Charles Lyell mused on the catastrophic labor crisis the planter faced when his workers struck for double pay in the middle of the harvest season. Lyell recorded that with neither additional laborers nor slaves to step into the breach, the planter lost his crop, valued at $10,000. Clearly aware of the dangers implicit in such a predicament and alive to the potential gains in productivity from regimented labor, one sugar planter declared that "by proper treatment and a judicious distribution of their work," the enterprising farmer could double the crop of the unreformed slave master. With exacting discipline and almost military regulation, one hundred slaves, Timothy Flint concluded in the 1830s, "will accomplish more on one plantation, than so many hired free men, acting at their own discretion." Sugar magnate Valcour Aime concurred, noting that the excellence of Louisiana production and the "advanced stage of our agriculture" owed a great deal to the superiority of its seasoned slaves over the newly imported African.[2]

Such superiority, however, did not derive entirely from the bondsman's physiognomy. As Aime fully realized, Louisiana slaves—unlike their African brethren—were accustomed to a punishing labor regime and habitualized to a managerial style that was almost militaristic in its organization. Planters usually referred to their work squads as gangs or crews, though they might just as readily have used the word "brigade" to describe workplace organization. Roaming through St. John the Baptist Parish, Amos Parker encapsulated the regimentation of field labor when he noted that "armies of negroes" advanced through the fields, cutting cane and hauling it to the mills. Parker was not alone in noting the martial order that dominated the sugar fields of south Louisiana. Twenty years later, Francis and Theresa Pulszky remarked on the military appearance of the slaves and overseers near New Orleans. Homer, an overseer who brandished a whip in one hand as he guided his Hungarian visitors through the labyrinthine maze of sugar fields, reminded the Pulszkys of

2. John H. Randolph to John R. Liddell, 22 March 1846, Moses and St. John Richardson Liddell Family Papers, LSU; Robert Russell, *North America, Its Agriculture and Climate: Containing Observations on the Agriculture and Climate of Canada, the United States, and the Island of Cuba* (Edinburgh, 1857), 249 (first quote); Charles Lyell, *A Second Visit to the United States of North America*, 2 vols. (New York, 1849), 1:127 (second quote); Timothy Flint, *The History and Geography of the Mississippi Valley*, 2 vols. (Cincinnati, Ohio, 1832), 1:244–45 (third quote); *De Bow's Review* 4 (November 1847): 385 (fourth quote).

those men who wielded the stick with equivalent ease in the Austrian and Russian armies.[3]

Homer's lash cracked on every plantation, impelling a work discipline that startled visitors with its ferocious efficiency. Like many of his contemporaries, Victor Tixier concluded that the slaves were "well regulated" on the estates he visited in St. James Parish. The term "regulation," of course, shrouded the slaves' suffering and cloaked the whip in distant terms. It did, however, candidly reflect the slaveholders' commitment to subordination, obedience, and antebellum business organization. While regulation proved alienating for the bondsperson, contemporary accounts suggest that "improved management," in the slaveholders' vernacular, yielded profitable fruit in disciplined work teams and efficient farming. Indeed, plantation slavery appeared so efficient that one West Indian visitor registered his surprise at the "industry of the slaves, even when the overseer was away." Perhaps these slaves quailed beneath the driver's whip, but their diligent labor convinced many observers that the sugar masters' unbending discipline wrought profitable returns. On a visit to Magnolia Plantation, one agricultural journalist marveled at the professionalism and competency of the slave workforce. Examining both the fields and improvements on the estate, he noted that "all work is done with regularity, and in an efficient manner." Others also lauded superior management practices, remarking that the Louisianans pursued greater economy than their Caribbean competitors, who, James Stirling concluded, lived self-assuredly in "stagnation and contented nonchalance."[4]

Just as the labor-specific demands of sugar farming dictated the demography of the cane world, sugar left its indelible stamp on plantation management. As the quintessential gang-labor crop, sugar—more than any other staple crop—lent itself to capitalist development. Densely planted, it also proved more labor-intensive than rice, coffee, or cotton. Whereas slaves on cotton plantations cultivated up to twenty-five acres per field hand, cane workers seldom farmed more than seven acres apiece. The intensity of sugar farming led to disciplined management, drilled gang work, and punishing management wherever it was planted. But Louisiana's

3. Amos A. Parker, *Trip to the West and Texas* (Concord, Mass., 1835), 188; Francis and Theresa Pulszky, *White, Red, Black: Sketches of American Society in the United States during the Visit of Their Guest*, 2 vols. (New York, 1853), 2:105–6.

4. Victor Tixier, *Tixier's Travels on the Osage Prairies*, trans. Albert J. Salvan, ed. John Francis McDermott (1844; reprint, Norman, Okla., 1940), 47 (first quote); *De Bow's Review* 15 (December 1853): 648 (second quote); *Southern Planter* 17 (August 1857): 484 (third quote); James Stirling, *Letters from the Slave States* (London, 1857), 124 (fourth quote).

climate-specific problems placed an additional premium on organization, speed, and tight supervision. Even in the most southerly parishes, every planter in the cane world faced a race against time. They wanted to mature their canes late into the year without risking serious frost damage. In most cases, planters compromised, ordering the first canes to be cut between mid- to late October and early November. In the following six to eight weeks, the entire crop had to be harvested and processed into raw sugar. This was no small task. The plantations stretched for miles, and harvesting eight-foot-high canes required back-breaking work. Once cut and stripped of their leaves, the canes had to be transported back to the mill for crushing and milling. Only then could the process of sugar making begin. These necessary procedures profoundly shaped the work culture of the cane world and placed disciplined supervision at a premium. Sugar planters in the Caribbean and Louisiana accordingly favored gang work as the most efficient means to plant and harvest the crop, and once the grinding season began, they valued discipline and speed. While the introduction of steam-powered sugar milling in the 1830s and 1840s increased the speed and extractive capacity of the mills, it also created new problems and more challenges. Whereas horse-drawn mills turned at the pace of a plodding mule, the new mills imposed a mechanical and industrial pace to work that set the tempo throughout the plantation. Slaves in the cane fields had to supply the mill with a constant flow of cane, and their counterparts in the mill house had to process the sugar in a mechanized and sequential manner. Planters could stop the mill and wait for the cane cutters to catch up, but for the most part they were disinclined to do so, especially if bad weather was in the offing. Instead, they kept the mills running and fashioned a labor order that toiled at the metered cadence of the steam age. In creating an industrialized labor force, they reconfigured work patterns, instituted assembly-line production, subdivided their laborers' tasks, introduced systematized shift work, and imposed a regimented order that proved both exacting and relentless. The combination of intense farming, steam power, and Louisiana's specific climatic conditions combined to create a punishing agro-industrial system that, at the very least, anticipated several aspects of modern industrialization. The sugar masters commanded this industry with brutal efficiency, and their penchant for industrialized management created one of the most rapacious and exploitative regimes in the American South.[5]

5. On gang work and industrial discipline, see Philip D. Morgan, "Task and Gang Systems: The Organization of Labor on New World Plantations," in *Work and Labor in Early America,* ed. Innes

For most of the 125,000 slaves who labored in Louisiana's cane fields prior to the Civil War, sugar production had been transformed within their lifetimes. The new steam-powered mills were frequently equipped with mechanical cane carriers that slaves fed with a steady supply of cane, and these constantly moving conveyor belts established an early form of assembly-line production. Mechanization forced operatives to keep astride of the mechanized process and ultimately to adapt their actions to its requirements. Slave laborers enmeshed within this transitional phase of assembly-line development toiled with relentless regularity as the ever-moving conveyor belts whisked canes through the mill at relentless speed. Throughout the late antebellum sugar country, leading planters sought to transform their slaves into industrial workers and to impose, in antebellum parlance, the "habits of industry." To coordinate their activities, planters altered the managerial structure of their enterprises and established a system of salaried managers, overseers, and assistant overseers, who directed the labor of the slaves with both incentives and the omnipresent lash. Those inferior to the task at hand faced swift demotion. Planter John Palfrey dismissed his estate manager and urged him to find a position as assistant overseer where he could acquire training in the "plantation business" before supervising "such an establishment as this." Beneath the cadre of estate managers came the slave drivers, who enforced workfloor discipline and served as a vital conduit between the slave community and the managerial structure over which the planter reigned. Within the sugarhouse, free and slave sugar makers, engineers, and skilled workers oversaw the discrete parts of the production process and pressed ahead at double-quick time. New technology permitted an enormous increase in the production of sugar, and during the grinding season the sugar planters advanced the business structure of their estates by subdividing the factory into distinct units while integrating mill operations through modern technology. With subunits functioning as separate though interdependent branches of production, the antebellum sugar mill stood within a transitional phase of industrial development, anchored in the social relations of slavery though mimicking aspects of northern capitalism. Local commentators applauded these

(Charlottesville, Va., 1988), 189–220, esp. 193, 210; Fogel, *Without Consent or Contract,* 25–26. On planters emulating the factory system and their early admiration of the industrial order, see Chaplin, *Anxious Pursuit,* 115; Dusinberre, *Them Dark Days,* 6; Mark M. Smith, "Old South Time in Comparative Perspective," *AHR* 101 (December 1996): 1432–69; Gerald David Jaynes, *Branches without Roots: Genesis of the Black Working Class in the American South, 1862–1882* (New York, 1986), 77.

administrative and supervisory improvements. As the *Franklin Planters' Banner* proudly trumpeted, regional success rested upon "good management on the improved principle adopted in Louisiana."[6]

In synchronizing the manual process of cutting canes with the mechanized production of sugar, estate managers in the cane world rationalized work and managerial practices to thread unfree labor into the fabric of business development. As historians Robert Fogel and Stanley Engerman contend, gang labor emerged as an unmerciful though efficient structure that provided the slaveholding elite with disciplined teams adept at intense work. Because of its dependence on gang labor, southern slave-based agriculture was approximately 35 percent more efficient than northern agriculture and as much as 28 percent more efficient than southern free farms. Disciplined, interdependent gangs could sustain an intense and constant rhythm of work; this in turn favored economies of scale and the efficiency of large-scale operations. Once equipped with an integrated order of established labor gangs, slaveholders could routinize labor while maintaining strict supervision over the slaves who toiled beneath the overseer's eye in the open field. The constant threat of the lash and its indiscriminate use drove the crews on, and throughout the sugar world planters balanced the whip with incentives to maintain effort and speed. To impel their workers to toil at full speed, estate managers additionally subdivided their labor crews according to task and

---

6. John Palfrey to William Palfrey, 19 January 1833, Palfrey Family Papers, LSU (second quote); *Franklin Planters' Banner*, 5 January 1854 (third quote). For the social impact of mechanization, see Thorstein Veblen, *The Instinct of Workmanship and the State of the Industrial Arts* (New York, 1914), 306–7; Siegfried Giedion, *Mechanization Takes Command: A Contribution to Anonymous History* (New York, 1948), 77. On the appointment of skilled sugar makers and assistant overseers, see Sitterson, "Hired Labor," 192–205; Sitterson, "Magnolia Plantation," 201; Sitterson, *Sugar Country*, 54–60; William Kauffman Scarborough, *The Overseer: Plantation Management in the Old South* (Baton Rouge, La., 1966). On the role of drivers, see William L. Van Deburg, *The Slave Drivers: Black Agricultural Labor Supervisors in the Antebellum South* (Westport, Conn., 1979); Robert L. Paquette, "The Drivers Shall Lead Them: Image and Reality in Slave Resistance," in *Slavery, Secession, and Southern History*, ed. Paquette and Ferleger (Charlottesville, Va., 2000), 31–58. On the parallel to factories in the fields, see Oakes, *Ruling Race*, 153–91. On the use of technology and the speed-up in northern factories, see Thomas Dublin, *Women at Work: The Transformation of Work and Community in Lowell, Massachusetts, 1826–1860* (New York, 1979), 71–72; Jonathan Prude, *The Coming of the Industrial Order: Town and Factory Life in Rural Massachusetts, 1810–1860* (Cambridge, 1983), 127–31; Barbara M. Tucker, "The Family and Industrial Discipline in Ante-Bellum New England," *Labor History* 21 (winter 1979–1980): 56 (first quote).

established a finely graded system of work specialization that required gang and labor interdependence.[7]

Like their Caribbean competitors, the sugar masters were not opposed to utilizing task work on occasion. For the primary cultivation and harvesting tasks, planters adopted the closely disciplined gang system and turned their fields into factories, but for plantation maintenance, ditching, scraping stubble cane, and—at times—the raising of corn and small grains, the sugar lords frequently adopted the more flexible task system, in which they could encourage rapid work with the promise of free time once the task was complete. Former slave Catherine Cornelius briefly alluded to the greater physical and psychological autonomy that task work afforded when she observed, "We all had certain tasks to do . . . if we finished dem ahead of time de rest of de day was ours." Tasking provided slaves with a modicum of flexibility over when and how they toiled during the workday or week, and it also allowed sugar growers to conduct their multifarious tasks and effectively manage the elaborate division of labor on a cane plantation. As Victor Tixier observed, tasks were "carefully proportioned" to the slave's age and strength, thus allowing the planter further to subdivide and routinize labor. Despite its appeal, task work never challenged the supremacy of gang labor, however. Although planters experimented with both systems, their predilection for tighter management and strict control ensured the prevalence of gang work, both under slavery and after emancipation. From the sugar masters' occasionally erroneous perspective, gang labor eradicated the slaves' ability to control the pace of their labor, enabling them rationally to allocate work according to gender, perceived strength, and even age. Sugar master William J. Minor, for instance, counseled his overseer to ensure that field hands "do a good days [*sic*] work according to their strength." Minor's stricture placed the onus of responsibility on his paid employee to gauge each slave's strength, his or her working potential, and any shortcoming in an individual's contribution to the crew's collective efforts. Slaveholders called their overseers swiftly to account when they failed in their duties. As one of the richest slave masters, Maunsell White, rebuked his estate managers, "I can't bear to see things done in a careless slovenly manner, in short, half-done. I want vigilance and care." Team and

---

7. On the economics and relative efficiency of gang work, see Fogel, *Without Consent or Contract,* 74–79; James R. Irwin "Exploring the Affinity of Wheat and Slavery in the Virginia Piedmont," *Explorations in Economic History* 25 (July 1988): 295–322; Jacob Metzer, "Rational Management, Modern Business Practices, and Economies of Scale in the Antebellum Southern Plantations," *Explorations in Economic History* 12 (April 1975): 139.

gang structures, planters firmly believed, would ensure that slaves worked vigilantly, efficiently, and "according to their strength." Gang labor, the *Southern Agriculturist* announced, would foster the "spirit of emulation . . . which makes them work with more cheerfulness, and . . . makes them execute the work better." Disciplined labor, moreover, carried an ideological thrust. For slaveholding planters, the regimented gangs symbolized their racial power and economic mastery over those in the fields.[8]

In Louisiana's sugar bowl, planters not only strove to perpetuate dependent social relations through gang labor, but they also discriminatingly specialized and subdivided work tasks in their quest for plantation success. Adapting this partly industrial and agricultural model to the exigencies of the steam age, sugar planters emulated the emerging factory system and crafted interdependent teams that would swiftly plant the crop in the New Year and efficiently harvest the canes come November or December. This practice exploited potential economies of scale and speed, as sugar cultivators organized their slaves into mutually dependent plowing and sowing crews that moved over the cane fields with military-like precision. On Robert Ruffin Barrow's Residence Plantation, estate manager Ephraim A. Knowlton carefully recorded the structured division of labor and gang work on Barrow's Terrebonne Parish farm. Taking charge on New Year's Day, 1857, Knowlton placed Barrow's slave driver, Andrew, in charge of the planting gangs and fourteen mules obtained from Barrow's Oak Grove estate in Lafourche Parish. Andrew soon found himself leading twenty-three other slaves. A week later, Andrew's gang counted thirty-five acres of cane planted. Meanwhile, Peter and Jerry, two other slaves given leadership responsibilities, led twelve of Barrow's slaves in

8. Interview with Catherine Cornelius, date unknown, WPA Ex-Slave Narrative Collection, LSU (first quote); Tixier, *Tixiers Travels,* 47 (second quote); Rules and Regulations on Governing Southdown and Hollywood Plantations, Plantation Diary, 1861–1868 (vol. 34), William J. Minor and Family Papers, LSU (third quote); Maunsell White to W. H. Haynes, 12 September 1848, Maunsell White Papers, UNC (fourth quote); *Southern Agriculturist* 7 (1834): 408, quoted in Joseph P. Reidy, "Obligation and Right: Patterns of Labor, Subsistence, and Exchange in the Cotton Belt of Georgia, 1790–1860," in *Cultivation and Culture,* ed. Berlin and Morgan (Charlottesville, Va., 1993), 147 (fifth quote). On task labor in sugar production, see William M. Polk, *Leonidas Polk: Bishop and General,* 2 vols. (London, 1915), 1:199; Joseph H. Parks, *General Leonidas Polk, C.S.A.: The Fighting Bishop* (Baton Rouge, La., 1990), 104. The literature on the economic and social significance of tasking, especially in the rice industry, is extensive, but see Coclanis, "How the Low Country Was Taken to Task," 59–78; Carney, *Black Rice,* 98–122; Betty Wood, *Women's Work, Men's Work: The Informal Slave Economy of Lowcountry Georgia* (Athens, Ga., 1995), 1–30.

repairing the levee, cleaning the canal, and rolling logs. To ensure the completion of all plantation duties, Knowlton further subdivided his labor crews, allotting five slaves to the hired white carpenter to work on repairing the slave quarters and cane house. Anxious that not a single hand remain idle, Knowlton additionally sent the young children, or suckling gang, to the fields to weed the canes and to engage in the perilous pursuit of burning logs. Others, aged eleven to fifteen, labored as water carriers, lugging fresh rainwater out to the field workers. Knowlton—and farm managers like him—thus optimized the performance and utilized the physical resources of all the slaves beneath their command. Slaves in their fifties and sixties who had survived the brutal conditions of the sugar country were not entirely liberated from labor but found themselves used as support workers across the estate. Samuel McCutchon, for instance, used older and "ruptured" slaves as cooks, stock minders, gardeners, carpenters, and nurses. No one, not even the fifty-seven-year-old paralytic jobber Paris Johnson, found respite from a never-ending life of disciplined and regimented labor. Like many of the antebellum sugar masters with fifty, eighty, or even a hundred working hands at their disposal, Knowlton and McCutchon were able to routinize work, subdivide their laborers' work and time, and separate tasks and responsibilities more efficiently than cotton growers, who often relied on no more than twenty or thirty hands. Equipped with slave crews that dwarfed those of the cotton southwest, the sugar masters possessed a considerable size advantage in allocating labor and in differentiating skilled and unskilled occupations on their large and labor-intensive estates.[9]

On many estates, planters divided the slaves into three or four separate but interdependent divisions. The first team, one contemporary noted, "consisting of

9. Residence Journal of R. R. Barrow, 1 January–13 February 1857, Robert Ruffin Barrow Papers, UNC; Inventory and Valuation of Slaves, Stock, and Farming Utensils of Oaklands Plantation, 1859, Samuel D. McCutchon Papers, LSU. On the stringent subdivision of labor, see Sitterson, "Magnolia Plantation," 202–3. On the Cuban and Jamaican sugar industries, see Knight, *Slave Society in Cuba*, 72–73; Richard S. Dunn, " 'Dreadful Idlers in the Cane Fields': The Slave Labor Pattern on a Jamaican Sugar Estate, 1762–1831," *JIH* 17 (spring 1987): 803–10. On the division of labor and the emerging factory system in the North, see Mary Blewett, *Men, Women, and Work: Class, Gender, and Protest in the New England Shoe Industry* (Urbana, Ill., 1988), 97–115; Prude, *Coming of the Industrial Order*, 129–32. Finally, on the differentiation of labor gangs in the cotton and sugar South, see Steven F. Miller, "Plantation Labor Organization and Slave Life on the Cotton Frontier: The Alabama-Mississippi Black Belt, 1815–1840," in *Cultivation and Culture*, ed. Berlin and Morgan (Charlottesville, Va., 1993), 163; John Hebron Moore, *The Emergence of the Cotton Kingdom in the Old Southwest: Mississippi, 1770–1860* (Baton Rouge, La., 1988), 108.

light hands, brought the cane by armfuls from the cart, and laid it by the side of the furrows; the second planted it, and the third covered it." Solomon Northup described drilled gang work from the slaves' perspective. Noting that three gangs operated in unison during planting, Northup recounted that the first gang drew canes from the stack and trimmed them of their leaves, another team laid the canes in the drill, and a third followed in their footsteps, drawing soil over the stalks to the depth of three inches. George W. Woodruff, an overseer on James Bowman's estate, additionally realized the advantages of gang labor during the planting and cultivating seasons, diligently recording that forty hands scraped the small sugar cane shoots in mid-March, while nine plow teams hoed the cane furrows, four slaves planted cotton, and fourteen hands planted corn. Woodruff divided the seventy-two-strong slave labor force into several teams that he could manage effectively; as a result, sugar operations at Frogmoor Plantation flourished, producing 167 hogsheads of sugar in 1857. Ever keen to improve his yield, Bowman had written his wife a year earlier, stating that he wished "to learn as much of planting as possible . . . [so] that here after I may better understand management and all unnecessary mistakes." Evidently Bowman learned his lessons well, as his Pointe Coupee estate surged in productivity in the following two years.[10]

Shrewdly calculating in questions of sexuality and brawn, planters also imposed a set of racial and gender assumptions that demoted women to the hoe gangs while stronger males staffed the plow teams. Plowing required the labor of sturdy or "likely" male laborers who could hold the plow firm and press the share point into the earth as the oxen marched forward. The hoe, in contrast, required significantly less gender specificity; all women and men could easily manage these simpler and lighter farm implements. Englishman William Howard Russell observed such a gendered division of labor during his sojourn in the sugar country. At five o'clock in the morning, when the daybreak sun was already excessively hot, Russell ventured into the cane fields, passing a cart laden with the slaves' uneaten breakfast provisions before gazing at the work scene that unfolded before him. "Three gangs of Negroes," Russell narrated, "were at work: one gang of men, with twenty mules and ploughs, were engaged in running through the furrows . . .

10. Olmsted, *Seaboard Slave States*, 2:324 (first quote); Northup, *Twelve Years a Slave*, 159 (second quote); Frogmoor Plantation Diary, 19 March 1857, and J. P. Bowman to Sarah Turnbull, 29 June 1856, Turnbull-Bowman-Lyons Family Papers, LSU (third quote); P. A. Champomier, *Statement of the Sugar Crop, 1857–1858*, 4; Champomier, *Statement of the Sugar Crop, 1858–1859*, 4.

another gang consisted of forty men . . . [while] the third gang of thirty-six women, were engaged in hoeing out the cane." The slaves, he added, were disciplined and driven "from point to point, like a *corps d'armée* of some despotic emperor maneuvering in the battle field." Russell returned to his host's mansion for breakfast convinced by this planter's "judicious employment of labor." He was not alone, however, in observing that planters divided their laborers by strength and sex. Thomas Bangs Thorpe similarly noted the use of age- and gender-defined work gangs. As he noted, a crew of the "most robust negroes" cleaned the myriad of drains and ditches on a plantation, while another, presumably less vigorous, prepared the fields for the plow. Such "judiciously applied labor," one commentator observed, lay at the axis of plantation success.[11]

In rendering an effective division of labor, planters often institutionalized their gender assumptions through dress. Watching a gang of slaves march home, Frederick Law Olmsted stood aghast. The first in line was an elderly driver, bearing in his hands the symbol of his authority—a whip. Next came forty of the "largest and strongest women I ever saw together," carrying themselves loftily, with hoes over their shoulders, walking "with a free, powerful swing, like *chasseurs* on the march." Without exception, they wore a uniform of a bluish checked cloth, but they ambled past him barefoot. Slightly to the rear came the "cavalry"—thirty strong in number, mainly men but including some women, two of whom rode the plow mules. The men, Olmsted recalled, wore small blue Scotch bonnets, while several of the women sported handkerchiefs worn turban fashion. At the rear of the column, strategically placed to urge the gang on and punish loafers, was a lean, wiry, and vigilant overseer, perpetually observant from atop a brisk pony. While these slaves

11. Russell, *My Diary North and South*, 179–80, 262, 271 (first quote); Thomas Bangs Thorpe, "Sugar and the Sugar Region of Louisiana," *Harper's New Monthly Magazine* 42 (November 1853): 756 (second quote); Solon Robinson, *Solon Robinson, Pioneer and Agriculturist: Selected Writings*, ed. Herbert Anthony Kellar (Bloomington, Ind., 1936), 186 (third quote). On the gendering of labor in southern economies, see Wood, *Women's Work, Men's Work*, 16–27; Carole Shammas, "Black Women's Work and the Evolution of Plantation Society in Virginia," *Labor History* 26 (winter 1985): 5–28. In the Caribbean, see Rhoda Reddock, "Women and the Slave Plantation Economy in the Caribbean," in *Retrieving Women's History*, ed. Kleinberg (Oxford, 1988), 107–10, 114–19. On the relationship between plowing and labor effort, see R. Douglas Hurt, *American Farm Tools: From Hand-Power to Steam-Power* (Manhattan, Kans., 1982), 11; Leo Rogin, *The Introduction of Farm Machinery in Its Relation to the Productivity of Labor in the Agriculture of the United States during the Nineteenth Century* (Berkeley, Calif., 1931), 3–36.

probably worked on a cotton plantation, slaves on neighboring sugar estates dressed and organized themselves in largely similar ways. Solon Robinson, who visited Ormond Plantation in 1848, similarly noted that thirty-six cane cutters labored on the plantation, all of whom were dressed identically in blue woolen shirts. This "uniform company" advanced so impressively, Robinson concluded, that "they might do the state some service in times of peril." The standardization of slave uniforms routinized labor discipline and imposed outward conformity to established plantation expectations. Later industrialists self-congratulatorily proclaimed that "fitting the worker to the job" and providing free uniforms reflected the caring welfare capitalism of corporate America; but in the cane fields of antebellum Louisiana, uniforms crudely marked the save laborer. Like the stripes of the convict, they served as an indelible badge of workplace exploitation and a degrading symbol of the slaves' bondage.[12]

As a process-centered industry, sugar production required rapid harvesting and prompt grinding. Vividly describing the furious activity of the grinding season to a national audience in 1853, Thomas Bangs Thorpe aptly used militaristic metaphors in his stirring—though excessively romanticized—account of the annual sugar harvest:

> And now may be seen the field hands, armed with huge cane knives, entering the harvest field. The cane is in the perfection of its beauty, and snaps and rattles its wiry-textured leaves, as if they were ribbons, and towers over the overseer as he rides between the rows on his good sized horse. Suddenly, you perceive an unusual motion among the foliage—a cracking noise, a blow—and the long rows of growing vegetation are broken, and every moment it disappears under the operation of the knife. The cane is stripped by the negroes of its leaves, decapitated of its unripe joints, and cut off from the root with a rapidity of execution that is almost marvelous. The stalks lie scattered along on the ground, soon to be gathered up and

---

12. Frederick Law Olmsted, *A Journey in the Back Country in the Winter of 1853–54* (1860; reprint, New York, 1907), 5–6 (first quote); Robinson, *Solon Robinson*, 167 (second quote). Also see Gerilyn Tandberg, "Field-Hand Clothing in Louisiana and Mississippi during the Ante-Bellum Period," *Dress* 6 (1980): 89–103. On welfare capitalism and the perceived degradation of uniforms, see Rick Halpern, "The Iron Fist and the Velvet Glove: Welfare Capitalism in Chicago's Packinghouses, 1921–1933," *Journal of American Studies* 26 (August 1992): 159–83; Walter Licht, *Working for the Railroad: The Organization of Work in Nineteenth-Century America* (Princeton, N.J., 1983), 271.

placed in the cane-wagons, which with their four gigantic mule-teams, have just come rattling on to the scene of action with a noise and manner that would do honor to a park of flying artillery.

Those who toiled at the frenetic pace of the grinding season might have gagged at Thorpe's purple prose, but they would have concurred with his militaristic description of the annual harvest. Agronomic commentator Solon Robinson declared that the uniformed crews, which were subdivided into mutually dependent teams that could advantageously exploit economies of speed, made a particularly imposing sight as they sallied forth with "their formidable-looking weapons, the cane knives" in hand. As the lead hand in a gang of fifty to a hundred slaves, Solomon Northup graphically described the interdependence of teamwork and the division of labor among the cane cutters in St. Mary Parish. Flanked on either side, the lead hand advanced slightly ahead of his compatriots, who formed the base of a triangle, which labored wholly in unison and at the pace of their squad leader. Armed with a razor-sharp knife, the lead hand sheared the cane from the ground, stripped the stalk of its flags, sliced off its top, and placed it behind him. Slightly behind their pacesetter, the two other cane cutters followed suit and laid their stripped canes upon the first stalk. A young slave who followed the squad gathered up the bundle and placed it in the cart that trailed behind the team. Once filled, the cart left for the sugarhouse, though it was quickly replaced by a second wagon, ensuring that the process of cutting, stripping, collecting, and loading the cane rarely ceased or slowed. Throughout the grinding season, this brutal yet highly efficient field labor regime supplied the voracious demand of the sugar mill from dawn to dusk.[13]

Cane cutters were at the vanguard of a plantation order characterized by the transformation of a raw material into a finished product. As an early prototype of the modern line-production system, sugar mills combined labor-saving techniques with production-raising methods. These innovations placed the sugar industry within a relatively small group of antebellum businesses that adopted experimental assembly-line systems. Flow production techniques—where the work was brought to a stationary worker by conveyor belts—evolved in the arms, clock, and pork-rendering industries, only advancing swiftly during the early twentieth century. Routinizing production through drill, regularity, and, above all, system,

13. Thorpe, "Sugar Region of Louisiana," 760 (first quote); Robinson, *Solon Robinson*, 167 (second quote); Northup, *Twelve Years a Slave*, 160 (third quote).

assembly lines ensured an efficient division of labor, rapid production, and a minimal role for the laborer, who toiled at the set speed of the chain. As one worker's wife declared about the Ford Motor Company, "The chain system you have is a *slave driver! My God!* . . . My husband has come home and thrown himself down and won't eat his supper—so done out!" Exhausted by the speeding-up and the punishing regimen of mass production, workers at Ford mirrored the sentiments of laborers who toiled on the line elsewhere. In the Carolina mill district, where labor-saving devices accelerated workfloor activity, the motif of wage slavery similarly echoed among workers, who likened the industrial conditions to "a sweatshop, slave place."[14]

While the antebellum sugar mill hardly matched the complexity of Ford's assembly line, the sugar masters found that flow production raised efficiency and increased production speed. With its conveyor belts and cane carriers, the mechanized mill linked the various stages of production and advanced with unbending regularity. Working at the methodical pace of the steam engine, mill hands staffed every part of the manufacturing process, from placing the cane onto a mechanized cane carrier to operating the mill, kettles, vacuum pumps, and all ancillary machinery. Ever astute to the nature of working conditions, Solomon Northup vividly portrayed the dynamics of mechanization. The mill, he recalled, was a vast brick building by antebellum standards, measuring some hundred feet in length and forty or fifty feet in width. The boiler was located outside the main building, but the main bulk of the machinery and engines stood on brick piers that were fifteen feet above the floor of the mill house. Connected to the machinery were two iron rollers, each two to three feet in diameter and six to eight feet in length. An "endless carrier" made of chain and wood led from the rollers through the entire length of the open shed. The carts full of freshly cut cane were unloaded just outside the main building, whereupon a team of slave children would snatch up the canes and place them upon the carrier that rattled endlessly and deafeningly for those within

---

14. On the use of flow production in the antebellum South, see Curtis J. Evans, *The Conquest of Labor: Daniel Pratt and Southern Industrialization* (Baton Rouge, La., 2001), 40. On flow production in antebellum industries, see Walter Licht, *Industrializing America: The Nineteenth Century* (Baltimore, 1995), 43. On Fordism, see David A. Hounshell, *From the American System to Mass Production, 1800–1932* (Baltimore, 1984), 216–61, esp. 259 (first quote); David Gartman, *Auto Slavery: The Labor Process in the American Automobile Industry, 1897–1950* (New Brunswick, N.J., 1989), 44–49. On chain production in the Carolina mills, see Sinclair Lewis, *Cheap and Contented Labor: The Picture of a Southern Mill Town in 1929* (New York, 1929), 3 (second quote).

earshot. From there, the canes were whisked high above the slaves on the ground toward the rollers, where they were crushed, the spent canes falling onto another belt that disappeared out of the building. The constant clattering of conveyor belts shuttling canes and semidry sucrose across the mill floor presented an alarming scene to visitors as they watched the workers literally sweat to keep time with the punishing order.[15]

Writing thirty years after the height of the slave regime, Attakapas sugar planter F. D. Richardson expressed his racial and class constructions of black labor in the context of the speed and tempo of slave work. The bondsman, Richardson recalled, appeared to be in his "native element" during the grinding season, and "his jokes and long-ringing laugh kept time with the rattle of the cane as he dashed it on the carrier and wheeled to get another turn." Richardson's lighthearted words, of course, belied the reality of mill work, in which cane carriers set an exacting and audible pace to the production line. To ensure unceasing production, agro-industrial plantations not only included rhythm-setting carriers, but they also operated late into the night. Joseph Ingraham observed the lengthening work schedules of sugar farming when he observed that slaves labored from eighteen to twenty hours during the grinding season. Likewise, Claude Robin remarked that the slaves' sole respite from the toil of sugar production came from a few hours sleep, snatched during the middle of the night.[16]

Whether in Cuba or Louisiana, nineteenth-century sugar growers faced a challenging disequilibrium between the manual aspects of the harvest (cutting the canes and transporting them to the cane shed) and the mechanized process of transforming the canes into sugar. The introduction of steam power accelerated the latter, but harvesting still advanced at the pace of the cane cutters. On some estates, however, elite planters sought to bring the mechanized tempo of the sugarhouse to the cane fields. Iron or wooden railroads with tracks radiating from the sugarhouse integrated the agricultural and industrial aspects of sugar production and quickened the pace of field work to match the incessant trundling speed of the locomotive. Field trains expanded the assembly-line structure of the sugar mill, synchronized field work with the industrial cadence of the mill, and impelled

15. Northup, *Twelve Years a Slave*, 161–62.

16. F. D. Richardson, "The Teche Country Fifty Years Ago," *Southern Bivouac* 4 (March 1886): 595 (first quote); Joseph H. Ingraham, *The South-West, by a Yankee*, 2 vols. (New York, 1835), 1:240; Claude C. Robin, *Voyage to Louisiana, 1803–1805*, trans. Stuart O. Landry Jr. (1807; reprint, New Orleans, 1966), 240.

greater urgency, stress, and risk on the workfloor. At Madewood Plantation, Thomas Pugh constructed a small iron railroad that brought freshly cut canes to the mill. From the carrier where the train of wagons arrived, conveyor belts whisked the cane shoots towards the mill for grinding. Keen to utilize the crushed canes as fuel, Pugh stationed additional carts at the base of the mill to catch the crushed canes as they fell from the mill. As editorialist Solon Robinson observed, after grinding, the freshly squeezed cane juice "runs to the vats .-.-. and thence to the kettles; thence to the coolers, and from there the sugar is carried upon railroad cars along lines of rails between the rows of hogsheads to the farther end of the building." While planters like Pugh and Leonidas Polk clearly favored mechanization, sugarhouse railroad transport, and assembly-line production, others similarly strove to build train tracks on their estates. Samuel Tillotson, for instance, spent $2,500 on a cedar railroad from his sugarhouse to the Mississippi River, where he built a small depot for storing sugar and goods until a steamboat arrived to carry his goods to market. Duncan Kenner appears to have gone one step further, purchasing portable tracks with light-gauge rails and metal cross-ties. These tracks could be easily moved around the plantation, allowing Kenner to focus on specific areas. Both Kenner's and Tillotson's railway cars linked the distinct units of the antebellum sugarhouse and created a chain between the various branches of production.

While the introduction of advanced machinery imposed the regimented order of the factory system, it also led to the further subdivision of labor and novel patterns of work organization. Mindful that speed defined the sugar harvest, planters carefully allocated tasks to ensure that crew productivity eclipsed the sum product of the individual team members. The New Orleans physician Dr. Samuel Cartwright observed that during the harvest season, "All of the laborers .-.-. are divided into two portions—one to labor in the field and to supply the mill house with cane; the other to manufacture the juice." Through such an occupational division of labor, "the negroes," Robert Russell observed, "are generally tasked up to their strength during the crushing system." On Robert Ruffin Barrow's Residence Plantation, overseer Ephraim Knowlton devised a roster that classified the occupational division of labor

17. Robinson, *Solon Robinson,* 162, 200 (first quote). Also see Jill-Karen Yakubik and Rosalinda Méndez, *Beyond the Great House: Archaeology at Ashland–Belle Helene Plantation* (Baton Rouge, La., 1996), 13; W. E. Butler, *Down among the Sugar Cane: The Story of Louisiana Sugar Plantations and Their Railroads* (Baton Rouge, La., 1981). For parallels to Cuba, see Moreno Fraginals, *The Sugar Mill,* 97, 144; Knight, *Slave Society in Cuba,* 32–33.

for the 1857 grinding season. Listing each slave's name below his or her expected task, Knowlton subdivided his labor force into a number of interdependent teams that worked on all tasks—from cooking a communal meal to operating the diverse and skilled functions of the industrialized sugar mill. On his Iberville Parish estate, planter Edward Gay echoed Knowlton's schedule, classifying his laborers according to chore and assignment. Preserving a small, all-male crew of kettle hands and engineers, Gay subdivided his labor force into a set of interconnected teams that supplied Jacob Lennox, his enslaved sugar maker, with canes.[18]

Anxious to meter the pace of work to the unbending regimen of the steam age, planters followed Caribbean precedent and established regular watches. Shift work of this type ensured that comparatively fresh hands were readily available to staff the machines and to conduct the complex art of sugar making, even in the early morning hours. John Hampden Randolph employed such a strategy when he instituted a system of watches for the 1857 grinding season. Dividing his slave force according to task and watch, Randolph established a revolving labor system in which he divided the working day and night into three eight-hour watches, of which most slaves worked two. Big Alfred, for instance, began his working day as a cart loader, following the cane cutters through the fields. Presumably Big Alfred rested in the late afternoon and evening, but in the early morning hours he entered the mill house, where he stood guard as the steam-engine fireman, controlling the fire beneath the sugar kettles. Weary from his night's labor, Big Alfred returned to the fields as a cart loader with the first morning light. At Shady Grove Plantation in Iberville Parish, Isaac Erwin instituted a similar watch system during November and December 1851. Following a strict division of labor, Erwin's interdependent gangs apparently operated in unison and efficiently enough to produce a strike every fifteen to twenty minutes. Driven by the inexorable timepiece, Erwin's teams hauled cane to the mill in "fine decent order," "cutting it rite" and producing "good sugar . . . pretty fast." To assure effective management and impose their own masterly imprint over the production process, planters not infrequently orchestrated nighttime operations and sometimes even established special dormitory rooms close to the mill where they or the slave might rest during harvest. Planter John Slack, for instance, spent sixty consecutive nights manning the night watch at Bay

---

18. *De Bow's Review* 13 (December 1852): 598 (first quote); Russell, *North America*, 273 (second quote); Plantation Record Book, 1849–1860 (vol. 36), Edward J. Gay and Family Papers, LSU; Residence Journal of R. R. Barrow, 3, 15, December 1857, Robert Ruffin Barrow Papers, UNC.

Farm on Bayou Grosse Tete. Charles D. Stewart, the master of Hog Point, also found his attendance in the sugar mill a necessity during the grinding season. Noting to his father that "Uncle Charley never leaves the sugar house when the mill is going," William B. Hamilton added that his kinsman served as a sugar maker and presumably as the plant manager on the Pointe Coupee Parish estate. The need to maintain personal oversight at all times appealed so significantly to Dr. H. G. Doyle, the owner and superintendent of Eureka Plantation in Iberville Parish, that he modified the mill to accommodate a platform or workstation from which he could monitor the workers. On a visit to his estate, William P. Bradburn of the *Plaquemine Sentinel* remarked that the "industrious proprietor" had a small though smart room overlooking the mill floor, where he retired after leaving his post at the sugar kettles. Functioning as the primary manager, engineer, sugar maker, and guard, Doyle clearly realized that in order to maximize productivity through the night, the planter's heavy hand and threatening whip proved invaluable.[19]

Doyle's attention to detail proved increasingly common among planters as they mechanized their equipment throughout the antebellum era. The constantly moving cane carrier and flywheels required highly attentive workers to keep an everwatchful eye on the turning belts and spinning gears that raced along at breakneck speed. Those who momentarily lost concentration found that the penalty for lax and careless work often proved excruciatingly painful, if not fatal. Jacob, a slave on William Palfrey's sugar estate, became caught on the steadily moving cane carrier, which dragged him up the conveyor belt toward the mill. Fortunately, he suffered only a dislocated collarbone and some severe bruises, but in an era of minimal safety protection, many were less fortunate. Writing to the Novelty Iron Works in New York, sugar master Maunsell White expressed his sorrow and anger at the death of a valuable female slave who died of complications following a grim accident in which she had caught her hand and arm in the moving parts of the mill while trying to unchoke it. The risk of industrial accidents during the sugarmaking process placed a premium on attention to detail and vigilance in the

19. Slave List, 1857, John H. Randolph Papers, LSU; Isaac Erwin Diary, 26 October–25 December 1851, LSU (first quote). Other examples of shift work include: John Slack to Henry Slack, 18 December 1854, Slack Family Papers, UNC; Franklin A. Hudson Diaries, 28 October 1853, UNC; Elu Landry Estate Plantation Diary and Ledger, 18 October 1849, LSU; William B. Hamilton to William S. Hamilton, 13 December 1857, William S. Hamilton Papers, LSU (second quote); *Plaquemine Sentinel,* 23 December 1857 (third quote).

workplace. Shift work partially resolved the imperative to maintain alert and efficient workers. Accidents slowed down productivity, but tiredness could prove more damaging if laborers tarried on the job, paid less attention to speed, or simply forgot to maintain an even temperature under the kettles. Under those circumstances, output materially declined, sugar production suffered, and plantation performance dwindled. By instituting shift work and close management, the sugar masters were able to guarantee that their skilled engine hands remained reasonably alert and capable of relentless manual labor hour upon hour, day after day.[20]

Shift work, like the commodification and division of labor, extended the masters' control and discipline by regimenting their slaves' daily activities. Although African Americans struggled courageously to maintain their dignity in this exploitative world, they were not able to overturn the routinization of their lives. Whether toiling in the cane fields or salvaging some brief moments of rest in the quarters, slaves were chained to the planter's regime and the inexorable rhythm of the ticking clock. Time was of the essence in sugar farming, and—like their efforts to fashion disciplined work gangs—planters similarly sought to instill clock-ordered discipline among the slaves. The sugar masters were not alone in sharing a commitment to formalized work rules and a clock-driven labor regime. Through ironclad factory schedules and the integration of flow production with the stability of shift labor, plant managers in the free North energetically strove to optimize workforce productivity through the drill and punctuality of industrial capitalism. The process of regulating workers by time management began in the eighteenth and early nineteenth centuries, when proto-industrialists began to impose clock discipline on their laborers. Finding the irregularity of the natural world incompatible with industrialization, reform-minded factory owners strove to enforce industrial discipline by measuring the worker's day with timesheets and by gauging their daily productivity. Ever anxious to enhance the intensity, reliability, and effectiveness of slave labor, southern planters similarly embraced the clock and its incumbent capitalist ethos, just as they had eagerly adopted shift work and industrial discipline. By timing their workers and introducing the clock as a component of plantation management, slaveholders, historian Mark Smith observes, acted as "clock-conscious

20. Plantation Diary, 1842–1859, 1867 (vol. 17), 21 November 1846, Palfrey Family Papers, LSU; Maunsell White to Messrs. Stillman, Allen, & Co., 1 December 1845, Letter Book, 1845–1850 (vol. 1), Maunsell White Papers, UNC.

capitalists . . . [who] in effect produced a time-based form of plantation capitalism" in the antebellum South.[21]

The sugar masters knew better than most that time is money and that their agricultural operations hinged on the careful assessment of weather, the projected duration of the harvest, and the length of time it would take their bondspeople to complete the various stages of the grinding season. As slaveholder Edward Butler observed, careless management, particularly during cold autumnal nights, "taught me the propriety and necessity of a sugar planter making the most of his time." Costly freezes impelled the sugar masters to maximize the productivity of the working day. Selectively drawing upon the improvement literature available in national and regional periodicals, cane planters increasingly sought to mirror the time-saving techniques of the scientific agrarians they read about. The steam-powered sugar mill stood as the greatest time-saving implement on the plantation, but draining machines, bagasse dryers, conveyor belts, and plantation railroads all gained more time for the slaveholders. Whether they utilized this extra time to mature their canes a little while longer or to gain a few more weeks for cane planting, the sugar masters suffused their lives with time management. In an ideal world, interdependent gang work optimized time, shifts codified a definitive working time, and the division of labor linked tasks and time still further. Industrial discipline and military regimentation in the cane fields and mill house saved minutes and hours over more lackadaisical management; even the central location of the slave quarters ensured that little time would be lost when the crews trudged out to the fields in the early morning light. In their estate journals, planters recorded when the planting commenced, when the canes suckered, and when the harvest began. Their harvest diaries detailed how much sugar was made, the duration of each sugar-granulating strike, and when each new shift came online. Estate managers observed their employers' fixation with record-keeping and kept plantation journals that listed their daily activities, the allocation of labor, and the division of time. Further reflecting the ascendancy of clock and calendar, the sugar masters scribbled down birth dates, recorded deaths, and noted the short birth intervals of those

21. On time management, see E. P. Thompson, "Time, Work-Discipline, and Industrial Capitalism," *Past and Present* 38 (December 1967): 56–97; Michael O'Malley, "Time Work, and Task Orientation: A Critique of American Historiography," *Time and Society* 1 (September 1992): 341–58. For the American South, see Mark M. Smith, "Time, Slavery, and Plantation Capitalism in the Ante-Bellum American South," *Past and Present* 150 (February 1996): 143 (first quote); Smith, *Mastered by the Clock*, 95–127, 144–46.

they called "good breeders." The regularly scheduled steamboats that carried them to New Orleans docked at a certain time and left planters with time enough to buy a slave or two before returning to the sugar country on the daily packet or scheduled railroad service. Once inside the New Orleans auction room, a planter gauged how many years of reproductive service might remain to the young woman on the rostrum block; once back on his domain, he might use a wet nurse to minimize lost time between pregnancies. Time defined the sugar regime, and most slave lords did not require an object lesson in its value or social function.[22]

The ticking clock and plantation almanac monitored the passage of time with unprecedented precision and enabled slaveholders to measure workfloor productivity in unnerving detail. Like other planters in the slave states, the sugar masters believed that the clock would serve as an ideal tool for disciplined, authoritarian control. The planters' time management added a gloss to their modern sensibilities, but this veneer proved wafer-thin when inculcating time awareness among their slaves. Planters feared that slaves who mastered the clock might soon divert time for their own activities, orchestrate slow-downs at an appointed hour, or even use the clock to time organized resistance. Telling the time proved a double-edged sword for both masters and slaves. By the late antebellum era, however, free and enslaved Louisianans shared a common time-awareness in which mechanically defined time merged with a belief in a naturally ordered clock.

Like most occidental people in the eighteenth century, West African slaves retained a belief in the natural order of time, in which the daily rhythm of farm work profoundly influenced the working day. Anthropologists contend that workers in preindustrial societies labored to the natural and rhythmic order of the agricultural year rather than internalizing the disciplined time-awareness of the industrial age. Sub-Saharan historians and anthropologists maintain that Africans remained bound to a natural order in which the sun and essential daily tasks shaped their concept of time. In contrast to time-conscious peoples, most nineteenth-century Africans conceived time as comprised of events. A day, month, or year is fundamentally the summation of events. There is no fixed time, anthropologists conclude, "which is independent of events, and which can be computed for its own sake." This traditional concept of time clashed with the emerging cult of rapidity that forcefully emerged during the market revolution and found its preeminent

22. E. G. W Butler to the Hon. Judge Thomas Butler, 2 December 1832, Thomas Butler and Family Papers, LSU.

form in New England. Europeans who visited the United States found the frenzied pace of life "decidedly faster, more frenetic than in the cities of Europe." As one American candidly concluded, "We are born in haste, . . . we finish our education on the run; we marry on the wing, we make a fortune at a stroke, and lose it in the same manner. . . . Our body is a locomotive, going at the rate of twenty-five miles an hour; our soul, a high-pressure engine." Aptly choosing the steam engine as a metaphor for his time-conscious culture, this nineteenth-century observer pointed to the emerging preoccupation with punctuality, time, and ordered industrial discipline. In northern factories, the capitalist principles of the industrial revolution rang out throughout the workplace as mill owners and plant managers established a clock-orientated labor regime that advanced, halted, and began afresh according to the hands on the factory clock. The white-faced clock with its slender black hands demarcated the working day. It timed labor shifts, relentlessly ticked away one's minutes of rest and leisure, and triggered the shrill peal of work bells that set labor and machinery in motion.[23]

Whether white- or black-faced, with opposing colored hands, the monochrome clock chimed orders throughout the sugar country and ticked in unison with the slaveholders' order. Aficionados of mechanization, the division of labor, and flow production, the sugar masters strove to substitute the natural pace of time with formalized work rules and the regimented discipline of the factory clock. Aping northern industrialists and drawing upon advice in the three principal southern agricultural journals, planters studied not only the value of time but, crucially, how to impose it. In a syndicated article that was reprinted at least six times between 1850 and 1855, one planter counseled fellow agrarians that the key to judicious slave management lay in the "saving of time." Calling upon slaveholders to

23. On the traditional concept of time among Africans, see Earl McKenzie, "Time in European and African Philosophy: A Comparison," *Caribbean Quarterly* 19 (September 1973): 82 (first quote); John S. Mbiti, *African Religions and Philosophy* (New York, 1969), 15–28; Ivor Wilks, "On Mentally Mapping Greater Asante: A Study of Time and Motion," *Journal of African History* 33 (spring 1992): 175–90. For work on the imposition of time discipline in the industrial North, see Carlene Stephens, "'The Most Reliable Time': William Bond, the New England Railroads, and Time Awareness in Nineteenth-Century America," *Technology and Culture* 30 (January 1989): 23 (second quote); Paul B. Hensley, "Time, Work, and Social Context in New England," *New England Quarterly* 65 (December 1992): 531–59; Martin Bruegel, "'Time That Can Be Relied Upon': The Evolution of Time Consciousness in the Mid-Hudson Valley, 1790–1860," *Journal of Social History* 28 (spring 1995): 547–64; Prude, *Coming of the Industrial Order*, 130–31; Anthony F. C. Wallace, *Rockdale: The Growth of an American Village in the Early Industrial Revolution* (New York, 1978), 177–79.

employ a central slave cook to prepare all meals for the bondspeople, the anonymous contributor calculated that prudent, time-conscious labor management of
this sort would yield hours of saved labor time that the planter could expropriate
for field work. With such broad dissemination, the reprinted article probably lay
on more than one sugar planter's desk and was closely read by those interested in
plantation supervision.[24]

Such time-saving advice about the preparation of meals unquestionably appealed to the sugar masters; during the grinding season, most planters selected one
or two cooks to prepare meals for all. Estimating that each slave family probably required two hours to cook, eat, and rest after their meal, planters realized that by
pooling their resources during the harvest, they could achieve a significant saving
in labor time. In 1859, for instance, Samuel McCutchon followed published wisdom and delegated three rather elderly and sick women to cook for all the field
hands. His cooks were plainly unsuited to field work, as they included Milly, who
was a perennial rheumatic, fifty-eight-year-old Beershiba, who was physically
handicapped, and asthma-suffering Betsey. McCutchon's kitchen staff undertook
to prepare meals for the 107 working adult slaves on his Plaquemine Parish estate.
By organizing a centralized refectory meal service, McCutchon appropriated the
precious working time of healthy, strong adults while also ensuring that his slaves
returned to their work adequately fed and freed from the chores of cooking and
washing. Thus fortified, slaves would be less likely to slow their pace as they labored
under the barking orders of the sugar masters. Overseer George Woodruff similarly
consolidated cooking operations prior to the rolling season at Frogmoor Plantation.
After discharging his daily duties, Woodruff wrote his plantation journal in a copy
of Thomas Affleck's *Sugar Plantation Record and Account Book*. Published primarily
for the improving planter, Affleck's register provided tabulated pages in which
overseers and managers could record every index of plantation performance. These
readily available ledgers routinized record keeping and plantation accountancy,
and they "followed the best and most advanced principles of efficient management." Beyond schooling the planter in the principles of double-entry bookkeeping

24. "The Management of Negroes," published successively in *Southern Cultivator* 8 (November
1850): 162–64; *Southern Planter* 2 (February 1851): 39–43; *De Bow's Review* 10 (March 1851): 326–28;
De Bow's Review 19 (September 1855): 358–63; and *Southern Cultivator* 13 (June 1855): 171–74. The four
journals—*De Bow's Review, Southern Cultivator, Southern Planter,* and *Southern Agriculturist*—while having a regional appeal, were purchased in large numbers by antebellum sugar planters. This is particularly true of *De Bow's Review*.

and reducing the overseers' records into a series of rows and charts—regulating the chaos of plantation accounts into apparent order—Affleck's account books included a two-page essay entitled "Duties of an Overseer." Urging the pursuit of judicious management, Affleck advised overseers to supply generous quantities of well-cooked food at regular intervals. Evidently following this stricture, Woodruff instituted centralized cooking on Monday, October 26, 1857, exactly one day before the start of the grinding season. Canteen-style food preparation further allowed the planter to regiment his slaves' diet as workloads peaked, while saving hours that could be appropriated to sugar production. Former slave Albert Patterson, who grew up on an estate in Plaquemines Parish, attested to the prevalence of centralized meal services when he recalled that a cook managed the large kitchen and prepared meals for the slaves while they worked. Characteristically, sugar lord William J. Minor required few lessons on the importance of regulating slave nutrition; he counseled his overseers to "see the various rations given out and that the food (particularly the bread and vegetables) be well cooked and delivered at proper hours to the hands." Like his compatriots, Minor understood that by centralizing operations and pursuing time-thrifty management, he might gain time for more pressing agricultural tasks and enhance plantation efficiency as a result.[25]

Other articles published on the optimization of slave labor also urged planters to teach their slaves that every minute belonged not to themselves, but to their masters. Publishing in both the *Southern Cultivator* and *De Bow's Review,* Robert Collins beseeched planters across his native South to ensure that their workers followed a rigidly structured day. According to Collins, slaves should begin work at the first light, break at eight A.M. for breakfast, and stop again at noon for a two-hour break during the heat of the early afternoon. They should return to their tools

25. Inventory and Valuation of Slaves, Stock, and Farming Utensils of Oaklands Plantation, Samuel D. McCutchon Papers, LSU; Sitterson, "William J. Minor Plantations," 66; Frogmoor Plantation Diary, 26 October 1857, Turnbull-Bowman-Lyons Family Papers, LSU; Interview with Albert Patterson, 22 May 1940, WPA Ex-Slave Narrative Collection, LSU; Rules and Regulations on Governing Southdown and Hollywood Plantations, Plantation Diary, 1861–1868 (vol. 34), William J. Minor and Family Papers, LSU (second quote). On the widespread acclaim for Affleck's *Plantation Record and Account Book,* see *American Agriculturist* 6 (November 1847): 356, and *Franklin Planters' Banner,* 14 January 1847. Also see Robert Williams, "Thomas Affleck: Missionary to the Planter, the Farmer, and the Gardener," *AH* 31 (July 1957): 46, quoted in Michael Mullin, *Africa in America: Slave Acculturation and Resistance in the American South and British Caribbean, 1736–1860* (Urbana, Ill, 1992), 120–21 (first quote).

at two o'clock and labor on until nightfall. To enforce such discipline, however, required the imposition of the mechanical clock and the sounding of bells to notify
slaves of the changing time. Other advice literature published in the *Farmer's Register* and subsequently in a prominent sugar cultivator's newspaper, the *Thibodaux
Minerva,* provided overseers with clear counsel to time the slaves' working day by
blowing a horn to appraise the bondspeople of their daily schedule. Leading agricultural journalists urged their readers to emulate their northern brethren and
classify daily tasks by time and productivity, using a horn or bell to time workers.
Just as surely as the levers and gears of industrial capitalism turned, so too did the
hands upon the overseer's pocketwatch or the clock on the mill-house wall. Writing in *De Bow's Review,* Joseph Acklen, a cotton planter in West Feliciana Parish, announced that the plantation manager rose at dawn each morning and proceeded to
ring a bell in the slave quarters to caution the slaves that roll call would occur in exactly twenty minutes. Acklen's rules for his plantation intimate that the slaves congregated within the allotted time and were quite conscious of the passing of time.
His "Rules in the Management of a Southern Estate" almost certainly shaped plantation supervision throughout the sugar country, where the sugar masters similarly utilized bells and horns to chime a variety of instructions—from the "appointed hour" for breast-feeding one's baby to the onset of the next shift.[26]

Embedding methodical and structured order within antebellum management
and seamlessly weaving time discipline into labor organization, the sugar masters
enforced a work discipline that advanced with clockwork punctuality. As early as
the 1820s, when Timothy Flint toured the sugar country, the contours of this exploitative regime had emerged. "There is in a large and respectable plantation,"
Flint observed, "as much precision in the rules, as much exactness in the times of

---

26. Robert Collins, "Essay on the Management of Slaves," *Southern Cultivator* 12 (July 1854):
205–6; *De Bow's Review* 17 (October 1854): 421–26; "Rules for Overseers," *Farmer's Register* 8 (April
1840): 230–31; *Thibodaux Minerva,* 30 June 1855; "Rules in the Management of a Southern Estate,"
De Bow's Review 22 (April 1857): 376. For other references to the use of horns and bells to institute
timed discipline and regular hours, see "Notions on the Management of Negroes," *Farmer's Register* 8
(December 1836): 494–95; "Management of Negroes," *Southern Cultivator* 9 (June 1851): 87–88;
"Management of Negroes," *De Bow's Review* 14 (February 1853): 176–78; Olmsted, *Seaboard Slave
States,* 2:314 (first quote). On the use of horns and bells as time signals, see Mark M. Smith, *Listen-
ing to Nineteenth-Century America* (Chapel Hill, N.C., 2001), 112–13; David Landes, *Revolution in Time:
Clocks and the Making of the Modern World* (Cambridge, Mass., 1983), 72–73; Moreno Fraginals, *Sugar
Mill,* 148.

going to sleep, awakening, going to labor, and resting before and after meals, as in a garrison under military discipline, or in a ship of war. A bell gives all the signals. Every slave at the assigned hour in the morning, is forthcoming to his labor, or his case is reported." Above all, system dictated plantation management, defining both estate efficiency and the successful production of sugar. As Flint noted, "All the process of agriculture are managed by system. Everything goes straightforward. There is no pulling down to-day the scheme of yesterday, and the whole amount of force is directed by the teaching of experience to the best result." Flint's detailed description of the links between time signals, discipline, and plantation productivity directly addressed the compatibility of slavery and the emergence of capitalist work rules in the sugar country. Thirty years after Flint made his observations, Bennet H. Barrow of Highland Plantation, who turned to sugar cultivation in the early 1850s, astutely counseled his overseers, "A plantation might be considered a piece of machinery, to operate successfully, all of its parts should be uniform and exact, and the impelling force regular and steady; and the master, if he pretended at all to attend to his business, should be their impelling force. . . . When a regular watch is established, each in turn performs his tour of duty, so that the most careless is at times, made to be observant and watchful—the very act of organizing a watch bespeaks a care and attention on the part of a master, which has due influence on the negro."[27]

Barrow's advice echoed throughout south Louisiana, where the emerging bourgeois obsession with time left its indelible print in the cane world. William Minor was only one of many planters who adopted the rubric of time management. Like his fellow agrarians, who counseled that "labor must be directed with an intelligent eye" and that plantation success rested on the "proper adaptation of the means to the end," Minor employed bells and established time signals to monitor his slaves' day. At nine o'clock in the evening, for instance, Minor's overseers rang the plantation bell warning all slaves that they must promptly return to their own houses. Minor's slaves undoubtedly fathomed the significance of both clock and time signals, especially when one's late arrival might prompt a brutal beating. Whether Minor physically measured or estimated the length of time it would take slaves to return to their quarters remains an open question, though he clearly stipulated that slaves

---

27. Flint, *History and Geography*, 1:244–45 (first quote); Edwin A. Davis, *Plantation Life in the Florida Parishes of Louisiana, 1836–1846, as Reflected in the Diary of Bennet H. Barrow* (New York, 1943), 409–10 (second quote).

had thirty minutes in the winter but only fifteen minutes in the spring and summer to return to their houses. The plantation bell commanded obedience, and, like Minor's strictures on the power of timed signals, sugar lords on other plantations strove to mimic the regulatory power of tolling bells. Much like contemporary industrialists and religious leaders before them, Louisiana's slaveholding elite strove to extend their authority over the clock and the plantation soundscape by aurally regulating their workers' time. Bells and whistles demarcated the working day and codified a precise, organized vision of labor in which discipline and regularity advanced at the inexorable pace of the clock and according to the slaveholder's behest. By aping northern business practices, planters announced their apparent modernity while simultaneously wielding a further weapon in their arsenal of disciplinary devices. Having mastered flow production in field and factory, planters adopted the regulatory power of time management and pierced the rural silence with clanging bells and chimed instructions. Although these time signals were specifically adopted to regiment the workplace, other signals, such as the peal of a steamboat bell tolling in the night, increasingly demarcated daily life in the southern states. Noting that the arrival and departure of scheduled steamboat services helped to instill order and regimen along the Mississippi Coast, Sir Charles Lyell observed that the "American captains are beginning to discipline the French proprietors into more punctual habits." Requiring no further lessons in promptness, slaves on the riverbank responded to a steamboat's imminent arrival by arriving exactly ten minutes after the captain sounded his bell. Whether slaves could tell time or merely heeded time according to bells, they were internalizing the metered pace of modern clock discipline, just as former slave Solomon Northup had done. Reflecting on his erstwhile owner's disciplined labor regimen, Northup noted that on William Turner's sugar estate, slaves entered and exited the mill house at the "proper time." Although such incidents might appear uneventful, African American slaves were responding to the disciplined time signals of the modern age and partaking of a clock-ordered regime that matched—and perhaps even surpassed—the activities of "some of the most *puissant* capitalists of the nineteenth-century world."[28]

28. *Southern Planter* 12 (June 1852): 163 (first quote); Rules and Regulations on Governing Southdown and Hollywood Plantations, Plantation Diary, 1861–1868 (vol. 34), William J. Minor and Family Papers, LSU; Lyell, *Second Visit to the United States,* 1:123 (second quote); Olmsted, *Seaboard Slave States,* 2:313; Northup, *Twelve Years a Slave,* 148 (third quote); Smith, *Debating Slavery,* 49 (fourth quote).

Few would have doubted such a conclusion in the antebellum sugar country. As early as 1819, Thomas Nutall firmly registered the powerful fusion between slavery and wealth formation during his visit to Wade Hampton's expansive sugar estate. Astutely observing that the slaves were the "engines of [Hampton's] wealth," Nutall concluded that bonded labor lay at the bedrock of an immense fortune that equaled "that of almost any English nobleman." Thirty years later, James Stirling similarly remarked on the efficacy of sugar work, noting that the Creole sugar master looked upon his slaves as "sugar machines." African Americans who toiled in this grim and forbidding land knew better than most that, as James Ramsey observed, the "discipline of a sugar plantation is exact as that of a regiment." Decades later, Ramsay's conclusions on West Indian slavery still rang true in the antebellum sugar country. Charles Stewart, a slave on Alexander Porter's Oak Lawn Plantation, echoed these sentiments from the workfloor when he observed that his master "wouldn't stand for no foolin' neither, I tell you. Things had to be jes' so . . . it was jes' stiddy management."[29]

Louisiana's sugar order left a brutal and exploitative imprint on those who worked the line and whose days were atomized, routinized, and divided by the ticking clock. Brigaded through the sugar fields by a predatory class of plantation sergeant majors, African Americans sweated through the heat of the summer on factory-like estates. Late antebellum sugar plantations hummed with the energy of the machine age and reverberated with the groans of exhausted slaves; field work and mill labor continued day in and day out with oppressive regularity and mind-numbing monotony. As ruthless and intrusive capitalists, the sugar masters modernized their immense agricultural enterprises and exploited the clock, the plantation layout, shift work, and the division of labor coldly and rationally to maximize slave labor. Whether in the 1780s or 1850s, the sugar regime gave an exacting rhythm to life in the sugar country. It dictated the amount of leisure time slaves fleetingly grasped, and it shaped the occupational pyramid within the plantation compound. And it was the demands of the Louisiana sugar economy that ultimately defined and dictated the master-slave relationship.

29. Thomas Nutall, *A Journey of Travels into the Arkansas Territory during the Year 1819* (Philadelphia, 1821), 239 (first quote); Stirling, *Letters from the Slave States,* 124 (second quote); James Ramsay, *An Essay on the Conversion of African Slaves in the British Sugar Colonies* (London, 1784), quoted in David Berry Gaspar, "Sugar Cultivation and Slave Life in the Americas," in *Cultivation and Culture,* ed. Berlin and Morgan, 114 (third quote); Charles Stewart, "My Life as a Slave," *Harper's New Monthly Magazine* 69 (October 1884): 738 (fourth quote).

four

"A VERY INGENIOUS AND MECHANICAL MAN"

J EAN DEBALLIÈVRE VALUED the excellence of slave workmanship. Firmly convinced of the compatibility of slavery with agro-industrial sugar production, Deballièvre not only surrendered the complex art of sugar making to his slaves, but he conceded his managerial accounts to the bondsman François, whom he purchased in 1829. A multitalented individual, François possessed the exceptional skills of reading and writing—faculties he clearly used, as his evaluation of $4,000 reflected his singular talent as an accountant. Deballièvre's shrewd interest in François was part of a broader purchasing drive that culminated in the acquisition of twelve highly trained and skilled slaves for a proposed plantation that he sought to establish near Baton Rouge. With bricklaying, coopering, and carpentry skills to match his accountancy talents, François's ability brought him to the attention of Achille Sigur d'Iberville, who offered to teach him the art of sugar making in the next grinding season. Hand-picked by Deballièvre as the crew leader, François was not alone in attracting this sugar planter's attention. Listed below the commander appeared the names of subcommander Nat, whose particular skills lay in carpentry and masonry, Jesse, a thirty-two-year-old mason and bricklayer, Jules, an excellent twenty-year-old bricklayer, and Peter, a young man of just sixteen, whose skills included shoveling, delving, plowing, and driving a wagon. Seven other craftsmen served to complete this exceptionally skilled slave crew.[1]

1. J. Deballièvre to John McDonogh, 24 March 1829, John McDonogh Papers, LSM. McDonagh ultimately sought to liberate his bondspeople in Liberia and thus had a particular interest in training slaves in specific skills. Although McDonagh was perhaps singular in teaching accountancy skills, planters throughout the sugar belt similarly trained their bondspeople. See Kendall, "New Orleans' 'Peculiar Institution,'" 21.

Deballièvre's chattel purchases made a mockery of antebellum racism and confounded definitions of southern backwardness. For almost a century, political economists from Adam Smith and David Ricardo to John Elliot Cairnes had condemned slavery as a deviant economic form that was doomed to drain the "very springs of human progress." Slavery, Emerson famously declared, "is no scholar, no improver; it does not love the whistle of the railroad . . . it does not improve the soil, everything goes to decay." As Adam Smith concluded, slaves were stripped of the right to the product of their labors and thus "have no other interest but to eat as much and labour as little as possible." With no recourse to wealth accumulation and the mutually reinforcing system of capital circulation and accretion, chattel labor proved ill-suited for efficiency, specialization, and an effective division of labor. But while François labored under compulsion and in the constant fear of violence, he—like many hundreds of skilled bondspeople in the sugar country—contested the damning appraisal of chattel slavery.[2]

Scholars of the American South have long grappled with John Elliot Cairnes's conclusions that slave labor was incompatible with work requiring care, forethought, or dexterity. The slave, Cairnes announced, "cannot be made to cooperate with machinery; he can only be trusted with the coarsest implements; he is incapable of all but the rudest forms of labor." The mid-nineteenth-century students who studied Cairnes's text and thumbed over his lecture notes in Queens' College, Galway, undoubtedly believed that the relative backwardness of the American South derived from its commitment to slavery. Perhaps some of them had been fortunate enough to see Frederick Douglass during his tour of Ireland or to hear Daniel O'Connell declare that slavery "has a natural, an inevitable tendency to brutalize every noble faculty of man." And perhaps they also knew that Professor Cairnes had mined Frederick Law Olmsted's *Journey in the Seaboard Slave States* to construct his damning indictment of the "Slave Power." Readers of Olmsted's

2. David Brion Davis, *Slavery and Human Progress* (New York, 1984), 80 (first quote); Ralph Waldo Emerson, "Address Delivered in Concord on the Anniversary of the Emancipation of the Negroes in the British West Indies, August 1, 1844," in ibid. (second quote), 110; Adam Smith, *An Inquiry into the Nature and Causes of the Wealth of Nations* (1776; reprint, Indianapolis, Ind., 1981), 388 (third quote). On the prevalence of contract and free-labor thought and slavery's anomalous position, see Amy Dru Stanley, *From Bondage to Contract: Wage Labor, Marriage, and the Market in the Age of Slave Emancipation* (Cambridge, 1998), 1–59; Eric Foner, "Free Labor and Nineteenth-Century Political Thought," in *The Market Revolution in America*, ed. Stokes and Conway (Charlottesville, Va., 1996), 99–127.

text, however, might have noted some inconsistencies in Cairnes's argument. The New Yorker not only described the division of labor and the prevalence of gang work in the slave South, but he also recorded the slaves' response to mechanized time signals, the introduction of costly machinery, yearly technological and agricultural improvements to the plantations, and the presence of slave mechanics. Olmsted rightly concluded that the enormity of capital, labor, and human lives expended in transforming the swamps into sugar plantations might have been better invested elsewhere, yet his portrait of the cane world presented an anomalous region in which slavery, industrialization, and conservatism commingled. Olmsted's evidence belied Cairnes's assertions and suggested that one part of the slave power was far from rudimentary. Indeed, it possessed some similarities to northern factory life. The speeding up and stretching out of the working day in Louisiana's cane world carried a slight echo of life in the New England mills, the timed bells on the sugar estate similarly tolled on the Philadelphia factory wall, and the incessant pace of machine work policed free and enslaved workers from Troy, New York, to Thibodaux, Louisiana. While there were many more differences than similarities between the Hudson Valley and the lower Mississippi Valley, Olmsted's sojourn in the sugar country indicated that a fracture existed in what he saw as a "bipolar" America.[3]

Historians have been trying to prize open that fracture for the past thirty years, and they have found multiple cases in which slavery and capitalism appeared to function as one. Whether in the Maryland Chemical Works, in the Tredegar iron foundry, or on the southern railroads, slave labor frequently proved highly proficient and exceptionally profitable. Capable of highly skilled work, slaves were not incompatible with industrialization or technical progress. Indeed, as Robert Fogel has contentiously argued, African American slaves embraced the Protestant work ethic and became "metaphoric clock punchers," who labored "under a regime that was more like a modern assembly line." Fogel's assumption that diligent slaves identified with capitalism proved problematic; but, as one antebellum

3. Fred Bateman and Thomas Weiss, *A Deplorable Scarcity: The Failure of Industrialization in the Slave Economy* (Chapel Hill, N.C., 1981), 80 (first quote); Olmsted, *Seaboard Slave States,* 2:322; Davis, *Slavery and Human Progress,* 253 (third quote). On Cairnes, see ibid., 244–45, 250–58. Daniel O'Connell quote in William Lloyd Garrison, "Preface," *Narrative of the Life of Frederick Douglass, An American Slave,* ed. Houston A. Baker Jr. (1845; reprint, New York, 1982), 37 (second quote). For an example of Olmsted's conclusion that the slave is suited only to the "simplest and rudest forms of labor," see his *Seaboard Slave States,* 2:239.

railroad booster declared, chattel labor was "not liable to strikes and riots and the consequent of tearing up rail and burning depots." Slave railway workers served in a variety of skilled roles, notably as brakemen, firemen, machine-shop mechanics, and even as locomotive drivers. The demand for slaves "as a measure of economy and good policy" created a significant demand for skilled hands in the southern railroad sector, where hundreds of bondspeople toiled on the line. Even the modestly sized Baton Rouge, Opelousas, and Gross Tete Railroad possessed skilled slaves, whose very presence on the tracks confirmed the sugar masters' convictions that slavery and mechanization were not a priori incompatible.[4]

Not all scholars, however, have been so convinced of the presence of this fault line. As Eugene Genovese argued in his outstanding study of American bondage, slave masters ultimately "presided over a plantation system that constituted a halfway house between peasant and factory culture." Bedeviled by a labor force that resisted the "regularity and routine which became the *sine qua non* for industrial society," Genovese asserted, slaves clung to a prebourgeois traditionalism. Although they derided laziness, they often proved careless with their implements and rejected routinized work, disciplined time management, and materialism in favor of a work ethic that shared aspects of African communalism and their own experiences as slaves. By downing tools, feigning illness, breaking hoes, and deliberately slowing their pace of work, slaves protected their culture and simultaneously minimized their labor for the slaveholding elite. Planters rationalized these episodes of black resistance, categorizing African Americans as lethargic workers who could be expected neither to work steadily nor to operate advanced machinery, *race*

---

4. The literature on industrial slavery is extensive. See Robert Starobin, *Industrial Slavery in the Old South* (New York, 1970), 230; T. Stephen Whitman, "Industrial Slavery at the Margin: The Maryland Chemical Works," *JSH* 59 (February 1993): 31–62; John Bezis-Selfa, "A Tale of Two Ironworks: Slavery, Free Labor, and Resistance in the Early Republic," *William and Mary Quarterly* 56 (October 1999): 677–700; Charles Dew, *Ironmaker to the Confederacy: Joseph R. Anderson and the Tredegar Iron Works* (New Haven, Conn., 1966), 22–37. On the notion of the slave as a "metaphoric clock puncher," see Fogel and Engerman, *Time on the Cross*, 208 (second quote); Fogel, *Without Consent or Contract*, 162 (first quote). On works challenging these assumptions, see Herbert Gutman and Richard Sutch, "Sambo Makes Good, or Were Slaves Imbued with the Protestant Work Ethic?" in *Reckoning with Slavery*, ed. David et al. (New York, 1976), 55–93; Herbert Gutman, "Enslaved Afro-Americans and the 'Protestant' Work Ethic," in *Power and Culture* (New York, 1987), 298–325. As Peter Kolchin observes, slaves accepted hard work, though it was their resentment of compulsion that defined their mentality. See Peter Kolchin, *Unfree Labor: American Slavery and Russian Serfdom* (Cambridge, Mass., 1987), 108–10. On railroads, see Licht, *Working for the Railroad*, 67–69, 224–25 (third quote, 68).

which called for care and attention. Southern industry and agriculture consequently languished in inertia as slaveholders rejected care-intensive industries for effort-intensive work. Even in the Chesapeake iron industry, foundry owners discovered that the institution of slavery imposed an overwhelmingly conservative force on mechanization and manufacturing. Skilled slave artisans, for instance, produced high-quality wrought iron by the old tilt-hammer process, as their owners shunned technical modernization. Southern industrialist Daniel Pratt also utilized slaves in his manufacturing plant, yet by the 1850s he increasingly consigned slave labor to the simplest tasks. Throughout the slave belt, planters' sporadic attempts at instituting clock time and factory discipline largely failed; they reverted to the whip and either chose the simplest tools for prouction or, convinced of their slaves' indifference to care-intensive work, declined to innovate. While this remains largely valid for much of the slave South, it does not hold true in the sugar country, where enslaved workers utilized sophisticated tools and implements, performed care-intensive work, accommodated the machine, and conducted skilled tasks. Detailing the slaves' tasks in the regimented columns of their plantation ledgers, planters recategorized black labor. Slaves like François toiled not as drudges, but as efficient and skilled workers who fused agro-industry with plantation slavery.[5]

Deballièvre, however, faced a peculiar dilemma. How could he motivate François to learn the art of sugar making and display his accountancy talents when the bondsman had nothing to gain by the quality of his own work? Deballièvre's ultimate goal was thus to design a plantation system that tempted the bondsman to work in his owner's interest. This chapter explores Deballièvre's dilemma, examining the prevalence of skilled labor in the sugar country and the use of incentives

---

5. Genovese, *Roll, Jordan, Roll,* 286 (first quote), 309 (second quote). As corroboration of Genovese's interpretation, the number of skilled hands to field workers in the cotton belt might have been as small as one in fifty. See Michael P. Johnson, "Work, Culture, and the Slave Community: Slave Occupations in the Cotton Belt in 1860," *Labor History* (1984): 325–55, esp. 331. On technological conservatism and the use of slaves for the simplest tasks, see Charles B. Dew, *Bond of Iron: Master and Slave at Buffalo Forge* (New York, 1994), 333; Dew, "Slavery and Technology in the Antebellum Iron Industry: The Case of Buffalo Forge," in *Science and Medicine in the Old South,* ed. Numbers and Savitt (Baton Rouge, La., 1989), 107–26; Charles G. Steffen, "The Pre-Industrial Slave: Northampton Iron Works, 1780–1820," *Labor History* 20 (winter 1979): 95. On Daniel Pratt, see Evans, *Conquest of Labor,* 53–54. For a parallel debate over slavery and technology in Cuba, see Moreno Fraginals, *Sugar Mill,* 112, 134–135; Scott, *Slave Emancipation in Cuba,* 26–28.

by François and his master. Through a network of compensation systems, slaves and slaveholders negotiated a series of compromises in which bondspeople accommodated the machine age and secured their labor stability in return for prerogatives that extended African American rights on the mill floor and in the plantation commissary. This mutualism, however, in no sense signified the sharing of collective values or peaceful compromise over the terms of labor. Far from remaining as mute factors of production in the planters' mechanized world, slaves adapted to the machine age and the disciplined nature of sugar work largely because the conditions of industrialization provided them with a range of potential opportunities to enhance their living conditions and extend their own pathways toward autonomy and independence.[6]

As accomplished craftsmen, teamsters, coopers, carpenters, masons, engineers, and sugar makers, bondspeople quickly assumed positions of technical and managerial importance on the sugar estates. Such an arrangement allowed planters like Deballièvre to develop a long-term skills base on their estates, maintain workplace stability as the grinding season advanced, and avoid the racially destabilizing presence of whites on the plantation. Whenever feasible, planters sought to exclude white masons and mechanics from the plantation world, believing that itinerant white laborers proved intrinsically detrimental to slave discipline. Moreover, planters felt that the occasionally slovenly disposition of white laborers might prompt the skilled slave population to question its own status. The use of slave mechanics, by contrast, had several compelling advantages. Their presence circumvented the need for white labor, their training was significantly cheaper than hiring free workers, and their success confirmed the vitality of bondage in the sugar country. In the later antebellum years, planters developed teams of skilled slave workers who replaced all or many of the technical posts formerly held by whites. All, of course, save the overseer, who remained a symbol of white authority. The sugar masters were thus able to expel many itinerant workers from the agricultural belt and to consolidate their holdings around a group of enslaved artisans and mechanics. By relying so heavily on slave labor,

6. How slaves rearticulated the labor process and wrought autonomy from the working environment has long received scholarly attention. See Ira Berlin and Philip D. Morgan, "Labor and the Shaping of Slave Life in the Americas," in *Cultivation and Culture,* ed. Berlin and Morgan (Charlottesville, Va., 1993), 1–45, and most recently, David S. Cecelski, *The Waterman's Song: Slavery and Freedom in Maritime North Carolina* (Chapel Hill, N.C., 2001).

however, planters left their economic success perilously at the behest of the slave population.[7]

Skilled slaves like François, Nat, Jesse, and Jules were not uncommon in the sugar country. During the 1840s and 1850s, planters pressed ahead with their use of skilled hands, despite dissent by white artisans and overseers, who saw their skills diminished and their racial authority challenged by a new cadre of workers. On Residence Plantation, Robert Ruffin Barrow placed so much confidence in his slaves' capacity to operate the steam engine that he named one of them, Jake, as the primary engineer during the 1857 grinding season. Complaining that the throttle valve and governor on the steam engine operated poorly, Barrow's plantation manager, Ephraim Knowlton, remarked significantly, if not a little coolly, "I fear Jake does not understand the engine well [but] Mr. B says he is willing to trust Jake, so I will say nothing." Equally confident in the abilities of his other slaves, Barrow relied extensively on his skilled slave mechanics, Ruben and Shell, to repair a loose cylinder bolt on the Myrtle Grove Plantation sugar mill. Both Ruben and Shell clearly possessed the technical competence that Barrow cherished and the engineering skills that were critical to plantation productivity. Ill-use of the engine placed unnecessary stress on the machinery, and few estate mangers could risk a shoddy repair that would delay resumption of grinding cane. Thus when Barrow assigned his bonded mechanics to run the engine and repair the cylinders towards the peak of the harvest, he displayed his trust in slave workmanship and a conscious decision to favor his chattel with significant and weighty responsibilities. Barrow's reliance on his skilled slave laborers for engineering and sugar-house work was hardly anomalous; on many plantations, a few highly skilled slaves were designated for mechanical work. Planter John Dymond, for instance, trained his male slaves to be carpenters and/or machinists, while Bernard Marigny assigned engineering tasks to two slaves. At Bayside Plantation in St. Mary Parish, planters Moses Liddell and Francis DuBose Richardson also placed the assistant engineer's post in the hands of Monday, a skilled slave, who continued to toil, several years later, on the first watch of the sugar-making roll for 1851. On neighboring Grand Cote, David Weeks also established a team of skilled slaves, including Peter Congo,

7. Olmsted, *Seaboard Slave States,* 2:333; Shugg, *Origins of Class Struggle,* 90–96. Also see Sitterson, "Hired Labor," 192–205. A very small number of free African American overseers (only twenty-five) resided in Louisiana in 1854, mostly working in the New Orleans area. Significantly, of this group, twenty-two were mulattoes, reinforcing white-only notions of this key managerial role. See Scarborough, *Overseer,* 19.

a forty-two-year-old sugar maker valued at $1,000 in 1853, Somerset, a sawyer in 1835 who within a decade held the driver's post, Isaac, a youthful engineer of twenty-five, and Frederick, the sugar-mill engineer, who at twenty-seven clearly possessed considerable talents.[8]

Across the sugar bowl in St. James Parish, William P. Welham also maintained a highly skilled labor force in which slaves occupied the primary positions of power and responsibility. Slave William Bias, for instance, served as the sugar maker on Welham's Mississippi River estate, producing a bumper crop of 371 hogsheads of sugar with his steam-powered mill in 1859. Welham rightly concluded that his talented bondsman had the skill and capacity for diligent and effective work. To sustain Bias in his production duties, Welham established a clearly defined division of labor on his estate, and he supplemented his force with forty-six-year-old Jesse as the plow foreman, Southern as the plantation blacksmith, Davy as a cooper, and forty-one-year-old Aleck Ross as the carpenter. In preparation for the next decade of bonded labor, Welham put young Jean Baptiste to work learning the carpenter's trade alongside Aleck and Davy; he also had seventeen-year-old René begin learning the crucial job of coopering. As field drivers, plow foremen, sugar makers, and engineers, slaves such as Ruben, Shell, Monday, and Jesse defined the pace and success of plantation productivity. By managing all ancillary tasks, such as carpentry and coopering, bondspeople could heavily influence all aspects of production, including the marketing of crop. René's ability to build water- or treacle-tight casks, for instance, would partially define marketing success. An ill-fitting cask with sugar or molasses oozing out of it would materially harm Welham's commercial reputation among commission merchants and trade handlers along the New Orleans levee. The quality of René's workmanship thus linked the small cooper's shop in rural Louisiana to regional and national sources of mercantile credit that flowed into the sugar country on the strength of a planter's standing within the network of local commission agents. René's cypress or hickory casks and the excellence of the sugar William Bias placed in them had a direct impact on Welham's

---

8. Residence Journal of R. R. Barrow, 30 November and 15 December 1857, Robert Ruffin Barrow Papers, UNC (quote); Moses Liddell to John R. Liddell, 29 November 1848, and Sugar Making Roll, 31 October 1851, Moses and St. John Richardson Liddell Family Papers, LSU; Copy of Act of Sale of Slaves and Plantation by Bernard Marigny and William and Haywood Stackhouse on 13 March 1852, dated 26 January 1872, Ross/Stackhouse Records, LSM; Interview with St. Ann Johnson, 8 February 1940, WPA Ex-Slave Narrative Collection, LSU; Inventory for 1835 and Inventory for 1846, David Weeks and Family Papers, LSU.

short-term profits and long-term business integrity. Few economically and socially astute planters willingly trafficked substandard produce, since the inevitable result was commercial and personal derision. René's coopering skills accordingly enmeshed him within a web of racialized production and consumption that linked the outbuildings at Welham's plantation with the slavemaster's standing in the community and his commercial reputation in New Orleans.[9]

Whether refined or not, finished sugar had traversed many disparate processes that converted six- to eight-foot-tall canes into crystalline granules. Dozens of individuals handled the produce during its manufacture, and the opportunities for commercial sabotage proved endless. Blunt knives slowed the cane cutting, a loosened wheel bolt brought the cane wagon crashing to one side, delays threw the flow procedure out of kilter, a well-placed kick lamed the mule, and too much heat scorched the sugar. But the risks only began there. Inattention to one's wood supply put out the fires beneath the kettles, alien objects or stones on the cane carrier brought the belts to a grinding halt, a misplaced spanner immobilized the steam engine, and shoddy fittings let hot steam out in shrill blasts that triggered alarm on the mill floor and anxious looks from planters and estate managers. And from there, the list of potential mishaps continued. Fires remained relatively infrequent, but a loose bung guaranteed that the planter's valuable produce would spill as laborers rolled out the casks and transported them down to the quay, while slippery hands either intentionally or unintentionally dropped the casks, and tired eyes watched sticky molasses seep out over the split barrel staves. Indeed, the risks continued until the hogsheads and casks were safely stowed away on board a steamboat or railroad car. Given the enormous risks involved in sugar production, it remains extraordinarily surprising that planters relied on a labor system that brazenly flaunted free labor ideology. Yet deep in St. James Parish, William Welham contested the critique that openly condemned slaveholders and their "peculiar institution" as doomed to economic and social backwardness.

Like Welham, Edward Gay found that skilled slave labor proved proficient for the administration of his estate in the late 1840s. Bondsmen William Saunders and Jacob Lennox served as Gay's sugar makers, while Bill Garner and a host of other skilled slaves built carts, made shoes, and coopered barrels on his plantation. Aware that skilled and experienced slaves were a necessity on every sugar estate,

9. List of Slaves Found on the Homestead of William P. Welham, 10 December 1860, William P. Welham Plantation Records, LSU; P. A. Champomier, *Statement of the Sugar Crop, 1859–1860*, 15.

Francis D. Richardson considered buying Colbert, a particularly dexterous slave, for his plantation on Bayou Teche. Whether the bondsman had been baptized Colbert or had been named after Jean Baptiste Colbert, one of Louis XIV's premier and most efficient administrators, he unquestionably possessed an array of talents and attributes that made him highly valuable. Colbert, William Winans wrote his nephew in the Attakapas district, "is a very ingenious and mechanical man" who had expertise in operating a steam engine, tanning leather, carpentry, wagon repairing, bricklaying, and blacksmithing. Skilled as a "practical engineer," Colbert was valued between $1,800 and $2,000. Despite his many attributes, Colbert proved to be somewhat slow at work, but as Winans declared, "His slowness is not the effect of indolence, but of an extreme desire to do his work well." Yet Colbert was industrious in his free time as well, for he had attempted to escape to the North on two occasions. Remarking that these incidents and Colbert's alleged lasciviousness proved problematic, Winans nonetheless urged his kinsman to strengthen his plantation slave crews with this particularly capable and skilled bondsman. Like François, William Bias, Ruben, and Shell, Colbert converted his technical expertise into autonomy and attained a position of power and responsibility in the sugar-production process. The proletarianization of workers that assured a downward trajectory in the white working experience operated in reverse for skilled slaves, who could translate their skills into upward advancement that carried payment, power, and a modicum of protection from the slaveholder's lash.[10]

Perhaps unsurprisingly, Louisiana labor practices mirrored those of the sugar-producing Caribbean. As in Louisiana, the planters' allocation of labor created an unofficial status system within the workplace that remained gender- and skill-defined. By the 1830s, slaves in Jamaica who had expectations of status within the plantation economy grasped the essence of freedom more readily than their less privileged counterparts, seeking semi-autonomous lives in skilled work. The experience of slaves like Colbert paralleled those in the Antilles, where a growing population of expert laborers converted their skills into status and their knowledge of the sugar economy into a powerful bargaining position that they exploited both before and after emancipation. Skilled labor, however, remained over- whelmingly a gendered phenomenon in both Louisiana and the Caribbean. Men dominated the technical roles and rose through the ranks from carters to drivers

10. Succession of Andrew Hynes, 12 April 1850, Edward J. Gay and Family Papers, LSU; William Winans to Francis D. Richardson, 15 April 1847, Simpson and Brumby Family Papers, UNC (quote).

and from coopers to sugar makers. Women, by contrast, carried out diverse field tasks and frequently staffed the cane carriers, but they rarely served as the sugar makers and engineers who toiled over the boiling kettles. Planters willingly appropriated the female body for their reproductive designs, but their gendered notions of work ultimately spared slave women from the worst excesses of the mill.[11]

The elevation of black skilled labor to a par with white work significantly challenged the racialized rubric of the antebellum South. On Eureka Plantation, the established paradigms of racial order collapsed as skilled slave workers proved integral in modernizing and improving the sugar works. Evidently impressed by the quality of the pure white sugar at Eureka, local visitors from neighboring Plaquemine scrutinized the estate's production facilities and examined the fusion of slavery and mechanization. Upon visiting James J. Hanna's Ardoyne Plantation in 1854, the editorial staff of the *Thibodaux Minerva* observed Hanna's slaves steadily managing a steam engine and two vacuum pans, producing a sugar "to rival the snow in whiteness, and the diamond in the sparkling brilliancy of its grain." Deeply impressed, the visitors remarked that Hanna "employed neither sugar maker nor sugar boiler, except the negroes who belonged to the plantation." Utilizing the latest technology with competence and dexterity, Hanna's bonded mill crews were highly successful, as annual production rose swiftly; by 1854, slaves produced far in excess of half a million pounds of sugar on his estate. Not all sugar makers were so convinced by the slaves' capacity to utilize the latest vacuum technology without the presence of white sugar makers, though on many estates where steam mills turned, relatively skilled slaves toiled at the kettles and throughout the sugarhouse.[12]

Ardoyne Plantation proved characteristic of many sugar estates in the late antebellum era. The widespread prevalence of skilled bondspeople—like Colbert, or

11. On the training of specialist Caribbean slaves and the necessity of planter-worker cooperation, see Michael Craton, *Searching for the Invisible Man: Slaves and Plantation Life in Jamaica* (Cambridge, Mass., 1978), 223–33; Mary Turner, "Slave Workers, Subsistence, and Labour Bargaining: Amity Hall, Jamaica, 1805–1832," in *The Slaves' Economy,* ed. Berlin and Morgan (London, 1991), 96. Lawrence T. McDonnell, Peter Way, and Fogel and Engerman allude to the presence of skilled slaves who "translated technical knowledge into a measure of autonomy." In terms of workplace organization, skilled slaves experienced an "upward movement," while a "downward convergence" occurred in the white work experience. See McDonnell, "Work, Culture, and Society in the Slave South, 1790–1861" in *Black and White Cultural Interaction in the Antebellum South,* ed. Ownby (Jackson, Miss., 1993), 133 (second quote), 137 (first quote); Peter Way, *Common Labor: Workers and the Digging of North American Canals, 1780–1860* (1993; reprint, Baltimore, 1997), 87; Fogel and Engerman, *Time on the Cross,* 149.

12. *Plaquemine Sentinel,* 23 December 1857; *Thibodaux Minerva,* 23 December 1854 (quote).

the slaves at Ardoyne—indicates that some planters consciously favored bonded labor and enthusiastically utilized slave hands even where free sugar makers plied
their trade. This process accelerated with the introduction of steam power in the
1840s, but even in the 1820s some sugar planters undertook the systematic training of their slaves in the art of making sugar. Noting that "all depends upon getting
one acquainted with graining and managing," farming periodicals apprized future
settlers that bondsmen required only a season to learn the sugar maker's trade adequately. Three decades later, training in the art of steam power and vacuum processing took somewhat longer. As Robert Russell observed, however, on his visit
to a large sugar estate in St. James Parish, the sugar maker and all "inferior
functionaries were negroes . . . who were, from the trust which was committed
to them, evidently possessed of considerable skill and intelligence." Even in the
most responsible jobs, accomplished slaves proved successful and influential.
Drawing attention to the sugar maker's signal role in boiling the cane juice,
Thomas Bangs Thorpe observed, "His commands, be he as black as midnight, are
attended to with a[n] unquestioning punctuality that shows how much is dependent upon his skill." Throughout the sugar country, these skilled sugar makers and
dexterous engineers challenged white Louisianans to rethink the terms of unfree
labor and to develop a slave regime that placed the skilled bondsperson at the axis
of business growth.[13]

Although some slaves parlayed their skills into positions of command and authority, the factory system that atomized slave work created many more unskilled
roles. On most sugar plantations, perhaps as much as 85 percent of the slave population occupied unskilled roles. For these slave workers, the shrill blasts of a
steam engine horn, the tolling of a plantation bell, and the booming commands of
an overseer subdivided work, fragmented the working day, and divorced production rates from human variables. These workers had limited control over the
means of production and were simultaneously bound to the machine and the monotonous pace of work. For most assembly-line production techniques, where
technology established a metered cadence to work, the training of slaves in low-
skilled tasks was a relatively quick process. In the sugar mill and out in the fields,
many duties merely required the bondsperson to perform repeated actions to the
pace of the machine or the cane-cutting or seeding gang. Yet the sugar masters ignored even the least skilled workers at their peril.[14]

13. *American Farmer* 10 (April 1828): 33 (first quote); Russell, *North America*, 275 (second quote);
Thorpe, "Sugar and the Sugar Region," 764–65 (third quote).

14. On the proportion of slave mechanics, see Gutman and Sutch, "Sambo Makes Good," 78–80.

Like most industrial sugar estates, Bowdon Plantation possessed a mechanical cane carrier that transported the freshly cut shoots to the mill. Eager to establish a work gang that could match the methodical drumming of his machine, Hore Browse Trist installed Aggy as the captain of the cane-carrier brigade on his Ascension Parish estate. Aggy's team served as an important link between the field and factory, as delays at this key bottleneck could disrupt the supply of cane to the expensive Rillieux apparatus. While the task of loading and sorting canes onto a conveyor belt required only minimum skills, Aggy's team actually possessed quite considerable leverage over the entire production process. Like their skilled brethren, they could use their collective power to paralyze the sugar regime. Few sugar masters took this—or Aggy—for granted. Obviously, recalcitrant workers could be replaced, but alternating and retraining a new crew entailed an aggravating delay in a very time-conscious industry. Sending Aggy's team to the fields additionally required training in new roles, and workers on the cane carrier might require extra supervision to ensure that they did not institute their own time delays or, worse still, commit sabotage. Planters understood the potential risks of go-slows and the additional supervisory costs involved in retraining even un- or semi-skilled workers. And perhaps most critically, they realized the fundamental necessity of negotiating and compromising with Aggy.[15]

Although the exigencies of sugar cultivation placed the field hand and mill-house operative in a potentially unparalleled position of power, few estates suffered from endemic workplace resistance. Paradoxically, slaves neither rejected the order of the industrial age nor eschewed the disciplined pace of the mechanized sugar mill. Rather, they staffed the steam engines despite exhausting conditions, long hours, and the feverish pace of the sugar harvest. And if the following accounts are to be even partially believed, they embraced the labor harvest with zeal. Northern contemporaries who traveled through the sugar districts naively observed that the grinding season was "certainly the gayest and happiest portion of the year." Closer to home, newspaper editors paralleled these puerile sentiments. One pronounced that "although the necessities of the crop demand almost incessant exertion, the happy blacks seemed to enjoy the fun, [and] they 'went at it' with much more ardor and zeal than at any other labor." Despite the absence of rest and recreation, African American bondsmen "prefer it to any other employment, and always look forward to the grinding season as a pleasant and exciting holyday." Some thirty years after Aggy toiled under the incessant strain of the agro-industrial order,

15. H. B. Trist to Bringier, 25 November 1854, Trist Wood Papers, UNC.

planter Francis DuBose Richardson recalled the sugar harvest as halcyon days of peace and stability. Drawing upon a highly selective version of his memories, Richardson remembered that despite night watches and a frenetic pace of work, the slaves worked to the cadence of a short-meter song "that fairly made the old cane-shed shake." Richardson's delight at the slaves' rhythmic accompaniment to their work echoed the sentiments of other slave masters, who encouraged singing because they believed it enhanced productivity. Songs carried very different and emotive meanings to the slaves themselves; but by imposing an African American rhythm onto the exploitative nature of Louisiana capitalism, the bondspeople set their own musical tempo to the repetitive beat of steam turbines and revolving flywheels.[16]

Frederick Law Olmsted proffered a more revealing explanation of why slaves appeared to work more diligently and cheerfully during the grinding season than at any other time of the year. "The reason for it," he wrote, "is that they are better paid; they have better and more varied food and stimulants than usual, but especially because they have a degree of freedom and social pleasure, and a variety of occupations which brings a recreation of mind, and to a certain degree gives them strength for, and pleasure in, their labor. Men of sense have discovered when they desire to get extraordinary exertions from their slaves, it is better to offer them rewards than to whip them; to encourage them rather than to drive them." While overestimating the possibilities of freedom and social pleasure during the grinding season and seriously misjudging the power of the lash, the New Yorker accurately portrayed the rich network of compromises and negotiations that defined the master-slave relationship in the sugar country. Revising the networks of bondage to accommodate their need for labor stability and the slaves' overwhelming desire for greater autonomy, masters employed incentives, rewards, and bonuses as a way to maximize production, while slaves utilized overtime work to bring the demands of both capital and labor into an uneasy balance.[17]

---

16. Ingraham, *South-West*, 1:238 (first quote); *New Orleans Weekly Delta*, 18 October 1847 (second quote); Richardson, "Teche Country Fifty Years Ago," 595 (third quote). On the use of songs to set the pace of work, see William Cullen Bryant, *Letters of a Traveller* (1850), quoted in John T. O'Brien, "Factory, Church, and Community: Blacks in Antebellum Richmond," *JSH* 44 (November 1978): 520.

17. Olmsted, *Seaboard Slave States*, 2:327 (quote); Sitterson, *Sugar Country*, 98. On incentives, see Stephen C. Crawford, "Punishments and Rewards," in *Without Consent or Contract (Technical Papers)*, vol. 2, *Conditions of Slave Life and the Transition to Freedom*, ed. Fogel and Engerman (New York, 1992), 536–50; Ronald Finley, "Slavery, Incentives, and Manumission: A Theoretical Model," *Journal of Political Economy* 83 (October 1975): 923–33; Stanley L. Engerman, "Slavery, Serfdom, and Other

Slaves and owners viewed this process of negotiation from very different perspectives. Planters conceived of incentives as a means to cajole recalcitrant bondspeople to work and entice the bonded workers to toil in their interest. Slaves, by contrast, viewed rewards, clothing, holidays, payment, and extra food allotments as due recompense for their labor. Either way, the dynamics of the marketplace ultimately defined the master-slave relationship in the cane world. At issue was the central question: what kind of pay for how much work? The definition of a bondsperson's wage proved malleable and included a vast array of what planters called incentives and what slaves deemed to be customary rights. Peering at the process of labor bargaining from opposite ends of a telescope, so to speak, slaves and slaveholders contested the terms of the master-slave relationship and derived very different benefits from the act of negotiation. From the sugar masters' economic perspective, negotiation and compromise established stability, improved morale, enabled them to introduce new technology and management practices, and fused the slaves' well-being with that of the plantation. Slaves considered the process of workplace bargaining as a way to derive the maximum possible return for their labor and to carve out a sphere of autonomy and independence. Unfortunately, labor's modest victories proved tantalizingly hollow. Although African Americans seized certain prerogatives from their masters, they dulled their radicalism with short-term material gains and ultimately accommodated not solely the machines but also the broader social, racial, and commercial system that exploited them.[18]

Slaveholders from the Carolinas to the Gulf of Mexico faced a set of almost identical problems, which revolved around one question: how could they create a plantation system that tempted the slave to work in his or her owner's interest? They knew from their extensive experience of paying African Americans for overtime work or the produce of their gardens that slaves seemed to work more diligently for

Forms of Coerced Labour: Similarities and Differences," in *Serfdom and Slavery*, ed. Bush (London, 1996), 31–32.

18. For incentives and bonuses under industrial and agricultural slavery, see Robert S. Starobin, "Disciplining Industrial Slaves in the Old South," *Journal of Negro History* 53 (April 1968): 115–26; Charles B. Dew, "Disciplining Slave Ironworkers in the Antebellum South: Coercion, Conciliation, and Accommodation," *AHR* 79 (April 1974): 393–418; T. Stephen Whitman, *The Price of Freedom: Slavery and Manumission in Baltimore and Early National Maryland* (London, 2001), 43–52. For antebellum northern factories and the use of paid overwork as a self-serving regimen, see Prude, *Coming of the Industrial Order*, 131, 226. On the use of bonuses and nonwage benefits in the postbellum textile mills, see Douglas Flamming, *Creating the Modern South: Masters and Millhands in Dalton, Georgia*,

themselves when the promise of wages, scrip, or credit lay in the offing. The challenge slaveholders confronted was to harness that energy for the plantation economy and to establish a compensation system of positive and negative incentives that would stimulate the slaves to toil even harder in the cane fields. But the challenge only began there. The sugar masters also had to persuade their bondspeople to put their expensive technology into action and to toil long hours through the harvest. The planters could and frequently did revert to the whip as the sole means for motivating their workers; yet, as economists contend, the lash was ill-suited for care-intensive work. Pain-driven incentives—like the whip, stocks, iron braces, and other ghastly tools in the planter's armory of regulatory devices—failed to generate the requisite attention for skilled work. Planters accordingly favored rewards and wage-driven incentives for complex and demanding tasks, and pain incentives—frequently whipping—for relatively simple, repetitive work. Southern masters fully grasped these distinctions and utilized rewards and incentives as part of a larger disciplinary structure that ranged from bonuses to gratuitous violence. But the spectrum of regulatory devices utilized by slave masters had more than merely economic value. Planters hoped that a combination of incentives and the lash would compel their workers to toil both conscientiously and at double-quick time, while also impressing the bondsmen or women with their mastery and its corollary, white power. Both aspects of this rule needed enforcement in order to create an overwhelming sense of submission to the planter's design. In reality, planters were not capable of exerting their mastery, and slaves certainly refused to submit to such a sweeping psychological and physical assault on their autonomy. Planters thus had to adapt their idealized model of slave management—to utilize force and incentive when necessary, but above all, to compromise.[19]

1884–1894 (Chapel Hill, N.C., 1992), 128–35. On the way slaves acted as "proto-peasants" and sold their labor in a marketplace, as well as for the coexistence of a free and wage labor market, see Mary Turner, "Introduction," *From Chattel Slaves to Wage Slaves,* ed. Turner (London, 1995), 2; Sidney Mintz, "Was the Plantation Slave a Proletarian?" in *An Expanding World,* vol. 18, *Plantation Societies in the Era of European Expansion,* ed. Bieber (Aldershot, 1997), 305–22; Bolland, "Proto-Proletarians?" 123–47; McDonald, *Economy and Material Culture of Slaves,* 59–80. For broader discussion of accommodating capitalism as self-expropriation, see Peter Way, "Labour's Love Lost: Observations on the Historiography of Class and Ethnicity in the Nineteenth Century," *Journal of American Studies* 28 (April 1994): 1–22.

19. On the intensity of slave work and cash payment, see Morris, *Becoming Southern,* 76; Schwalm, *Hard Fight,* 62. On punishment and skilled labor, see Stefano Fenoaltea, "Slavery and Supervision in Comparative Perspective: A Model," *JEH* 44 (September 1984): 635–68.

Evidence from the West Indies and other parts of the American South underscores the singular importance of negotiation between master and slave. As many Americans were undoubtedly aware, violence and rebellion defined Caribbean slavery during the first half of the nineteenth century. The rise of the western Cuba plantation economy, for instance, triggered intense resistance. Vast sugar estates utilizing steam-powered grinding facilities dotted the landscape east of Havana, and by the mid-1840s monocultural sugar production defined the plantation economy. The slaves responded to this agro-industrial transformation by leveraging their collective strength against their masters and the punishing regime in a series of large uprisings. Cuba's experience between 1825 and 1845 paralleled that of the British West Indies, where vast slave insurrections occurred in Barbados, Demerera, and Jamaica as planters implemented agro-industrial sugar production. The Jamaican revolt, in particular, occurred because planters changed the length of the Christmas holiday and altered established ground rules without warning. Louisiana's sugar elite did not make the same mistake as their Jamaican counterparts. They established and then held to compensation systems and reward structures that balanced the demands and requirements of masters and slaves.[20]

Events in Louisiana's recent past also reminded planters of the benefits accrued from negotiation and compromise. The largest slave revolt in the history of the United States had occurred in January 1811, when some two hundred African Americans marched through the sugar plantation belt before meeting overwhelming white resistance. The memory of 1811, the haunting specter of Haitian revolutionary Toussaint L'Overture and his marauding cane hands, and the frightening ghosts of Gabriel and Nat Turner preyed on the slaveholder's mind. Southern planters exorcized their misgivings by strengthening slave codes, reinforcing patrols, and redoubling their belief in an idealized master-slave relationship. Fearful of the "germ of revolt" and desperate to harvest their crops, the sugar masters forged a negotiated settlement of inducements and incentives that established relative stability in Louisiana's cane world from the 1830s until the destabilization of the Civil War.[21]

20. Paquette, "Drivers Shall Lead Them," 45–47. On Cuba's expansion, see Laird W. Bergad, *Cuban Rural Society in the Nineteenth Century: The Social and Economic History of Monoculture in Matanzas* (Princeton, N.J., 1990). On Jamaica, see Mullin, *Africa in America,* 254.

21. For Joseph Pontalba's response to the 1796 Pointe Coupee plot, see Lachance, "The Politics of Fear," 168 (quote); Herbert Aptheker, *American Negro Slave Revolts* (New York, 1943), 249–51.

While events in the Caribbean steered planters toward socially stabilizing compromise, the pressing crop and meteorological constraints of Louisiana's grinding season placed a premium on the efficacy of these inducements. From mid-October until Christmas, the sugar masters simply could not afford any delays. On the one hand, this forced planters to negotiate with their slaves, while on the other hand, it encouraged the lash. Hunton Love, a former slave driver on Bayou Lafourche, revealed the axiomatic relationship between force, discipline, and economic success when he recalled, "I had to whip 'em, I had to show 'em I was boss, or the plantation would be wrecked." Like Love, William Howard Russell underscored the symbiosis of force and slave supervision, remarking that the "anxieties attending the cultivation of sugar are great and so much depends upon the judicious employment of labour, it is scarcely possible to exaggerate the importance of experience in directing it, and of the power to insist on its application." Slave driver Solomon Northup clarified the centrality of the lash and rigorous supervision in facilitating agro-industrial sugar production. As he recalled, "The whip was given to me with directions to use it upon anyone who was caught standing idle. If I failed to the letter, there was another one for my own back." Northup's duties extended beyond physical coercion to time management and calling on and off the "different gangs at the proper time." Most planters, however, combined brutality with effective compensation systems that induced the slaves to work fourteen- and eighteen-hour shifts and, crucially, to apply the masters' technology.[22]

The plantocracy was not alone in attempting to negotiate with its enslaved workers. In various industries throughout the slave states, planters faced an analogous dilemma. They had to orchestrate a labor system and reward structure that would encourage slaves to apply their knowledge and experience of sugar farming. But in so doing, slaveholders faced two primary problems. They could not completely monitor their slaves' actions, and they occasionally knew less about the technical art of sugar making than their slaves. To be sure, the planters clearly understood what they wanted accomplished, but their skilled slaves increasingly had an edge over the estate managers and overseers in terms of both knowledge and experience. Furthermore, given the huge number

22. Interview with Hunton Love, date unknown, WPA Ex-Slave Narrative Collection, LSU (first quote); Russell, *My Diary North and South*, 263 (second quote); Northup, *Twelve Years a Slave*, 148 (third quote).

of slaves working on a sugar estate, planters could not always be sure that a slave had utilized his or her skills in a way that best served the slaveholder's interest. The planters' increasing reliance on skilled slaves only enhanced those difficulties. The time pressures of agro-industrial sugar production in Louisiana and the rise of skilled slave artisans in the 1850s placed a further premium on compromise, and during the late antebellum era slaves and slave-holders negotiated a plethora of compensation systems that underpinned business success.[23]

Although their work experience gave bondspeople an effective bargaining tool, not all sugar masters were willing to cede complete control of sugar production to their slaves. Most planters were residential landlords, and while several of them owned two or three estates, the majority of Louisiana's planting elite lived in the rural parishes and knew the art of cane farming and sugar making. If they did not, or they wanted to augment their own expertise, they could draw upon a pool of white laborers ranging from overseers and sugar makers to mechanics and engineers. Experience, however, proved costly. As early as 1835, a competent sugar maker commanded $450 per season, while an engineer (who serviced the mill) could earn up to $150 per month. Assistant overseers earned $300 per annum, while farm managers on small estates received between $500 and $800 a year in the 1850s. Incomes on larger operations were predictably higher. Overseers who man-aged large or multiple estates commanded up to $1,200. Not surprisingly, estate managers who possessed managerial and technical skills commanded significant annual salaries of between $2,000 and $3,000 to work on the largest estates. Expenses of several thousand dollars for technically skilled labor and managerial assistance chipped into annual profit returns and convinced many planters of the efficacy of slave labor. To be sure, the sugar masters sought to give themselves the upper hand in management, but every planter knew that the success of the annual

23. On the economics of information and the principal-agent relationship, see Coclanis, "How the Low Country Was Taken to Task," 66; Charles Kahn, "An Agency Approach to Slave Punish-ments and Rewards," in *Without Consent or Contract (Technical Papers*, vol. 2, *Conditions of Slave Life and the Transition to Freedom)*, ed. Fogel and Engerman (New York, 1992), 551–65. On compromise as a management tool between principals and agents, see Samuel Bowles and Herbert Gintis, "Con-tested Exchange: New Microfoundations for the Political Economy of Capitalism," *Politics and Society* 18 (June 1990): 165–222. On management difficulties and compromise, see also John W. Pratt and Richard J. Zeckhauser, "Principals and Agents: An Overview," in *Principals and Agents*, ed. Pratt and Zeckhauser (Boston, 1985), 1–35.

crop ultimately hinged on the quality of the slaves' work and their readiness to act in the master's interest.[24]

Skilled slaves possessed a further advantage over their owners or their representatives. Most overseers in the cane-producing parishes were not native to Louisiana and were frequently the sons of small planters. Whether hailing from Tennessee or the lowlands of Scotland, they entered the sugar industry with basic management skills, which they honed on the sugar estates. Just thirty-four years old, the average overseer was a font of neither technical knowledge nor practical experience. While some had served as agents and sub-overseers in the sugar country, they had by no means mastered every aspect of the trade. Most had no extensive training in the art of subtropical cane agronomy and sugar production. As a result, they often failed to live up to their employers' expectations. Their failure to remain attentive and diligent in business and slave management, in particular, produced constant irritation and frequent dismissal. Some planters, like William Minor, itemized the overseer's responsibilities, but most made almost impossible requests of their employees. Minor himself was guilty of this, replacing his overseers frequently and maintaining them on monthly and yearly contracts. Such job insecurity bred poor relations between planters and overseers, giving little incentive to the ablest employees; in all probability, it perpetuated managerial incompetence. In an industry where experience overwhelmingly trumped book-learning, the average overseer knew little more—and often less—than the skilled, semiskilled and field hands whom he supervised. The thirty-year-old overseer, who had recently arrived and would soon be dismissed, might have been older than the youthful slaves who toiled in the sugar country, but he was ill-prepared to address the experience and information gap between himself, his employer, and the slaves. And given the short annual contracts on offer, he had little motivation to master the complications of the industry. The gulf between technical and estate-base experience thus tended to expand rather than contract, especially on

24. On wage rates, see Sitterson's three articles: "Hired Labor," 198; "William J. Minor Plantations," 70; and "Magnolia Plantation," 201. See Flanders, "Experiment in Louisiana Sugar," 160; Scarborough, *Overseer*, 28, 29, 36; Shugg, *Origins of Class Struggle*, 92; Olmsted, *Seaboard Slave States*, 2:348. Also see Nathan M. Cochran to A. J. Conrad, 14 August 1835, David Weeks and Family Papers, LSU; Elu Landry Estate Plantation Diary and Ledger, 1 February 1849, 8 February 1850, 2 January 1851, LSU; Franklin A. Hudson Diaries, 23 December 1856, UNC; Expense Book, 1847–1853 (vol. 5), 7 and 10 January, 21 March, 20 August, 15 October, 18 and 28 November 1847, John H. Randolph Papers, LSU; Residence Journal of R. R. Barrow, 21 April 1858, Robert Ruffin Barrow Papers, UNC.

plantations where farm managers faced frequent dismissal and slaves had accumulated years of familiarity with the peculiarities of sugar and their own masters' ways. In such circumstances, the possibilities for conflict were many. Inexperienced overseers and seasoned slaves accordingly squared off over the terms of compensation and the practice of sugar farming. Worse still, the overseer's sexual licentiousness undermined plantation stability and led to further divisive conflict. Little wonder that Moore Rawls, a native North Carolinian who had moved to his employer's sugar plantation in Rapides Parish, announced just a year after arriving in the sugar belt that "men of the right stamp to manage negros are like Angels visits[,] few and far between."[25]

In south Louisiana, angels came as rarely as Rawls predicted. Having trained a cadre of skilled slaves, the sugar masters saved hundreds of dollars each year by using their own workers. Yet in so doing, they surrendered power and information to a group of individuals who could paralyze the harvest or any other part of the cultivation and production process. Training Ruben, Shell, François, and René thus proved a double-edged sword. The planters obtained high-quality sugar workers, yet they created several overlapping problems. Not only could Colbert choose to apply his knowledge to render the steam engine inoperable, but Bill Garner might ignore telltale signs of extreme wear and tear on his mill, Aggy could immobilize the cane carrier, and Peter Congo might burn his sugar. And what could David Weeks, his kin, or any other planter do about it? They could not possibly monitor every aspect of slaves' work, and even if they could, their overseers often lacked detailed expertise of sugar making. Well and truly hoisted upon their own petard, the sugar masters had to compromise if they sought workfloor stability as the harvest approached.

Sugar masters faced what economists describe as hidden action and hidden information cases. In hidden action scenarios, the effort and work rate of the slave

---

25. Rawls quote in Sitterson, *Sugar Country,* 56; Scarborough, *Overseer,* 170–71. On average, overseers in the sugar country were no older than those in other southern economies, but in sharp contrast to other regions, where native-born overseers predominated, only 36 percent of sugar country overseers were born in Louisiana. See ibid, 57–61. Sitterson likewise concludes that "many overseers were only one step above the average Negro drivers in their knowledge of plantation affairs." This may, however, be a little unfair to the driver, who likely possessed vital expertise on planting, cultivating, and harvesting cane sugar. Sitterson, *Sugar Country,* 55 (quote); Sitterson, "William J. Minor Plantations," 62–63. For examples of planter-overseer conflict relating to sexual licentiousness, see William F. Weeks to Mary Clara Conrad, n.d., 1854, David Weeks and Family Papers, LSU.

[agent] made a direct impact upon the business success of the planter [principal]. The size of the estate, the multiplicity of tasks, and the overseer's at best marginal advantage in planting expertise all complicated the task of monitoring the slaves' work in the field and mill. Although regimented gang labor partially resolved this problem, it did not do so entirely. Struggling to oversee every loiterer, planter Andrew Durnford bitterly complained that he had severely threatened his "rascally" kettle hands to do their duty. "I cannot be everywhere," Durnford exclaimed, in what was probably a common complaint. The second scenario builds upon the increasing expertise of the slave community. In hidden information problems, the principal must motivate the agent to put his or her technical expertise or information into action, even if both parties are aware that the principal lacks the knowledge or capacity to judge whether the agent is working in his or her interest. In the sugar country, most masters and overseers knew when slaves were openly acting against the planter's affairs, yet they nonetheless confronted the typical principal's predicament of motivating those individuals. The institution of slavery made this quandary even more complex, as the typical rewards and compensation of a free labor system were unavailable to them. Or were they?[26]

In the cane world, planters negotiated a set of incentives in which skilled slaves received direct payment for work conducted after the end of the working day. Others were remunerated for piecework, and proficient gangs and individuals obtained rewards for rapid work. These incentives probably induced a self-selection process in which the strongest or most financially desperate individuals toiled feverishly in the evenings to gain money or minimize their debt at the plantation commissary. In addition to these direct or in-kind payments, planters awarded post-harvest bonuses at the close of the year. Frequently, the scale of these Christmas handouts reflected the size of the crop, ensuring that the amount received was proportional to the work effort. Incentives of this type provided slaves with a modest income while encouraging self-policing and the chastisement of shirkers. The effectiveness of such arrangements should not be underestimated. In other locations and eras, sugar planters also divided payments among the members of a gang in order to force

26. Andrew Durnford to John McDonogh, 30 January 1843, John McDonogh Papers, TUL (quote). On the hidden action and information problem, see Kenneth J. Arrow, "The Economics of Agency," in *Principals and Agents*, ed. Pratt and Zeckhauser (Boston, 1985), 37–51. For an overview of incentives, see David E. M. Sappington, "Incentives in Principal-Agent Relationships," *Journal of Economic Perspectives* 5 (spring 1991): 45–66; Robert Gibbons, "Incentives in Organizations," *Journal of Economic Perspectives* 12 (fall 1998): 117, 121–23.

workers to monitor each other's contribution to the collective output. And by transferring some of the regulatory burden from the overseer to the slave community itself, sugar masters were better able to cajole and monitor their enslaved workers. In addition to various types of payment, slaves and slavemasters negotiated specific holidays, improved housing, and the provision of whiskey and other goods for postharvest celebrations. Each of these incentives served as part of a broader compensation system designed to maximize effort through the harvest season.[27]

Planters did not stop with these efforts to dupe their enslaved laborers into complicity with the owners of production. Throughout the slave states and the Caribbean, bondspeople utilized the paid overtime system as a means to secure financial remuneration for chopping wood, growing corn, or trading moss, poultry, and livestock. And in doing so, they carved out a meaningful sphere of both self and communal identity through the trading of diverse goods. The market for slave produce, however, emerged as an erratic and conflicted space. Wage-earning slaves grasped the essence of liberty through independent production, but paradoxically they strengthened the regime by working on plantation business in their own time. In Louisiana's sugar country, slaves quickly deemed it a customary right to receive payment for their wood and corn just prior to the grinding season, and on most plantations slaves entered the harvest with their demand for disbursable income at least partially satiated.[28]

The commodification and sale of slaves' overtime work maximized the appalling expropriation of labor in the sugar country and exploited their desire for

27. Expense Book, 1847–1853 (vol. 5), and Expense Book, 1853–1863 (vol. 6), John H. Randolph Papers, LSU; Olmsted, *Seaboard Slave States,* 2:317. Incentives fell into three separate categories: short-term prizes for meeting performance goals; rewards designed to influence behavior over periods of intermediate duration; and long-term rewards that might take years before they paid off. Fogel and Engerman, *Time on the Cross,* 148–49. These types of incentives were designed specifically to resolve problems of information asymmetry—a problem regularly faced by both modern businessmen and slaveholders with skilled workers. See Paul Milgrom and John Roberts, "An Economic Approach to Influence Activities in Organizations," *American Journal of Sociology* 94 (1988): S154–S179. On incentives in other gang labor–sugar economies, see Ralph Shlomowitz, "Team Work and Incentives: The Origins and Development of the Butty Gang System in Queensland's Sugar Industry, 1891–1913," *Journal of Comparative Economics* 3 (1979): 41–55.

28. The literature on the slaves' internal economy is extensive, but see McDonald, *Economy and Material Culture,* chap. 2; Hudson *To Have and To Hold,* 32–78; Lawrence T. McDonnell, "Money Knows No Master: Market Relations and the American Slave Community," in *Developing Dixie,* ed. Moore (Westport, Conn., 1988), 31–44.

pecuniary gain to optimize production. With their earnings, slaves were able to purchase food, clothing, household goods, and tobacco; yet by accommodating these compensation systems, they traded short-term benefits for the exploitation of the steam age. Slaves materially gained from these compensation systems, which in essence were highly manipulative and encouraged accommodation over resistance and rebellion. This was the crux of the sugar masters' success. To be fair, the slaves had little choice but to accept this arrangement. Escape from the sugar country was almost impossible, and the power of the local militia and whites could not be surmounted, as slaves along the Mississippi knew only too well following the brutal repression of the 1811 revolt. Patrols prowled the cane-producing parishes with fearsome regularity, and the criminal code was punitive toward slaves. Through their extra labor, African Americans worked even longer hours for the plantocracy in return for very modest wages; yet by engaging in acts of independent production and by utilizing their power as workers, they eked out a better material existence even as they exposed the central flaw in slavery. Ironically, planters had to adopt aspects of the free labor economy and pay wages for diligent work, long hours, and the operation of their machines.[29]

If overtime work and incentives channeled slave labor into profitable ends for the planters, the slaves viewed them (and other bonuses as well) as due recompense for their work. Wages in specie or in kind laid bare the need to use the tools of free labor in a slaveholding society and suggests that slaves and slaveholders in Louisiana's sugar country were effectively combining aspects of free labor in a slave-based economy. For the slaves, this provided due compensation for the sale of their labor and skills beyond customary levels, while for slaveholders it offered greater flexibility and even more ways to extract maximum labor for a minimal cost. Wage labor accordingly fortified the institution of slavery. Once combined with rewards and other compensation systems, it enhanced workfloor stability in the sugar country and provided a robust avenue through which to introduce technology. And when payment or incentives failed to maintain work speed, planters retreated to more customary models of antebellum labor discipline and firmly grasped the whip.

29. The payment of wages in cash and wages in kind to slaves began a process of commodification of their labor. This is a process by which "formerly non-market mediated activities come to take commodity form." Christopher Chase-Dunn, *Global Formation: Structures of the World Economy* (Oxford, 1989), 18, quoted in Bolland, "Proto-Proletarians?" 126.

In recounting this episode of exploitation, we must guard against leaving all the power in the principal's hands, for agents also possess agency. The very money or scrip-payment that slaves earned and expended vividly denied the planters' claim to all their labors. It commodified their laboring power and blazed a trail that would lead to self-employment and independence. Selling their labor on Sundays or in the evenings, slaves collectively undermined the concept of bondage and grasped the essence of wage work. Not unlike free laborers in the antebellum North, who viewed wage work as a fleeting step on the road to independent proprietorship, slaves in the sugar country grasped at the promise of a self-owning, self-producing, and self-governing life. When quizzed about his vision of the future, the slave William observed that he would work for a year, "get some money for my self," buy some land, and "raise things in my garden, and take 'em to New Orleans and sell 'em dar." Plainly aroused by the whiff of freedom contained in the questions being put to him, he added that his fellow slaves would hire themselves out once they were freed and that "den, wen dey work for demselves, dey work harder dan dey so now to get more wages—a heap harder." The postwar history of the sugar belt ushered in wage work, not independent proprietorship, for African Americans, but William and those around him were quickly learning that "Money is Power."[30]

The collective power of the work gangs similarly provided slaves with a weapon for potential resistance and was a well-honed tool in their arsenal. The narrow, strip-like estates of the sugar country furthered the possibility for workfloor alliances and silent sabotage as the tightly knit gangs ranged over the fields, placing pressure on the sugar regime where and when they could. Slaves occasionally lashed out at the economic system that held them so cruelly in bondage, but black activism never completely undermined the plantation order or brought the steam engines to a crashing halt. Despite occasional episodes of violence and resistance, the slaves rarely trumped the colossal and overwhelming power of the planter class. Slaveholders were keenly aware of the potential of slave resistance, and they accordingly amended the structure of bondage to minimize that risk and evade workplace confrontations as the harvest season approached.[31]

---

30. Olmsted, *Seaboard Slave States*, 2:343–44 (first quote); James O. Breeden, ed., *Advice among Masters: The Ideal in Slave Management in the Old South* (Westport, Conn., 1980), 274 (second quote). On notions of wage work, see Lawrence B. Glickman, *A Living Wage: American Workers and the Making of Consumer Society* (Ithaca, N.Y., 1997), 12; Stanley, *From Bondage to Contract*, 60–97.

31. On silent sabotage, see Kolchin, *Unfree Labor*, 241–44; Jim Scott, "Everyday Forms of Peasant Resistance," *Journal of Peasant Studies* 13 (January 1986): 5–35. Roderick McDonald states that slaves

African Americans unquestionably grasped the significance of their position at the axis of production and the power it implicitly entailed. Occasionally they exercised this leverage through acts of individual and communal resistance, but ultimately the slaveholders' power, their prodigious use of the lash, their beady-eyed regulation of the slaves' working environment, and their grossly exploitative compensation systems ensured that worker power and resistance was channeled away from unified class action. Louisiana slaves adapted to mechanization not because they had imbibed the Protestant work ethic and had become metaphorical clock punchers, but rather because the compensation systems of incentives and rewards provided them with just enough inducements to apply their expertise to the planters' advantage. Bribed into accepting the machine age, slaves in the sugar country were not willing converts to the agro-industrial order, but through their influence at the axis of harvest production they were able to wrest some new economic and social space from the slaveholders' regime. In sum, they were realists, carving out some very basic rights for themselves and their families from the wretched life of the cane world.

Irrespective of the newly defined terms of the compensation systems, slaves in the sugar region pressed firmly upon the limits of white authority; while not halting the annual harvest, their active resistance proved a piercing shock for slave and planter alike. Violent murder and smoldering flames upended the planters' idealized master-slave relationship, while runaways exposed the vanity of the slaveholders' charade. In local newspapers, the simple woodcut of a runaway slave reminded slaveholders that accommodation and acquiescence might easily give way to escape and flight. By denying their labor in totality, escapee slaves brazenly inverted proslavery logic, defied the chains of the plantation order, and offered a stimulating role model for fellow African Americans within the cane belt. Planters remained alert to the potentially destabilizing and subversive role of fugitive slaves and reinforced their policing and regulatory mechanisms both on and off the plantations. Thus when the slave Richard temporarily escaped the prowling slave dogs before being rearrested in Pointe Coupee Parish, John Hamilton reported to his kinsman that he had given Richard "a good flogging and am having him ironed at the shop this morning." In the rich Attakapas sugar region, sporadic episodes of

---

"exercised power by affecting productivity," but his richly researched volume does not detail how the slaves methodically wrung concessions from their owners. McDonald, *Economy and Material Culture of Slaves*, 50.

slave resistance alarmed local residents, who advocated murdering slaves "if it can be done legally" once the edifice of slavery came under attack. Slaves like Bob, who deliberately overslept and fled before the lash tore his young back, and Hector, who prized open the jail to seize a firearm, buckshot, and a gourd of powder, dangerously threatened the plantation order. Their individual acts against slavery forced local whites to return to the most primitive and oppressive forms of labor control. Stripping the veneer from the terms of compensation, St. Mary Parish planters—like their brethren elsewhere in the sugar country—struck out violently at the slave who threatened the delicate balance of regional sugar production and sought to arrest him before his absence inspired the other hands. On Tally-Ho Plantation, overseer L. Hewitt expressed such concerns when he reported to landlord John D. Murrell that as preparations advanced for the 1848 rolling season, bondsman Lew remained at large, "beating about the place every night or two." Presumably gaining sustenance from friends and family within the Tally-Ho slave community, Lew defied the bounty hunters who pursued Hewitt's $25 reward; his continued absence convinced the overseer that "when he is caught, it will be best to put [him] in a chain gang . . . and let the yellow fever kill him if it can." On Grand Cote Island, the slave Linzy likewise bedeviled plantation overseer John Merriman. Runaway slaves like Linzy had little chance of securing their freedom in this dismal and "remote corner of the Globe," but they could flagrantly deny their labor for the grinding season. On November 11, 1840, Linzy escaped from Grand Cote with his wife and child, as well as weapons for his defense. Merriman reported that Linzy was armed with his butcher's knife and a double-barreled gun and that he intended to join a group of fugitive slaves who had stolen a boat for their use in the swamps surrounding Cypremort Bay. Whether these slaves just sought to abscond while the rolling season monotonously advanced remains their secret, yet the impact of their absence reverberated throughout the Weeks estate. Merriman detailed "a great difference in the behavior of several Blacks on the Island," whom he suspected were in regular contact with the dissidents. Sounding the alarm, Merriman counseled fellow whites to flush out the fugitives before they brought down "miseries upon [our] heads." While undoubtedly inspiring to their fellow slaves, rebels like Linzy or Lew did not shatter the plantation order, even though their actions denied the slaveholders the product of their labor. During the grinding season, pinpoint acts of individual and collective resistance provided a timely reminder to whites that their success ultimately rested on an

runaways

unwilling labor force that would strike for freedom whenever the opportunity beckoned.[32]

Like Linzy and his fellow fugitives in the Attakapas swamps, slaves were most likely to resist or run away during November, in the middle of the grinding season. Indeed, many estates were afflicted by various forms of resistance during the harvesting season. On Moses Liddell's estate, for instance, the mill was brought to a sudden halt at the height of the harvest when one slave—either deliberately or otherwise—left a nine-inch-long steel bolt on the cane carrier, which "caused the shaft that connects the engine and mill to give way with a great crash" once the metal embedded itself within the engine. Liddell responded promptly, inserting a reserve shaft that had the mill turning once again by daylight. Perhaps more alarming to Liddell than the accident itself was the fact that "10 or 12 at a time are moping about with hands swollen in stings and unable to do [their] duty." Just over a week later, several bondspeople continued to evade work on account of sore fingers and hands, while others suffered from headaches and diverse pains. Fuming over the incapacity or unwillingness of his crews to return to work, Liddell reduced his medical treatment to doses of salts mixed with rhubarb and seethed that his slaves' medical ailments "really annoy me extremely." His decision to "give up the practice of medicine" proved unwise. A week later, he wrote, his slave force continued to be plagued by sore fingers and "are useless in the work before us." To further confound the planter, the mill broke for a third occasion that season,

32. John A. Hamilton to William S. Hamilton, 15 November 1859, William S. Hamilton Papers, LSU (first quote); Boyd Smith to William F. Weeks, 20 and 27 August 1834, and W. H. White to John Moore, 15 December 1858, David Weeks and Family Papers, LSU (second quote); L. Hewitt to John D. Murrell, 3 September 1848, John D. Murrell Papers, LSU (third quote); A. T. Conrad to David Weeks, 16 December 1827, David Weeks and Family Papers, LSU (fourth quote); John Merriman to Francis D. Richardson, 11 November 1840, David Weeks and Family Papers, LSU (fifth quote). The narrative is also recounted by John Hope Franklin and Loren Schweninger, *Runaway Slaves: Rebels on the Plantation* (New York, 1999), 50, 168. In attempting to "count the cost" of runaways, Franklin and Schweninger point to Homestead Plantation in St. James Parish and suggest that 4 percent of the slaves aged thirteen and older had absconded. Ibid., 279–94. On the iconography of slave runaways, see Schafer, "New Orleans Slavery in 1850," 33–56. Occasional episodes of murder and arson are reported in planters' correspondence; see, for example, H. B. Trist to Willie Trist, 25 December 1854, and H. B. Trist to Browse and Bringier, 5 March 1855, Trist Wood Papers, UNC. On the structures of runaways' capture, arrest, and costs to the planter, see Taylor, *Negro Slavery in Louisiana*, 169–75; Moody, *Slavery on Louisiana Sugar Plantations*, 24–27; Schafer, *Slavery, the Civil Law, and the Supreme Court of Louisiana*, 98–104, 115–26.

though it turned again after a short delay and some makeshift repairs. Liddell's prudent supply of spare parts and his engineering wizardry finally failed on Christmas Day, 1851, when the rudimentary repair gave way and the mill came to a complete stop before local ironworkers riveted some new iron cogs into the engine. Liddell did not openly blame the bondspeople for the tumultuous harvest, though his repeated adversities suggest that they played a role in his bad fortune. While the slaves might not have placed the steel bar on the cane carrier, no one removed it; and although they might have been innocent in the subsequent breakdowns, several of them took full advantage of the occasion to minimize their labor and perhaps to exaggerate, if not feign, injury. The actions of Liddell's slaves, however, in no sense matched the destructiveness of Old Pleasant, a slave at Bowdon Plantation in Ascension Parish whose transparent act of sabotage brought the mill house to an alarming halt. Recounting the incident a week later, H. B. Trist described Old Pleasant's "villainy" at length:

> He suffered the water in the boilers to get so low that there was scarcely any left in them, and when informed by some of the hands that there was something wrong, told them to mind their own business. The engineer . . . made his appearance about this time, and on going to the boilers found them heated almost to redness—He gave the alarm and bid all in the neighborhood run for their lives—but Pleasant instead of showing any concern . . . went and seated himself very coolly on one of the boilers . . . shortly after I arrived another alarm was given; steam escaped with violence from the top of one of the boilers and made the ashes and brick fly about . . . a bolt that had been put in to stop a hole in the boiler had been heated so that the gasket under the washer had burned out so as to allow steam to escape. This leak being stopped, steam was again raised when to our dismay it rushed out from other places and on cooling down and examining, we found several rents in the first boilers—it took five days to repair damages . . . but the syrup in the tanks and filters got sour and we made sugar of inferior quality for several days.[33]

---

33. Moses Liddell to John Liddell, 7, 10, and 25 December 1851, and Moses Liddell to Mary Liddell, 23 November, 1 December 1851, Moses and St. John Richardson Liddell Family Papers, LSU (first five quotes); H. B. Trist to Bringier, 25 November 1854, Trist Wood Papers, UNC (sixth quote). The conclusion on the seasonality of slave resistance rests on thirty incidents of reported resistance on sugar estates: June, 0 incidents; March, April, and May, 1 incident; July and October, 2 incidents; January, February, September, and December, 3 incidents; August, 4 incidents; November, 7 incidents.

Old Pleasant, like Aggy, clearly enjoyed significant responsibilities in the mill house and used his power and influence to halt the harvest and damage the commercial value of the crop. But as Trist's account indicates, Old Pleasant acted alone, as the other slaves either raised the alarm or escaped from the building. Fleeing for their lives, these slaves might not have tampered with the boilers, but—just as when Aggy's team raced out of the sugarhouse and disappeared down the lane— they undoubtedly took full advantage of the temporary halt in sugar making.

Theft similarly harmed plantation operations by impeding day-to-day business, eroding profit margins, and inverting plantocratic definitions of property and ownership. Occasionally slaves unscrewed and dismantled costly equipment and sold expensive items to petty traders along the rivers and bayous. Others expressed their resentment at the plantation order by stealing and subsequently selling the produce of their labor. By doing so, slaves comprehensively rejected the master's property rights over his sugar and molasses and redefined the notion of plantation profit. Theft accordingly became part of the active negotiation of the master-slave relationship, as bondspeople manipulated the contours of sugar production to further their own autonomy. Maunsell White witnessed the combined effrontery of theft and slave flight when he apprehended one of his bondsmen for stealing molasses and selling it to river traders for cash, produce, or services rendered. Concealing himself near the riverbank, White observed his bondsman and the boatmen stowing the molasses before he seized the culprits and searched the boat, only to uncover a fugitive slave who had absconded from his master two and a half months earlier. During this time, Leander—the escapee—had worked in New Orleans and had evidently flourished, as he possessed a silver watch and over $35 in cash when White placed him under custody. Swiftly addressing his neighbor and fellow sugar master George Lanaux, White urged him to receive Leander with clemency and "propriety," as "I have myself made it a rule to pardon 'Runaways' and found it good policy." One can only imagine that Leander's time-consciousness impressed his owner, even if it directly challenged his control of the ticking clock.[34]

Like Maunsell White, other slaveholders endured persistent theft and frustrating damage on their estates. Ellen McCollam, for instance, reported that chicken theft remained a perennial problem on Ellendale Plantation. Between March and

---

34. Maunsell White to Mr. Laneau(x), 15 April 1859, George Lanaux and Family Papers, LSU (quote); Olmsted, *Seaboard Slave States,* 2:334. On slave theft and loitering, see Alex Lichtenstein, "'That Disposition to Theft, with which They Have Been Branded': Moral Economy, Slave Management, and the Law," *Journal of Social History* 21 (1988): 413–40; Genovese, *Roll, Jordan, Roll,* 599–612.

August 1847, eight hens vanished from their plantation coops, for which Old Armistead received a whipping. McCollam's problems would only continue. Just three days after her hens disappeared, a newly purchased ham was unaccounted for, and a month later the carriage horse perished at the hands of the slave George. Such irritations were hardly unusual. Chicken theft had previously menaced Ellen's small business in 1845, and a surprise fire in February 1847 almost burned down the sugarhouse just as Andrew McCollam was preparing to leave for New Orleans to purchase new castings for the sawmill that had fortuitously broken that day. On this occasion, just the woodpile was set ablaze, but the four hundred smoldering cords of wood represented extra work, for which slaves on many plantations received overtime payments. Whether McCollam's slaves sabotaged operations and committed arson remains an open question, but their incessant (if small) acts of day-to-day resistance chipped away at the plantation edifice and left the estate manager with no doubt about the potential for highly destructive slave resistance. On Grand Cote, the slave Charity similarly expressed her contempt for the Weeks's regime by rejecting the paper clothes patterns her mistress provided for her. By cutting and trimming the jackets "carelessly," Charity exasperated the Weeks family, who lost time, materials, and their tempers with the headstrong and unyielding bondswoman. Charity's small-scale—but telling—act tested the patience and resolve of her masters just as assuredly as Linzy's escape, reminding those in the family mansion on Bayou Teche that plantation work would advance seamlessly only at the slaves' sufferance. Like many slavemasters, William F. Weeks grudgingly and wearily accepted these subversive episodes as constant irritations in the daily management of his land and labor holdings. Remembering his days on Lasco Plantation, former slave Albert Patterson emphasized the slaves' communal resistance and the planters' tolerance of runaways when he recounted how fugitives would return to the plantation and break into the meat house. Explaining how the runaways evaded the ever-present dogs, Patterson recollected that they "put bay leaves on de bottom of their feet and shoes" before walking in manure in order to disguise their scent and mislead the hounds. That way, the former bondsman reported, a fugitive man could come and see his family and obtain a meal come nightfall. Runaways on Lasco Plantation, however, utilized the planter's labor shortfall to erase their litany of crimes and benefit from any payment due them. According to Patterson, "When it's time to get busy to gather the crops," plantation miscreants and fugitives reappeared from the canebrakes and offered their services, stating point blank, "I'll take de crops off if you'll let me." The planter had lost a little meat and the labor of a few individuals during the lay-by period, but the harvest call to arms succeeded in bringing fugitive slaves out of the woods and back into the fields.[35]

For every stolen valve and broken cog, however, one could recount literally thousands of hours when plantation operations ran relatively smoothly. Likewise, for every Leander or Old Pleasant, there were many more who trudged through their daily tasks and labored in the field or on the sugar-production line. This does not minimize the dignity of those who stood up against an appalling regime. Yet within the context of a plantation order that grew and thrived over half a century, the occasional loosened gasket did little more than momentarily halt the power of the planters or check the industry. Perhaps more remarkable than these sporadic outbursts of individual and communal resistance was the relative efficiency of the harvest. To be sure, machinery broke down, but the engineering of the mid-nineteenth century was hardly robust. Nor were the newly designed vacuum pans and steam pumps perpetually reliable. For the sugar masters, the perils and hazards of this new machinery presented numerous problems. When compelled to run flat-out for weeks, cast-iron teeth cracked, primitive gaskets gave way, and the technological revolution shuddered to a halt. Yet despite these unforeseen delays, the sugar industry was characterized by relative efficiency and labor stability even during the most exacting and demanding time of the year. How the sugar masters achieved such a notable level of business success with a slave labor force bewildered contemporaries, who were left questioning their assumptions about the incompatibility of slavery with progress.

In the final two decades of the antebellum era, slavery in the sugar country evolved as a hybrid form of wage and slave labor in which planters and slaves compromised over the terms of payment. The nascent labor market was still a shadowy fraction of the postbellum era, but by the eve of the Civil War neither planters nor slaves were novices in the principles of wage work, reward structures, or the sale of labor. The transition from slave to wage work did not therefore occur overnight when General Benjamin Butler occupied the sugar country in April 1862.

35. Diary of E. E. McCollam of Ellendale Plantation (1847–1851), 29 October 1845, 20 February, 15 August, and 28 September 1847, Andrew McCollam Papers, UNC; William F. Weeks to John Moore, 5 August 1855, David Weeks and Family Papers, LSU (first quote); Interview with Albert Patterson, 22 May 1940, WPA Ex-Slave Narrative Collection, LSU (second quote).

For twenty years before the invasion of the Union Army, slaveholders in the lower Mississippi Valley had been combining free and coerced labor for profit. This was the essence of capitalism. It was the most effective way for slaveholders to extract, the "largest share of the surplus value being produced by productive workers." The sugar masters often viewed their actions through a paternalist lens, yet their use of distinct forms of labor ensured profit, stability, and above all, a satisfactory way to induce slaves to act in the planters' interest.[36]

36. Immanuel Wallerstein, *The Capitalist World-Economy: Essays* (Cambridge, 1979), 220 (quote); Mintz, "Was the Plantation Slave a Proletarian?" 311.

five

"MEN OF SENSE"

WHEN FREDERICK LAW Olmsted observed that "men of sense have discovered that when they desire to get extraordinary exertions from their slaves, it is better to offer them rewards than to whip them; to encourage them rather than to drive them," he commented on a labor system that blended aspects of wage work and contract labor with one of the oldest and most savage forms of human exploitation, slavery. Bondspeople in the cane world endured their brutal working conditions because they received payment, benefitted from improved food rations, and carried out a variety of tasks. Plantation success in the antebellum sugar country, Olmsted discovered, rested foursquare upon the bargaining and contestation of the master-slave relationship.[1]

Ever attentive to the theatrics of southern life, Olmsted watched the unfolding script as master and slave disputed and negotiated the terms of "Christmas money" one autumn morning. Both parties, as we shall see, pressured one another to maximize their own interests and revealed the sometimes subtle and occasionally transparent art of workplace bargaining. The eleven o'clock bell had sounded when the visiting northerner joined his host in a walking tour of the plantation. The sugar master seemed "energetic and humane," but for the past few weeks he had been confined to his residence due to a bout of illness. Sallying forth into the fields, the slavemaster encountered some young bondswomen returning to their infants in the plantation nursery and inquired into their health and that of their children, his troubled brow and anxious questions about Sukey's poorly child revealing his patriarchal humanitarianism. The planter laced his self-serving questions with seeming affection for his family, black and white. The slave women

1. Olmsted, *Seaboard Slave States,* 2:327.

were not fooled by the sympathetic munificence of their paterfamilias or his ardent efforts to further his authority through face-to-face contact. They evaded his questions, obliquely sidestepped confrontation, and assumed a mien of dissembling deference. Content with his affecting display of kindness and benevolence, the slaveholder continued his perambulations, perhaps musing on his humane management and the reciprocal duties that bound master and slave as one. His understanding of reciprocity, however, was about to be tested. A few minutes later, a young African American drew up to his master and with practiced reverence doffed his hat and asked, "How is 'ou master?" The sugar master performed his part to the hilt. "I'm getting well you see. If I don't get about, and look after you, I'm afraid we shan't have much of a crop." Coyly playing the patenalist's card, the slaveholder exposed his market-oriented personality before closing with the rhetorical question, "I don't know what you will do for Christmas money." Pointedly, his refrain both articulated his beliefs about white authority and linked seasonal incentives directly to the size of the crop and the amount of work completed. The cart driver launched a quick riposte, hammering home the slaves' achievements and informing his master, however subtly, that he would indeed pay Christmas money (and presumably a fair amount of it). The young man was clearly schooled in the art of allusion, concealment, and verbal trickery; he hazarded a joke with the slavemaster: "Ha!—look heah, massa!—you jus' go right straight on de ways you's goin'; see suthin' make you laugh, ha! ha!" The slave had seized the upper hand on this occasion, and with the promise of a good crop ahead, he reminded his owner of his side of the bargain and the expectation of due payment.[2]

The planter's charade of benevolence and kindly paternalism was an act, but one in which he believed. Southern slaveholders could damn their chattel as lazy beggars to be "licked like blazes," watch their slaves perish in frightening numbers, and brutally exploit their bondswomens' bodies, but they still believed themselves to be paternalists bound to their slaves by mutual obligations and reciprocal duties. African Americans keenly understood the extent of this fiction. Having watched young men grow old and their infants die, slaves in the sugar country found the paternalistic claims of the master class to be as empty as their own stomachs. Far

2. Ibid., 2:315. On the social ethic of a shared black and white family, see Genovese, "'Our Family, White and Black,'" 69–87. On the concept of dissemblance and duality in slave performance, see Hine, "Rape and the Inner Lives of Black Women," 912–20; Bertram Wyatt-Brown, "The Mask of Obedience: Male Slave Psychology in the Old South," *AHR* 93 (December 1988): 1228–52.

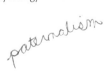

more important to them was the recognition of reciprocity and mutualism as the defining features of nineteenth-century social relationships, which provided them with opportunities to transfer perquisites gained on the mill floor or cane fields into customary rights. Those "rights" provided money and income for the slave community, but they proved sadly profitable for the sugar masters, who rewove the mutual obligations of paternalism to exploit their slaves still further.[3]

Paternalism evolved in the late eighteenth century, when the ideology of patriarchal subordination gave way to reciprocity in the individual family and society at large. Drawing upon a transatlantic wave of bourgeois, revolutionary, and humanitarian thought, southern planters articulated a new defense of slavery by arguing that they governed over an organic society in which mutualism and domesticity ensured moral progress for all. They believed that they possessed a reciprocal relationship not only with their wives and families, but with their slaves as well. Both master and slave, planters believed, lived by a code of obligations and duties that bound the well-being of the workforce to that of the master or mistress. Slaveholding paternalism, however, carried the indelible stamp of antebellum racism. Unlike their predecessors, nineteenth-century planters no longer viewed African Americans as exotic, alien "others," but rather as wayward children who required guidance and occasionally disciplinary control. The prevailing logic of slave-holding paternalism reached its apogee in proslavery rhetoric, in which racial bondage appeared as an altruistic exercise in white benevolence and humanitarianism.[4]

3. John A. Hamilton to William S. Hamilton, 5 November 1859, William S. Hamilton and Family Papers, LSU. On Genovese's interpretation of paternalism as an "organic relationship" linking master and slave, its impact on the creation of planter hegemony, and its effect on slave resistance, see *Roll, Jordan, Roll,* esp. 1–97, 285–324, 661–63. On Genovese and paternalism, also see Smith, *Debating Slavery,* 20–22, 44–46; Kolchin, *Unfree Labor,* 128–40.

4. Anne C. Rose, *Victorian America and the Civil War* (Cambridge, 1992), 147. Rose suggests that patriarchy was a system of deference, while paternalism evolved out of "customs based on reciprocal sentiments." On the late eighteenth- and early nineteenth-century transition from patriarchal to paternalist relations, see Willie Lee Rose, "The Domestication of Domestic Slavery," in *Slavery and Freedom,* ed. Freehling (New York, 1982), 18–36; Young, *Domesticating Slavery,* 40–53; Peter W. Bardaglio, *Reconstructing the Household: Families, Sex, and the Law in the Nineteenth-Century South* (Chapel Hill, N.C., 1995), 25–27. On the religious impulses of eighteenth-century society (particularly the Great Awakening) that shaped the emergence of paternalism, see Allan Gallay, "The Origins of Slaveholders' Paternalism: George Whitefield, the Bryan Family, and the Great Awakening in the South," *JSH* 53 (August 1987): 369–94. On paternalism within southern families, see Stowe, *Intimacy and Power*

Recent scholarly interpretations of the prevalence of paternalistic thought within planter ideology suggest that paternalism ultimately enabled slaveholders to perceive of themselves as Christian stewards whose enlightened mastery ensured a perfect world in which devoted slaves esteemed and honored their masters. The reality was, of course, very different, but this did not preclude slaveholders from believing in paternalism as an ideal relationship. As historian Jeffrey Young contends, they embraced paternalism as a romantic, "fictive idea" through which they could vigorously defend the institution of chattel slavery and miraculously exorcize its gross realities. Paternalism thus evolved as a cohesive, if blinkered, ideology that deluded planters into believing their own idealized notions of mastery. Yet it offered still more advantages to slavemasters. By casting themselves as Christian stewards who stood in loco parentis to the slave, planters could legitimize their authority and consign the African American to perpetual dependence. Plantation management accordingly hinged on the slaveholders' efforts to maximize this dependence, be it in doling out provisions, awarding prizes, or orchestrating health care. From the slaveholders' perspective, paternalism not only made these power relationships moral, but it also served as a vital means of stabilizing social hierarchies and creating relative harmony on their estates. According to Samuel R. Walker of Elia Plantation near New Orleans, slavery was a "good and wise despotism" and a model relationship. "Slavery," he continued, "is from its very nature eminently patriarchal and altogether agricultural."[5]

---

in the Old South, 164–68, 190–91; Joan E. Cashin, "The Structure of Antebellum Planter Families: 'The Ties That Bound Us Was Strong,'" *JSH* 56 (February 1990): 55–75. Paternalism shaped the perception of the family, as planters created narrow gender roles that relegated women to domesticity and a socially constrained private sphere. Some women challenged these assumptions, while others developed a culture of resignation to address the double bind of paternalism and gender iniquity. For an excellent overview, see Joan E. Cashin, "Introduction: Culture of Resignation," in *Our Common Affairs*, ed. Cashin (Baltimore, 1996), 1–41; Victoria E. Bynum, *Unruly Women: The Politics of Social and Sexual Control in the Old South* (Chapel Hill, N.C., 1992), 59–110; Elizabeth Varon, *We Mean To Be Counted: White Women and Politics in Antebellum Virginia* (Chapel Hill, N.C., 1998). On racist constructions of African Americans as childlike and the impact of paternalism on these patterns of thought, see George M. Frederickson, *The Black Image in the White Mind: The Debate on Afro-American Character and Destiny, 1817–1914* (New York, 1971), 43–70; Larry E. Tise, *Proslavery: A History of the Defense of Slavery in America, 1701–1840* (Athens, Ga., 1988), 5–6, 11–14.

5. Blake Touchstone, "Planters and Slave Religion in the Deep South," in *Masters and Slaves in the House of the Lord*, ed. Boles (Lexington, Ky., 1988), 108–9 (second quote). On the notion of

If paternalism morally justified exploitation, at least in the slaveholders' eyes, it also codified their class power. Through kindness and mutualism, violence and threat, planters could mediate conflicts of interest and foster personal ties that en- *class* couraged the slave to identify with his or her master. While slaveholding paternalism evolved most particularly on the demographically stable estates in the American South, it might appear logical to argue that the deathly conditions of the sugar country would have blasted any remnants of identification that might have occurred between slave and master. Irrespective of the cold facts, however, slaveholders permanently residing in Louisiana's cane world sought to inculcate just such dependence. Yet beneath its comforting promise of idealized mastery, paternalism was a façade for exploitation and a convenient tool for managing labor come the grinding season. The notion of mutualism within the domestic family seldom carried economic overtones, but when applied to a workforce, plantation lords and nineteenth-century industrialists discovered that reciprocity was a compelling motivational tool for workforce management.[6]

Olmsted was not alone in concluding that "men of sense" found Christmas rewards a profitable way to create powerful economic and social ties between the enslaved and the enslavers. In parts of the South both before and after emancipation, a system of market paternalism served to motivate workers, reduce monitoring costs, and reaffirm the planters' class power. Market paternalism, one historian argues, "must be built upon an illusion which conceals the commercially oriented

---

paternalism as a well-orchestrated theatrical act in which the planter sought to impress the enslaved audience, see Drew Gilpin Faust, *James Henry Hammond and the Old South: A Design for Mastery* (Baton Rouge, La., 1982), 72–104; Young, *Domesticating Slavery*, 133 (first quote), 168, 170, 228.

6. Michael Tadman and William Freehling have critiqued the paternalist thesis, arguing that the constant threat and use of the interregional slave trade in the southwest left paternalism as an empty shell. Mutual distrust, fear, and racism shot through the paternalistic visage and left both groups in permanent hostility. See Tadman, "The Persistent Myth of Paternalism: Historians and the Nature of Master-Slave Relations in the American South," *Sage Race Relations Abstracts* 23 (February 1998): 7–23; Freehling, *Road to Disunion*, 59–76, 229–30. For examples of how paternalism was utilized by industrialists, see Prude, *Coming of Industrial Order*, 110–31; Philip Scranton, "Varieties of Paternalism: Industrial Structures and the Social Relations of Production in American Textiles," *American Quarterly* 36 (1985): 235–57; Howard Newby, "Paternalism and Capitalism," in *Industrial Society*, ed. Scase (London, 1977), 59–73. As Genovese notes, however, industrialists who used paternalism were not espousing the same hegemonic values as southern planters, who sought not only the subordination of a class and race, but of individuals, to the master. Genovese, *Roll, Jordan, Roll*, 662.

nature of the employment contract." Slaveholders accordingly had to veil their economically focused incentives beneath a façade of paternal reciprocity that seemingly bound the slave and his or her interests to the goodwill of the master. Such masking allowed slavemasters to disguise their incentives—and the prevailing market dynamics of the master-slave relationship—with the sheen of paternalism. In order to do this, planters had to "handcraft [their] social authority link by link" through gifts, Christmas rewards, acts of apparent charity, and both practical and symbolic rituals. The scene played out before Olmsted's eyes reflected such hardnosed market paternalism even as it underscored the planter's exploitative, though nonetheless authentic, view of himself as the master and steward of his sugar demesne. Herein lay the essence of the sugar masters' success. The reciprocity of paternalism provided business-conscious planters with an ideological vocabulary for negotiating a contractual relationship with the slaves that aided plantation productivity. Planters' compromises, however, proved highly self-serving, as slaves ceded labor stability in return for cash and increased rights under bondage. The sugar masters espoused market paternalism and utilized the motif of reciprocity as an effective management tool, though they conceived of themselves as idealized stewards who benevolently catered for the needs of their entire households. The slaveholders' pastiche was an act of denial, but it defined their self-image and articulated their objectionable, though bona fide world view.[7]

When Olmsted sauntered into the fields with his host, he encountered the dynamics of market paternalism at work. The sugar master had boldly played the patrician's card when he inquired into the health of some slave women and a valuable infant, mixing apparent benevolence with an eye to the reproductive potential of his bondwomen. Upon meeting the young cart driver, the planter again donned his paternalistic visor, underscoring his inclination to pay a hefty Christmas bonus—though only if the required work had been completed. By weighting the bonus to the harvest yield, this sugar master profitably fused paternalism and business

---

7. On market paternalism, see Jaynes, *Branches without Roots,* 107–8 (first quote); James Scott, *Weapons of the Weak: Everyday Forms of Peasant Resistance* (New Haven, Conn., 1985), 308 (second quote). On the economics of paternalism, see Lee J. Alston and Joseph P. Ferrie, *Southern Paternalism and the American Welfare State: Economics, Politics, and Institutions in the South, 1865–1965* (Cambridge, 1999), 23–28. On reconciling the problem of reconciling productivity with paternalism/patriarchy, see David Brion Davis, "Looking at Slavery from Broader Perspectives," 465; Robert Olwell, "'A Reckoning of Accounts': Patriarchy, Market Relations, and Control on Henry Lauren's Lowcountry Plantations, 1762–1785," in *Working toward Freedom,* ed. Hudson (Rochester, N.Y., 1994), 33–52.

success. The slave, however, was nonplussed. For him, reciprocity and compromise did not carry the ideological baggage of paternalism and mastery; rather, it was a convenient way to put money in his pocket. Yet by affirming the slaves' diligent work, the bondsman unwittingly revealed the reality of bondage. He and his fellow slaves would gain some relatively small recompense, but the field had been planted, cultivated with attention, and harvested with care, promising considerably more money to the planter. The scene might well have been carefully orchestrated to impress the visitor with the slaveholder's benevolence; but even here, the New York correspondent unmasked the dynamics of market paternalism. In his perambulations, Olmsted learned that at Christmas the slaves collectively received $1 for each hogshead they had produced. Fusing the apparent munificence of a yuletide gift with an unabashed economic incentive for hard work, the sugar master had spread his seasonal cheer far and wide during the previous year, when this sum had amounted to over $2 a head. The slaveholder could gratify himself with the thought that by sharing the profits of the estate with, in the planters' vernacular, "his people," his mutualism and reciprocity brought stability and harmony to plantation life. But perhaps more important, he could reconcile such paternalistic values with hard profit motivations. Olmsted further noted that both master and bondspeople enforced a community-wide work discipline, and that "if any [slave] had been particularly careless or lazy, it was remembered at this Christmas dole. . . . The effect of this arrangement was to give the laborers a direct interest in the economical direction of their labor: the advantage of [which] is said to be very evident." For the sake of a few hundred dollars—a pittance in the annual costs of making sugar, hiring extra hands, or purchasing new slaves—the sugar master succeeded in creating a highly effective labor regime. Slackers received no payment at all, and the slaves self-policed themselves by exerting pressure on laggards. Most pernicious of all, the slaves now possessed a stake in the success of the plantation. Better still for the planters, this type of incentive created a new type of dependency, that of the slave to the harvest yield.[8]

Perennially calculating in the presentation of his economic and social mastery to slaves and planters alike, John Hampden Randolph fused plantation performance with Christmas bonuses to encourage maximum output from his bonded slave crews. At Forest Home, Randolph paid his slaves $175 as a Christmas bonus in December 1851. One year later, he increased his bonus by $25. In January 1854 he

8. Olmsted, *Seaboard Slave States,* 2:317.

rewarded his slaves with $300—a significant increase, marking the signal success of the new vacuum pans that had produced 680 hogsheads of sugar. Randolph's slaves, who controlled the expensive and complex pans with considerable aplomb, found mechanization financially advantageous, as each adult gained approximately $5 by accepting the new machinery. By scaling the amount of the Christmas rewards to the volume of the crop, Randolph likely triggered communal pressure in the slave quarters against those who loitered in the sugarhouse but who drew an equal share of the Christmas bonus. Furthermore, Randolph could retire to his residence confident that his slaves considered the vacuum facilities not as their enemy but as an additional way of improving their lifestyle. With their bonus money, slaves could purchase consumables available from river hucksters, town traders, and provision stores on the estates. Some slaves were undoubtedly aware that the Christmas bonuses merely served Randolph's interest, just as others must have realized that such rewards duped the workers into accepting mechanization. In truth, there was little yuletide merriment on offer. The Christmas bonus was merely an attempt to make the slaves work harder, police themselves, and forge multiple dependencies that linked the bondspeoples' monetary interest to the planter, the plantation, and the pumping pistons of the steam engine. And by doling out plantation profits on the Lord's birthday, Randolph was able to mask his exploitative ruse with seeming benevolence and Christian generosity.[9]

Having paid his chattel, Randolph could muse over the designs for his Italianate palace that he named after the Virginia county from which his antecedents hailed. Nottoway Plantation possessed sixty-five rooms and projected Randolph's wealth and mastery to all who visited or spotted the residence from a riverboat. Both practical and representational, the mansion served as a stage upon which Randolph could portray a hierarchical social structure and conspicuously perform his role as master of sugar, slaves, and households. It embodied the monumental sense of his personal prosperity and displayed his persona to everyone in the community, be they black or white, male or female, rich or poor, native or visitor. The classicism of the house reflected Randolph's claims to master all with virtue, intelligence, and reserve; it publicly displayed Randolph's private emotions and exposed his desires to be honored and respected by all beneath him. Nottoway's stately columns and

9. Expense Book, 1847–1853 (vol. 5), and Expense Book, 1853–1863 (vol. 6), John H. Randolph Papers, LSU; P. A. Champomier, *Statement of the Sugar Crop, 1851–1853*. On the size of the slave population at Forest Home, see Paul Everett Postell, "John Hampden Randolph, A Louisiana Planter," *LHQ* 25 (January 1942): 192–93.

ornamental ironwork impressed travelers, and those who ventured within discovered the most up-to-date conveniences, including running water, indoor bathrooms on every floor, and fireplaces that burned gas manufactured on the plantation. In Nottoway, as in his sugarhouse, Randolph was thoroughly modern in coercing labor to employ his machinery or in heating his rooms, yet simultaneously he strove to be a lord who exercised authority over "his" people. Planters like Randolph found no disjuncture between slavery and modernity, and throughout all aspects of plantation management they reconciled profit-centered measures with market paternalism.[10]

The dynamics of slaveholding paternalism proved highly seasonal. Indeed, it remained every bit as season-specific as the agricultural calendar. During the harvest months, for example, bondspeople received an adequate diet that met their caloric requirements; during the rest of the year, however, they consumed at best the minimum number of calories required for their daily work. As in other slaveholding societies, where death, hunger, and deprivation stalked the plantation quarters, southern planters remained singularly myopic about the effects of their slave management. The elite sugar masters were no different. They did not agonize in self-doubt or wallow in self-criticism over their treatment of their slaves, and they certainly did not conceive of themselves as monstrous fiends starving their dependents. Even if reality pointed in a different direction, they still believed themselves to be good masters. But they knew that to secure workplace stability and induce their slaves to operate their machinery during the harvest, they would have to compensate their laborers with direct payment, rewards, and non-wage incentives, including food. Compensation provided the slaveholders with an ideal opportunity to add a paternalistic polish to a network of coldly rational economic incentives and to dress exploitation in patriarchal garb. They distributed their Christmas bonuses with apparent largesse, granted post-harvest holidays, and

10. On Nottoway, see Postell, "John Hampden Randolph," 210–11; Rehder, *Delta Sugar*, 86–87. On buildings as emblematic symbols of power and identity, see D. J. Olsen, *The City as a Work of Art: London, Paris, Vienna* (New Haven, Conn., 1986), 287; Enrico Dal Lago, "Southern Elites: A Comparative Study of the Landed Aristocracies of the American South and the Italian South, 1815–1860" (Ph.D. diss., University College London, 2000), 238–59; Rhys Isaac, *The Transformation of Virginia, 1740–1790* (New York, 1982), 34–39, 305–8; Coclanis, *Shadow of a Dream*, 3–11. On the shifting architectural style of Anglo-American plantation mansions, see Barbara SoRelle Bacot, "The Plantation," in *Louisiana Buildings, 1720–1940*, ed. Poesch and Bacot (Baton Rouge, La., 1997), 107–25.

orchestrated the post-harvest and Christmas feasts in order to enhance their own power as masters who permitted such revelry.[11]

Planters accordingly used incentives as both a motivational tool to maximize work and as a way of strengthening their social authority and ritualizing their mastery. For this reason, paternalism could whiplash between relative extremes. When a planter's slaves had performed their half of the bargain, he might extend his munificence; if they failed to labor at an appropriate speed or produce an adequate crop, however, the same planter might strip away the harvest incentives or wield the lash with uncompromising brutality. In reality, most sugar masters chose a middle course, attempting to manipulate the reciprocal ties of the master-slave relationship and make their power complete with payment, incentives, and food.[12]

As the grinding season approached, slaves throughout the region saw their weekly rations increased. During the agricultural year, the average slave diet was comprised of 1 to 1.5 pecks of corn and 3.5 to 4 pounds of pork meat per week per hand. Occasionally, enterprising planters increased the rations of pork, corn, molasses, yams, and other vegetables to ensure that field hands would remain healthy and strong. Yet, as the peripatetic agricultural journalist Solon Robinson observed, enhanced rations frequently cloaked raw self-interest. Recounting in the *American Agriculturist* his visit to Thomas Pugh's Madewood estate, Robinson informed his readers that working hands drew 1.5 pecks of corn, 5.25 pounds of mess pork, and extra vegetables as their weekly ration. Pugh's generosity, however, was far from simple largesse, for the improved rations paralleled his introduction of time- and labor-saving technology. Toiling to supply the railroad that brought the freshly cut canes to the mill and to feed the conveyor belts that whisked the canes

11. On the seasonality of paternalism, also see Russell, "Cultural Conflicts and Common Interests," 297, 339–40. As Drew Faust and Jeffrey Young indicate, however, paternalism as self-image continued even when surrounded by death and natural population decrease. See Faust, *James Henry Hammond*, 77; Young, "Ideology and Death on a Savannah Rice River Plantation, 1833–1867: Paternalism amidst a 'Good Supply of Disease and Pain,'" *JSH* 59 (November 1993): 673–706.

12. On ritualizing mastery through the provision of clothes, food, housing, etc., see Joseph P. Reidy, *From Slavery to Agrarian Capitalism in the Cotton Plantation South: Central Georgia, 1800–1880* (Chapel Hill, N.C., 1992), 43–44; Genovese, *Roll, Jordan, Roll*, 550–84. Although suspicious of the utility of paternalism, Dusinberre notes that rice planters sought to orchestrate their "privilege" by ritualizing food and clothing distribution; see *Them Dark Days*, 183. On the duality and interactivity of ritual, see also Charles Joyner, *Shared Traditions: Southern History and Folk Culture* (Urbana, Ill., 1999), 93–102; Wayne K. Durrill, "Routines of Seasons: Labour Regimes and Social Ritual in an Antebellum Plantation Community," *Slavery and Abolition* 16 (August 1995): 161–63.

throughout the sugarhouse, slaves at Madewood needed every ounce of energy they could get. Others were less fortunate. Even Robinson observed that most slaves drew less than on Madewood; on Bishop Leonidas Polk's plantation, for example, the daily ration extended to half a pound of pork—a quantity somewhat closer to the standard slave ration. Elizabeth Hines, who grew up on S. G. Laycock's farm near Baton Rouge, frankly remarked that the slaves "et greens and pickled pork . . . pickled pork and corn bread!" Carlyle Stewart might have concurred, noting that his master in Jeanerette proved singularly less generous than those planters encountered by Solon Robinson. "We had grits and cornmeal and sometimes side meat and 'lasses," Stewart recalled about the cheap dry goods and fatty cuts doled out in the weekly ration. Not all slaves lived on such a meager diet, but for most of the year estate managers barely provided the caloric requirements necessary for strenuous farm labor.[13]

Once harvest approached, planters provided liberal quantities of energy-rich food to sustain their workforce during the grinding season. Ever alert to regional variations during his sojourn in Louisiana, Olmsted recorded that the field slaves acquired extra rations of flour and molasses at harvesttime, while those working in the sugarhouse received a plentiful supply of coffee, tobacco, and hot molasses or syrup. This sweet and sticky juice allegedly proved particularly beneficial to slaves, who "revive and become robust and healthy" upon its consumption. Explaining this phenomenon, Dr. Samuel Cartwright wrote that the fattening qualities of sugar emerged most markedly among boiling house workers, who enjoyed ready access to the calorie-rich cane juice. Not satisfied with relying upon the slaves' own inclination to drink the product of their labors, planters additionally distributed "a generous allowance of *sirop*, or molasses . . . every week during the winter and early summer." The effect of such additional allowances was palpable. Visitor Matilda Houstoun, for instance, naively observed that after four or five

13. Solon Robinson, "Agricultural Tour South and West, No. 4" *American Agriculturist* 8 (April 1849): 118; "Agricultural Tour South and West, No. 9," ibid. 8 (September 1849): 283; "Agricultural Tour South and West, No. 10," ibid. 8 (October 1849): 315; "Agricultural Tour South and West, No. 11," ibid. 8 (November 1849): 337; Sitterson, "William J. Minor Plantations," 66; Interview with Elizabeth Hines, WPA Slave Narrative Project, Arkansas Narratives, vol. 2, pt. 3, Federal Writers' Project, U.S. Work Projects Administration (USWPA), Manuscript Division, Library of Congress. See "Born in Slavery: Slave Narratives from the Federal Writers' Project, 1936–1938," http://memory.loc .gov/ammem/snhtml/snhome.html (first quote; accessed 30 April 2004); Interview with Carlyle Stewart, 3 May 1940, WPA Ex-Slave Narrative Collection, LSU (second quote).

days of feverish work, the slaves appeared "as cheerful and merry a set of people as I ever saw." As in Louisiana's sugar country, Jamaican planters also gave their slaves unrestricted access to the cane juice and prided themselves that the bonds-people appeared healthy and cheerful during the grinding season. But, as John Masterson Waddell observed, those smiles were merely derived from full bellies. "'Ere the season closed," the Presbyterian missionary continued, "they began to suffer, were fagged and sickly, from excessive toil and want of food." Waddell's observations rang true throughout the circum-Caribbean, where planters offered stimulants and production-focused incentives to their primary work crews. In short, the introduction of harvest perquisites heralded the onset of the grinding season, as planters inscribed their authority over the slaves' diet. Providing one's bondsmen with the fruits of their labors made good nutritional sense, but it also linked their well-being to the harvest and the master's apparent munificence.[14]

Although planters masked self-interest with benevolence during the harvest, ritualism and paternalism reached their apogee in the Christmas and post-harvest vacation, when the cane lords extended their gifts, paid their debts, and sought to tie their bondspeople both materially and psychologically to the plantation regime. The slaves did not subscribe to the paternalistic charade, and the harvest holiday served as a contested terrain in which African Americans sought to gain key rights and compensation for their harvest work. Whether receiving payment for extra work or due recompense for their labor, slaves in the sugar country greeted the close of the grinding season and onset of the vacation with considerable delight. Predictably, the cutting of the final cane was an orchestrated ritual, designed to impress the slaves with the planter's power and authority. Celebrations included ceremonial cane cutting and symbolic mule- and oxen-driven processions that led to the big house. The following description centers on the drama and display of the harvest season and focuses upon the interplay between white authority and the slaves' apparent accommodation to the planter's work schedule.

> When the hands reached the last rows left standings, the foreman (*le com-mandeur*) chose the tallest cane, and the best laborer (*le meilleur coteau*)

---

14. Olmsted, *Seaboard Slave States*, 2:317; Ingraham, *South-West*, 1:240 (first quote); *De Bow's Review* 13 (December 1852): 598–99 (second quote); *The Plow* 1 (November 1852): 352; Matilda C. Houstoun, *Hesperos, or Travels in the West* (London, 1850), 155 (third quote); Rev. Hope Masterton Waddell, *Twenty-Nine Years in the West Indies and Central Africa: A Review of Missionary Work and Adventure*, quoted in Sheridan, *Doctors and Slaves*, 153 (fourth quote).

came to the cane chosen, which was the only one left in the field uncut. Then the whole gang congregated around the spot, with the overseer and the foreman, and the latter taking a blue ribbon, tied it to the cane, and, brandishing the knife in the air, sang to the cane as if it were a person, and danced around it several times before cutting it. When this was done, all the laborers, men, women, and children, mounted in the empty carts, carrying the last cane in triumph, waving colored handkerchiefs in the air, and singing as loud as they could. The procession went to the house of the master, who gave a drink to every Negro, and the day ended with a ball, amid general rejoicing.

Fittingly enough, the spectacle honored the best laborer and terminated with a well-planned and ostentatious display on the steps of the plantation mansion. The black foreman was accorded due prestige, and the estate manager oversaw the performance of this brief, coordinated tableau. Whether this staged pageant impressed the slaves remains unknown. Others, however, were plainly not convinced. On his visit to the sugar country in the 1830s, Joseph Ingraham overheard slaves singing as the harvest ceased and the holiday began. The song, "which is improvised by one of the gang, the rest all joining in a prolonged and unintelligible chorus," Ingraham added, "now breaks night and day upon the ear, in notes 'most musical, most melancholy.'" Unintelligible to Ingraham and perhaps other whites as well, the slaves celebrated the end of the harvest with a painful lament that reflected their collective response to the exploitation of the grinding season.[15]

Like monetary incentives, post-harvest vacations and holiday meals became standard practice throughout the sugar country and served at least three overlapping functions. They fitted within the grid of duties and reciprocal obligations inherent in paternalistic slaveholding, they augmented the master's power as provider and protector, and they imparted a veneer of benevolence to the grinding season. As with material rewards, holidays took on giftlike proportions, for which planters expected due gratitude in terms of stability and work. And to further the slaves' sense of obligation and debt to the plantocracy, many planters harmonized the harvest celebration with Christmas. By conflating Christ's holy day with the slaves' holidays, planters could preen their devout egos and convince themselves that the sugar regime was truly a civilizing influence upon the cursed seed of Cham.

---

15. R. A. Wilkinson, "The Sugar Fields of Louisiana," *Southern Bivouac* 2 (June 1886): 18, reprinted in Moody, *Slavery on Louisiana Sugar Plantations*, 88; Ingraham, *South-West*, 1:241.

Indeed, on Charles Slack's plantation, Christmas assumed a multiplicity of meanings. Slack himself recalled that the slaves tested his nerves on Christmas Eve, but he also remembered the "merriment and confusion" of the early morning, with "a crowd of darkies rushing in to wish me a merry Christmas" and laying claim to their Christmas presents. Slack mused on his slaves' seeming loyalty, but even the most blinkered of optimists must have realized that African Americans used Christmas to extract benefits, in the shape of presents, from their slavemasters. Slave and slaveholder might well have construed the events of Christmas morning from opposing perspectives, though the repeated chanting and ostentatious act of bestowal confirmed to Slack the moral rectitude of racial bondage and its peculiar benevolence in the sugar country. But as Charles Slack and fellow planter William F. Weeks soon discovered, the slaves' holidays often proved taxing and irritating for the master class. As the twenty-three-year-old estate manager for the vast family plantation on Vermillion Bay, Weeks moaned to his father-in-law that after finishing the grinding season in mid-January, the "negroes have holidays now, and appear to enjoy themselves much more than I do, as it is extremely dull to be here without occupation." Weeks's holiday sojourn on Grand Cote Island nonetheless proved necessary if the young master was to impress the bondspeople with his authority over their leisure time.[16]

Other estate managers proved more miserly, utilizing the vacation as a punishment when the crop disappointed. When Ephraim Knowlton growled bitterly that "there is something wrong in the management of this plantation," slavemaster Robert Ruffin Barron expressed his displeasure by giving his slaves a miserably short three-day Christmas vacation, during which they cooked a freshly killed cow. More than a little disappointed that his seventy-five hands had produced a paltry 175 hogsheads of sugar, the master of Residence Plantation evidently saw little reason to share either harvest or seasonal cheer with them. While he could not risk abandoning a tradition that slaves clearly viewed as one of their collective rights, Barrow's reduction of the slaves' vacation offered a clear stimulus for the following year and a warning against inadequate labor. Paternalism of this variety left Barrow's slaves with an object lesson in harvest productivity and a painful reminder of their master's power and sway. But Barrow's miserly approach to the

16. Charles A. Slack to John A. Slack, 25 December 1846, Slack Family Papers, UNC (first quote); William F. Weeks to John Moore, 23 January 1848, David Weeks and Family Papers, LSU (second quote).

slaves' vacation served as only one side of the double-edged sword of slaveholding paternalism. Planters who expressed their anger at slave behavior in one season could swiftly reverse themselves in a self-serving display of slaveholding generosity. In contrast to Barrow's meanspiritedness, Harriet Meade produced a large and sumptuous meal for her slaves after they labored particularly efficiently during the 1858 harvest season. Meade had obvious grounds for praising her slaves, and they had every reason to expect more for their efforts; after the installation of a steam-powered apparatus on the Bayou Teche plantation earlier in the year, Meade's slave crews had increased production by 60 percent. With paternalistic largesse and condescension, Meade observed that despite the wet weather, her slaves would enjoy a "frolic" in the sugarhouse, where a supper and dance awaited them. Furnishing her bondspeople with ten geese and a hog, she promised a plentiful supply of cakes and pies as well. "Poor creatures," she observed on her recent act of bestowal, "they deserve it[,] for they had a long and tedious sugar making." Both reward and incentive, the promise of a frolic or the threat of its withdrawal served to codify the slavemistress's authority and underscore the market orientation of the harvest supper.[17]

While most planters were neither as uncharitable as Barrow nor as generous as Meade, the vast majority of sugar masters provided their slaves with a special Christmas or harvest meal that frequently included extra rations of pork, rice, sugar, potatoes, coffee, and flour. The ramifications of these events were manifold. First and foremost, the meal served as a customary right that slaves expected in return for their harvest labor. Simultaneously, the theatrics of slaveholding entailed a showy display of white humanity that seemingly linked master and slave in a unique personal bond. Harriet Meade, for instance, reinforced those ties by—in an ironic role reversal—baking cakes and pies for the slave crews. Former slave Caroline Wright, who grew up on Warren Wortham's plantation, likewise recalled that at Christmas, "The white folks allus give us presents and plenty to eat." Musing on Wortham's festive benevolence, she added that the slaves had a big dance "five or six times a year" and "de young missus learned us our ABC's." Caroline's focus on her master's generosity underscored the intensity of the paternalist bonds, which turned the meal and gifts into acts of bestowal. Similar acts of apparently thoughtful consideration added further gloss to the planter's self-image, reinforcing the

17. Residence Journal of R. R. Barrow, 11, 15, and 16 January 1858, Robert Ruffin Barrow Papers, UNC (first quote); H. W. Meade to Mary C. Moore, 28 January 1859, David Weeks and Family Papers, LSU (second quote); P. A. Champomier, *Statement of the Sugar Crop, 1856–1858.*

sense of stability, community, and mutualism that slaveholders strove to instill. Former slave Catherine Cornelius espoused precisely these values when remembering life in the antebellum sugar country. "We had gud times," Catherine averred. "Shucks, we didn't need no money in dem days. We got everything we wanted to eat. We had plenty of clothes. . . . We had everything de white folks had." Beyond this, Catherine received fifty cents at Christmas, holidays, and when the showboats came to neighboring Baton Rouge. Frances Doby likewise recalled her master's carefully orchestrated charade of slaveholding paternalism on New Year's Day. Blowing his trumpet from the mansion gallery, Master DeGruy beckoned the slave children forward and presented each one with a picayune or nickel. While perhaps masking their true histories from the white interviewers, both Catherine and Frances nevertheless dwelled on the strength of the paternalist bonds, even if those bonds remained skin-deep.[18]

The occasional wad of tobacco or a fattened goose were relatively harmless items to distribute, but alcohol was a wholly different matter. Drunken slaves might threaten social stability and dangerously challenge white authority. At the very least, whiskey might enhance slaves' self-consciousness, prompting many to question their condition and voice their resentment. Slaves were thus frequently banned from consuming alcohol. Despite its legal prohibition, however, many bondspeople clearly developed quite a taste for whiskey, which they purchased with money gained from overtime work payments or by trading with river hucksters. The social and economic implications of illicit slave drinking were far-ranging, especially when bondspeople stole pieces of machinery and sold them for a fraction of their value—chiefly, planters complained, for whiskey. Anxious to evade capture, the slave wisely consumed the proceeds of his thievery at once and left his master to fume against those "rascals" who bought, sold, and stole near the riverbank. The allure of whiskey not only provoked theft, but it was a relatively undetectable means of converting the planter's capital into the slaves' recreation—a suitable inversion of the plantation norm. As abolitionist Theodore Weld noted, trafficking in pilfered goods with river traders was so prevalent that "there are very

---

18. Interview with Caroline Wright, WPA Slave Narrative Project, Arkansas Narratives, vol. 2, pt. 3, Federal Writers' Project, U.S. Work Projects Administration (USWPA), Manuscript Division, Library of Congress. See "Born in Slavery: Slave Narratives from the Federal Writers' Project, 1936–1938," http://memory.loc.gov/ammem/snhtml/snhome.html (first quote; accessed 30 April 2004); Interviews with Catherine Cornelius, date unknown (second quote), and Frances Doby, 6 December 1938 (third quote), WPA Ex-Slave Narrative Collection, LSU. On the use of Christmas gifts, see Genovese, *Roll, Jordan, Roll,* 575.

many masters on 'the coast' who will not suffer their slaves to come to the boats, because they steal molasses to barter for meat; indeed they generally trade more or less with stolen property. But it is impossible to find out what and when, as their articles of barter are of such trifling importance." Slaves who snatched a few cents or a drink from the river trader could find alcohol elsewhere as well, most notably in neighboring towns and villages on Sundays. Plainly, slave drinking was a significant irritation to the white community and threatened their emotional and physical security. The *Franklin Planters' Banner* announced uneasily, "It is no uncommon site [*sic*] to see staggering drunken negroes in our streets at 9 or 10 o'clock, and hear them cursing like madmen." Clearly neither the slaves' words nor their demeanors were a welcome sight to white ears and eyes. Mad, these slaves might have been, but their threatening speech and alarming looks were likely grounded in their status as enslaved beings within a forbidding regime.[19]

Whiskey was an undeniable release from the horrors of bondage, offering a temporary form of escapism and camaraderie. Under its heady influence, African Americans could recast themselves as autonomous individuals or dream of inverting the racial pyramid. On both emotional and psychological levels, a whiskey-induced reverie enabled fathers and husbands to salvage some dignity and fellowship with other enslaved men who watched the white erosion of African American patriarchy. A drink hardly healed the physical and mental wounds of those women raped or forced into early sex to satiate the planter's desire for demographic growth, but it did offer a temporary respite from the many and multiple evils of the sugar country. In the hands and bellies of tired sugar workers, alcohol channeled resentment into comparatively risk-free expressions of release. Whiskey-induced bluster enabled black Louisianans to display their independence,

19. On theft and alcohol, see *Franklin Planters' Banner*, 24 January 1850, also quoted in Taylor, *Negro Slavery in Louisiana*, 127 (third quote); Moody, *Slavery on Louisiana Sugar Plantations*, 68–69; McDonald, *Economy and Material Culture of Slaves*, 70, 81; Sitterson, "William J. Minor Plantations," 70. Olmsted, *Seaboard Slave States*, 2:334 (first quote); Weld, *American Slavery as It Is*, 29 (second quote). On the slaveholders' disapproval of granting alcohol, see Breeden, ed., *Advice among Masters*, 250–56. Nonetheless, in almost every slave society, planters provided their slaves with whiskey during specific harvest celebrations and holidays or as a designated reward for work accomplished. See Genovese, *Roll, Jordan, Roll*, 577; Joyner, *Down by the Riverside*, 102. On the way planters attempted to stamp out late-night revelry with curfews, see William L. Richter, "Slavery in Baton Rouge, 1820–1860," in *Plantation, Town, and County*, ed. Miller and Genovese (Urbana, Ill., 1974), 389. On the illicit white-black alcohol trade in Georgia, see Timothy James Lockley, "Partners in Crime: African-Americans and Non-Slaveholding Whites in Antebellum Georgia," in *White Trash*, ed. Wray and Newitz (London, 1997), 57–72.

amplify their achievements and sexuality, defend their autonomy in a world that denied it, and, through bragging and hyperbole, seize power and control from their masters. Given the peculiar significance of alcohol within the slave community, white masters vigorously attempted to stamp out its availability. Despite its legal prohibition, slave drinking nevertheless flourished. On many estates whiskey emerged as a bargaining chip in the master-slave relationship. In return for their harvest labor, slaves translated their desire for alcohol into expectations upon which planters delivered during Christmas and end-of-season festivities. From the planter's perspective, the temporary easing of Louisiana bondage with a barrel of whiskey aided his self-image as a beneficent father figure whose largesse extended to assuaging the slaves' taste buds. At Frogmoor Plantation, George W. Woodruff accordingly gave the slaves four hogs, potatoes, molasses, and five gallons of whiskey. Even if watered down, this was enough alcohol for the merriment—if not inebriation—of the seventy-two-strong slave force. By limiting the amount of whiskey, Woodruff checked the extent to which the slaves' festivities could develop revolutionary or dangerous overtones that might threaten the edifice of racial bondage. Whiskey consumed in moderation added luster to the festivities and enhanced the slaveholder's paternalism, without potentially rebellious side effects.[20]

Beyond its undeniable value to enslaved African Americans, alcohol heightened the dependence relationship on which white hegemony thrived. Frederick Douglass observed the degradation slaves endured when forced to drink to inebriation. As the former bondsman and abolitionist remarked, planters conspired "to disgust their slaves with freedom, by plunging them into the lowest depths of dissipation" and deceiving them with a "dose of vicious dissipation, artfully labelled with the name of liberty." When the holidays ended, he concluded, "We staggered up from the filth of our wallowing . . . feeling, upon the whole, rather glad to go, from what our master had deceived us into a belief was freedom, back to the arms of slavery."[21]

Although Douglass perhaps exaggerated the baneful effects of whiskey, sugar planters gained in multiple ways by fueling slave drinking. They could use alcohol as wages in kind and display their apparent benevolence in supplying these drugs. Yet alcohol also weakened the community, leading to marital stress and a

20. Frogmoor Plantation Diary, 26 December 1857, Turnbull-Bowman-Lyons Family Papers, LSU. On the watering down of slaves' whiskey, see William J. Rorabaugh, *The Alcoholic Republic: An American Tradition* (New York, 1979), 13–14. On drink, autonomy, and working-class identity, see Roy Rozenweig, *Eight Hours for What We Will: Workers and Leisure in an Industrial City* (Cambridge, 1983), 61–64.

21. Douglass, *Narrative of the Life of Frederick Douglass,* 115–16.

depressing cycle of dependency. Whiskey rations pacified workers, leading them to conduct their day-to-day tasks thinking less of escape but more of their nightly toddy on the levee crest. And finally, once slaves had become dependent on the bottle, planters could withhold alcohol rations for inadequate labor or offer it as an incentive for extra work. In both cases, planters could thus exercise control over their young male workforce. Although some African Americans might have shared Douglass's particularly fatalistic vision of holidays as conductors "to carry off the rebellious spirit of enslaved humanity," others agreed with Solomon Northup's conclusion that the vacation transformed dour-faced workers and stirred repressed personalities. "They are different beings from what they are in the field," Northup remarked, "the temporary relaxation, the brief deliverance from fear, and from the lash, producing an entire metamorphosis in their appearance and demeanor." Former bondsman Hunton Love similarly recalled the release and pleasure associated with alcohol. "We had a nice time frolicing," Love explained, "We sang and danced and drank anisette," an easily made liqueur on a sugar plantation. But in his temporary contentment, Love was unwittingly playing into the slaveholders' hands. As one Alabama planter observed, "Frequent holidays are given[,] and as much interest as possible thrown them[,] to make them look forward to them with pleasure as seasons of enjoyment, where they can revel in the fun and frolic." Exposing the incalculable worth of slave frolics and harvest dinners, he added, the bondsman, "whose life is never illuminated by the cheerful beams of Hope, is devoid of any inducements to praiseworthy actions, and must be driven to discharge the duties of every day, solely by the fear of punishment." Such an individual, the Alabamian continued, "will be an eye-servant, without pride of character, and only fit to be constantly under the eye of the overseer. To beget in mine the essential principle, the hope of reward is constantly held out as an inducement." Hunton Love might have differed from this planter's observations about pride of character, but almost every cane planter agreed that holidays and meals proved invaluable inducements with which to inculcate the essential principles of hard work and diligent labor, while also dissipating resentment in the agro-industrial sugarhouse.[22]

22. Northup, *Twelve Years a Slave*, 169 (first quote); Interview with Hunton Love, date unknown, WPA Ex-Slave Narrative Collection, LSU (second quote); Rusticus, "Plantation Management and Practice," *American Cotton Planter and Soil of the South* 1 (December 1857): 372–75, quoted in Breeden, ed., *Advice among Masters*, 260–61 (third quote). On the social impact of drinking in working-class culture and its manipulation by workplace managers, see Way, *Common Labor*, 181–87. On alcohol as a cause of slave marital stress, Stevenson, *Life in Black and White*, 255.

If slavemasters mused on the potentially profitable returns that a holiday or dinner might bring, the "happy looking mortals" described by Northup and Love used the mask of contentment to advance African American rights, including the acquisition and consumption of alcohol. The slaves on Bay Farm evidently attached signal importance to their whiskey ration and expected planter John Slack to follow established convention and supply them with an adequate quantity of liquor for their Christmas holiday. To ensure that he did so, they crowded their master, swept him off his feet, and proceeded to carry him on their shoulders until he promised them plenty of whiskey. Although he later recounted the story with considerable mirth, Slack clearly perceived the whiskey reward as an acceptable and well-established precedent, upon which he—like other planters—had to make good. The sugar master interpreted his slaves' recent act of apparent devotion as a touching pastiche of slaveholding paternalism. Yet by hoisting him upon their shoulders and performing a scene in the liturgy of accommodation, John Slack's human property inverted his claim to master all and seized their rights. Most planters and estate managers seldom reached the stage of being jostled, swept up, and precariously raised onto men's shoulders. In such circumstances, goodwill could easily turn from badgering into bullying. For Slack, the effect of several large and strong men pressing him physically and mentally for whiskey must have been potentially alarming, and it also carried broader ramifications. Here, black Louisianans negated white stereotypes of the obsequious stammering "sambo" and presented themselves as activist, vaguely threatening figures who claimed their rights by physical force if necessary. In a society where brute strength lay in the arms, legs, and backs of the workers but power and authority rested in the overseers' hearts, minds, and whips, slave agency cut directly across established precedent and left white power structures temporarily confused by black contestation. By grabbing their master and manhandling him upon their shoulders, the slaves at Bay Farm had simultaneously crowned yet dethroned their master. By prodding him, grasping his legs, and grappling with his writhing body, Slack's slaves had effectively turned the auction block upon its head.[23]

Slack's pyrrhic victory masked the incongruities within slavery and underscored the occasionally physical negotiation of the master-slave relationship. Few planters tested their slaves' patience or goodwill to such an extent. Most followed the example of George Marsh, who realized that after a grueling three-month-long

23. John Slack to Henry Slack, 18 December 1856, Slack Family Papers, UNC.

grinding season, the slaves on his Petite Anse Plantation deserved a week-long vacation, during which Marsh orchestrated an evening of dancing and eating. Ellen Betts, a former slave on Bayou Teche, likewise recalled that her "marse sure good to them gals and buks what cutting the cane." After the termination of sugar making, planter William Tolas Parsons offered a drink called "Peach and Honey" to Betts and the other female slaves and whiskey and brandy to the men. In doing so, Parsons was neither unique nor especially novel in his slave management. Like many other "men of sense," he had discovered the potential profit to be derived from encouraging those in the cane fields with the promise of drinks and a well-earned holiday.[24]

The sugar masters also used the post-harvest vacation as a time to distribute clothing and to pay the slaves for their extra work. As one white contemporary recalled, "It is the season of enjoyment and festivity, and the time for settling up their outstanding accounts with each other, and the master and mistress of the plantation. The long running account for chickens, eggs, and vegetables, is liquidated by the good housewife; and the master pays for innumerable things, which have been provided by the slave, without interfering with his accustomed labors." By paying her slaves during the Christmas break, the plantation mistress translated her debts into gifts, and her husband similarly converted his market transactions into acts of generosity. On Shady Grove Plantation, Isaac Erwin adopted a comparable program of Christmas incentives, although he combined it with an additional schedule of holidays and vacations prior to the grinding season. In his plantation journal, Erwin noted that in late September the slaves received a two-day pre-harvest holiday, which the bondspeople put to profitable use by digging and storing their crop of potatoes. During this mini-break, Erwin furnished the slaves with their fall clothing. This included two pairs of pantaloons, a coat, and a shirt for the men, and two slips and a dress for the women. At Christmas, Erwin supplied his slaves with freshly slaughtered meat and with sufficient coffee, rice, potatoes, flour, and whiskey "to make a big ball." By timing his rewards and incentives to coincide with the beginning and end of the grinding season, Erwin provided two complementary stimuli to effective plantation work during the harvest period. Perhaps somewhat contented by Erwin's pre-harvest mini-holiday, his slaves efficiently kept pace with their master's regimented labor system—one that

---

24. George Marsh to Sarah Craig Marsh, 1 February 1840, Avery Family Papers, UNC; Botkin, ed., *Lay My Burden Down,* 126 (quote).

included disciplined watches and timed shifts in which Erwin measured his slaves' productivity with timepiece in hand. Despite a hard frost in mid-December and the inconvenience of changing his overseer in mid-harvest, Erwin recorded no discipline problems with his slaves, who produced 260 hogsheads of sugar before receiving their post-harvest rewards on Christmas Day.[25]

Erwin was not alone in utilizing market paternalism or a multiplicity of stimuli to impress upon his slaves the benefits of work and his own authority as the provider for all. Like Erwin, Elu Landry and Alexander Franklin Pugh also employed incentives both before and after the harvest to goad their slaves into particularly productive work during the grinding season. Landry, for instance, distributed shoes just two days before the commencement of the rolling season; in mid-November, after the completion of an unseasonably early sugar harvest, he issued a three-day holiday in lieu of Sunday work. To supplement this brief respite, he also gave his slaves a Christmas and New Year vacation. Not infrequently, planters orchestrated midsummer festivities featuring a communal meal and dance with slaves from neighboring plantations, and in which hats and tobacco were distributed just prior to the hay, corn, and cane harvesting seasons (or after laying by the cane crop in June and July). Maunsell White added Fourth of July festivities to his schedule of vacations and celebrated Independence Day in style. The heady republican rhetoric must have carried a hollow ring for those in fetters, but former Albert Patterson recalled that July 4 meant killing a cow and roasting it on a large barbecue. National holidays aside, Independence Day brought temporary relief from the crushing cycle of "no pleasures, jes' work" at Lasco Plantation. Patterson's experience paralleled that of slaves throughout the South, who doubtless endured the national flag-waving with indifference. As one Georgian slaveholder advised, Fourth of July festivities should avoid the "platitudes of the Declaration of Independence" and evade any "mischief-making discontent . . . that they are *white* men with black skins and that all men are really 'born free and equal.'" A far better course was to convert Independence Day into a "*negro* and *family* affair," in which the dinner "is a mere gratuity to be given or withheld according to merit or demerit." Such arrangements would have a "controlling power in the management of negroes," who would "be greatly animated and encouraged in their labors . . . [by] the fear of losing it." Fife and drum sounded in July the Fourth celebrations

25. Thorpe, "Sugar and the Sugar Region," 767 (first quote); Plantation Diary (1848–1868), 28 September 1849, 25 December 1851, Isaac Erwin Diary, LSU (second quote).

along the Mississippi, but at Lasco Plantation, dependence, not independence, marked the festivities.[26]

Christmas holidays and midyear vacations provided White, Landry, and Erwin with valuable incentives and a showy opportunity to display their beneficence, even if few of the enslaved audiences believed the charade. But when paternalistic negotiation and compromise failed to ensure plantation efficiency, planters swiftly reverted to the whip, exposing the fact that however useful incentives proved, slavery remained a relationship built on force and coercion. Indeed, the omnipresent lash gave compromise and negotiation extra force. By cracking the whip, slaveholders could remind workers of the expediency of compromising with their demands and the gory consequences of inadequate work. Whipping served as an appalling though integral part of slaveholding; in Louisiana's cane belt, the agronomic and environmental pressures that led to negotiation also promoted the lash. When crews failed to keep pace with the pumping pistons of agro-industrial capitalism, planters unrolled the bullwhip and showed that market paternalism also wore an iron mask.

Excessive driving hinged on the brutality and intelligence of the plantation overseer—a figure who had the unhappy task of marrying the planter's economic goals with his call for fair treatment, while supervising a labor force that tested a new arrival's wits, nerve, and resolve. Too easily, the overseer was neither firm nor gentle, as plantation registers advised him to be; instead he relied on violence or the threat of it when arbitration turned icy cold. Aptly portrayed by John Benwell as "embruted creatures with a ruffienly mien, prowling sulkily about, watching every motion of the bondsmen," overseers throughout the cane belt sharply emphasized that reciprocity lay in the whip. The visiting Englishman noted that the estate manager ensured rapid work by "quickening the steps of a loiterer by a word, or threatening with his whip" any bondsman or woman who slowed their labor. With his short-handled whip, weighted at the butt and with a lash four to five times the length of the staff, the overseer embodied the strict plantation supervision within the mill house and cane fields. John A. Hamilton, who whipped

26. Elu Landry Estate Plantation Diary and Ledger, 21 October and 21 November 1849, LSU; Alexander Franklin Pugh Diary, 28 October 1859, Alexander Franklin Pugh Papers, LSU; Frogmoor Plantation Diary, 19 July 1857, Turnbull-Bowman-Lyons Family Papers, LSU; Diary of E. E. McCollam of Ellendale Plantation (1842–1846), 7 July 1845, Andrew McCollam Papers, UNC; Interview with Albert Patterson, 22 May 1940, WPA Ex-Slave Narrative Collection, LSU (first quote); Breeden, ed., *Advice among Masters*, 263–265 (second quote).

his house slaves "until there is no place to whip" and believed fervently in a "good flogging," applied the lash with venom, complaining that his hands seemed "determined to out do me" in the midst of the rolling season. Hamilton suffered from heart palpitations as he flogged those who defied his authority, later writing exasperatedly, "To fight negroes nearly kills me—damn their skins I wish they were all in Africa." Overseer Vinson on the Palfrey family estate might well have concurred with Hamilton's sentiments. Described by his employer as "intelligent" when initially hired in February 1833, Vinson soon showed his true colors, flying into violent rages and wielding a club against the bondspeope. Only when the slave Anderson was brutally clubbed over his head and shoulders did John Palfrey dismiss his overseer, declaring him "a man of violent and unimaginable temper and of a jealous, suspicious, and vindictive disposition." Like Vinson, others whipped with gusto and suffered few lasting consequences for their actions. When the lash proved inadequate to the task at hand, overseers like William C. Riley employed the paddle with such ferocity that one man's skin was almost stripped from his back. "Passionate about punishing negroes," Riley humiliated his victims, whipping the driver for not chastising more severely the slaves under his command and leaving others to nurse their wounds while lying naked under the midday sun. At times, one observer noted, Riley would pick up sugar canes or a stick and beat the slaves over the head. More often, he resorted to his fists, punching their faces and striking their sides, heads, and bodies.[27]

Riley was not alone in his despicable acts. Former slaves particularly recalled the brutalizing nature of plantation discipline and the vicious form of physical punishment that they or their fellows endured at the hands of their overseers. Recollecting the ferocity of one estate manager in Rapides Parish, Peter Hill asserted that when his sister proved recalcitrant at work, the overseer punished her by staking her out and breaking her legs. Hunton Love, a former driver who wielded the lash against his fellow bondspeople, recognized the managerial compulsion to whip, but he was also horrified by the brutality of plantation discipline in south

---

27. J. Benwell, *An Englishman's Travels in America: His Observations of Life and Manners in the Free and Slave States* (London, 1853), 106 (first and second quote); Ingraham, *South-West,* 1 : 237; John A. Hamilton to William Hamilton, 5 and 15 November 1859, William S. Hamilton and Family Papers, LSU (third quote); John Palfrey to William Palfrey, 6 February and 3 November 1833, Palfrey Family Papers, LSU (fourth quote). The Riley incident is reported in Franklin and Schweninger, *Runaway Slaves,* 46, 351 n. 73 (fifth quote). Riley was not alone in using farm implements to beat his slaves. See Scarborough, *Overseer,* 96.

Louisiana. Recounting how the overseer tied one woman face down in a bed of ants with a heavy weight on her back "so she couldn't budge," Love described how the bondswoman visibly suffered and declared that she "was tortured awfully." As the bondsman recalled, "It wuz neces'ry sometimes" to apply the whip, but "these overseers . . . wuz brutal." Hunton Love's experience was unexceptional. On Valsin Mermillion's sugar plantation, one slave remembered that "one of his [master's] cruelties was to place a disobedient slave, standing in a box, in which there were nails placed in such a manner that the poor creature was unable to move. He was powerless even to chase the flies, or sometimes ants crawling on some parts of his body." Mermillion further proved his heartlessness when he purchased a young slave who had no experience of field work and had been raised with his previous master's children. When the youth refused to plow, Mermillion shot him dead, watching him fall into the hole that he had just dug. Brooking no challenge to his authority, Mermillion preferred to sustain a financial loss than have his supremacy questioned. Nor was he alone. When other slaves failed to measure up to the incessant pace of work, planters roughly beat and occasionally tortured them. Daffney Johnson, for instance, recalled that slaves were stripped to the waist before cats were released to claw and scratch "de blood out of our backs." Others had their ears cropped and faced brutal disfigurement. One individual who slowed at his labor received a hideous retribution from his master, who cut off one of his ears, lashed his body (leaving telling scars), and branded him with a hot iron. The slave had been marked a miscreant for life, and the slaveholder had exposed the transparency of his stewardship.[28]

The snarling crack of the whip added force to the planter's pretense at benevolent tyranny. Gracie Stafford remembered the powerful psychological impact of whipping, noting that the "ol' folks used to say that the master was hard on the slaves and had 'em whipped until the blood sometime stained the groun'." Lashed with a rawhide strap that channeled the overseer's strength and violence into stinging torture, the whip cut savagely into flesh and flaying open wounds. Even rope whips packed a frightening punch; their knotted ends tore open skin and left the victim nursing a lacerated back for weeks on end. When the slaveholder sought to inflict pain without disfigurement, he employed the paddle. Measuring about

28. Interviews with Peter Hill, 18 May 1937, Hunton Love, date unknown (first quote), Unidentified Ex-Slave, 17 August 1940 (second quote), and Daffney Johnson, date unknown (third quote), WPA Ex-Slave Narrative Collection, LSU; John Roach v. H. Holland, 20 February 1817, First Judicial District Court, Orleans Parish, Louisiana, cited in Franklin and Schweninger, *Runaway Slaves*, 194.

sixteen inches in length, the wooden paddle was covered with leather and was applied between sixteen and thirty times against the slave's naked buttocks. The emotional and psychological impact of being whipped or beaten with a paddle left a searing impression on ex-slave Ceceil George. Seventy-five years after slavery ended, she lamented, "Down here, dey strip yo' down naked, an' two men hold yo' down an' whip yo' till de blood come—Creuel [sic], Oh, Lawd." As the collective memory of the African American population attests, violence punctuated the working day. Elizabeth Ross Hite recalled the sugar master's regime with characteristic precision: "Dey slaves was punished fo' fights, being late fo' work, lying, runnin away and stealin'. . . . Dey would put ya in a stark [stock]. Yo hands and foots was buckled up and ya stayed dere fo months. No, dey did not hang ya. . . . Ya jest got a whippin. De mean master would tie de slaves to a tree and beat dem to death." To illustrate the horrors of plantation discipline, Hite recounted the story of "Old lady Cater," a runaway slave with six children who escaped and "built a home in de ground" to deceive her pursuants. When the driver caught her, Hite reminisced, he "whipped her to death. He beat her until her skin fell off an she died." Untying his victim from the tree that doubled as a makeshift whipping post, the driver buried her in front of the quarters, presumably for the maximum theatrical and menacing effect.[29]

The scars of a whip stigmatized the slave in the eyes of the plantation elite for the course of his or her natural life and warned others against violating the planter's regime. Theoretically, the whip awed the slave with the planter's superior physical force. As sugar lord William Minor reminded his overseers, in order to maximize the whip's psychological impact, punishment should be inflicted "in a serious, firm, and gentlemanly manner." Minor frowned on frenzied brutality; rather, the overseer must "endeavor to impress the culprit that he is punished for his bad conduct only and not for revenge or passion." Thomas Affleck's *Sugar Plantation Record and Account Book* contained similar advice on slave management. "Be firm and at the same time gentle in your control," Affleck warned. "Never display yourself before them in a passion, and even if inflicting the severest punishment,

---

29. Interviews with Gracie Stafford, 25 October 1940 (first quote), Ceceil George, 15 February 1940 (second quote), and Elizabeth Ross Hite, date unknown (third quote), WPA Ex-Slave Narrative Collection, LSU. On the paddle, see "Testimony of a New Orleans Free Man of Color before the American Freedmen's Inquiry Commission," in *Freedom*, ser. 1, vol. 3, *The Wartime Genesis of Free Labor: The Lower South*, ed Berlin et al., 522. On whipping, see Freehling, *Road to Disunion*, 62, 69; Kolchin, *Unfree Labor*, 120–26.

do so in a mild, cool manner, and it will produce a tenfold effect." Cold, calculating, and ruthlessly methodical in theory, if not in practice, whipping joined the
panoply of estate management tools as planters strove to emulate disciplinarians
of Minor and Affleck's stripe. Composed individuals who applied the lash with unnerving indifference vividly revealed their social authority, their command over
emotion, and their masterly intelligence. Experienced slave masters like Walter
Brashear quickly applied these tactics. By warning his overseer that no more than
five lashes should be applied without his permission, Brashear left both his slaves
and his overseer in no doubt about who possessed the ultimate authority to award
clemency or double the punishment. Rationally and coolly delivered punishment
convinced the enslaver—if not the enslaved—of the veracity of the master-slave relationship, in which steward-like planters cared for, pardoned, and chastised those
in their paternalistic care. Screaming insults and bludgeoning recklessly would
achieve little, planters believed, either in correcting the slave or in parading their
own idealized notions of mastery. "Indiscriminate, constant, and excessive use of
the whip is altogether unnecessary and inexcusable," Affleck advised, adding that
"when it can be done without a too great loss of time, the stocks offer a means of
punishment greatly to be preferred." Apparently Colonel Maunsell White agreed
with this advice. Former slave Albert Patterson remembered that his childhood
master "was not cruel, he wouldn't whip, he'd punish." Conflating cruelty with the
lash and punishment with fetters, Patterson recalled that White used shackles and
an iron neck brace to maintain discipline on his sugar estates. The former bondsman testified that after placing ankle cuffs on him, White had forged an iron band
around his neck with bars curving to the front, back, and sides. Noting that it was
impossible to lay down with the brace on, Patterson added that he had to pad the
iron band so that its collective weight would not cut into his skin. With manacles
riveted to his ankles and iron round his neck, Patterson nonetheless stoically endured this primeval form of punishment.[30]

30. Rules and Regulations on Governing Southdown and Hollywood Plantations, Plantation
Diary, 1861–1868 (vol. 34), William J. Minor and Family Papers, LSU (first quote); Thomas Affleck,
The Sugar Plantation Record and Account Book No. 2, Suitable for a Force of 120 Hands or Under. 4th ed.
(New Orleans, 1854), Turnbull-Bowman-Lyons Family Papers, LSU (second quote); Walter Brashear
to Robert Brashear, 7 February 1836, Brashear and Lawrence Family Papers, UNC; Interview with
Albert Patterson, 22 May 1940, WPA Ex-Slave Narrative Collection, LSU (third quote). On the need
for "calm and deliberate" flogging, see Greenberg, *Masters and Statesmen*, 22; Genovese, *Roll, Jordan,
Roll*, 64–67; Faust, *James Henry Hammond*, 100.

Not unlike White's fearsome collar, the stocks aimed both to correct and shame the miscreant and remind others of the wisdom of submission. Gracie Stafford, a slave on Myrtle Grove Plantation, recollected that her aunt "was put in stocks 'cause she wouldn't give in." In Terrebonne Parish, Ellen McCollam reported that after returning to the plantation, runaway slave Kit received a severe whipping and was placed in stocks for a week. Physical retribution of this type aimed to stigmatize the culprit, instill feelings of inferiority among the punished, and elevate the planter as the fount of all justice. No other act exalted the planter's power more clearly than his ability to grant clemency and his right to deny it. Albert Patterson, saved from the cruelty of a whipping by the punishment of chains, touched on the power of clemency when he described Maunsell White in terms of surprising and perhaps orchestrated respect, noting that his master would not allow overseers to beat his fugitive slave father "'cause . . . he was a good man." In these small acts of leniency, planters specifically aimed at enhancing their power and reminding the slave of his or her dependence upon their mercy. Subsequent acts of benevolence fortified the singularly personal bond forged in violence and strengthened the paternalistic ties. Advice to overseers accordingly focused on aloof and deliberate control. The cardinal virtue an overseer must possess, the *Capitolian Vis-à-Vis* proclaimed, was humanity and the attendant ability to manage without the lash. Those who relied too heavily on the whip, Joseph Acklen wrote in *De Bow's Review,* deserved prompt dismissal. "Negroes," Acklen counseled slaveholders in the sugar parishes, "must be kept under strict discipline, which can be accomplished by talking to them, and punishing moderately, but promptly and certainly." William Whitmel Pugh, who found the "management of negroes exceedingly disagreeable," concurred, noting that while "some severity is absolutely necessary at the start," wanton cruelty was to be avoided before "milder measures may be adopted a short time hence."[31]

Inevitably, the sugar masters failed to live up to their ideal. As Elizabeth Ross Hite, Hunton Love, and Gracie Stafford testified, planters in the cane world frequently reached for the whip before experimenting with mild measures or restraint. In the heat of the moment, disciplined moderation was shelved as

31. Interviews with Gracie Stafford, 25 October 1940 (first quote), and Albert Patterson, 22 May 1940 (second quote), WPA Ex-Slave Narrative Collection, LSU; Diary of E. E. McCollam of Ellendale Plantation (1842–1846), 5 February 1845, Andrew McCollam Papers, UNC; *Port Allen Capitolian Vis-à-Vis,* 13 September 1854; *De Bow's Review* 22 (April 1857): 379 (third quote); William Whitmell Pugh to Josephine Pugh, 17 February 1846, Pugh Family Papers, TEX (fourth quote).

tempers flared. Julia Woodrich remembered how slaves were dragged onto the levee crest and lacerated with a bullwhip in plain sight of their neighbors and observing whites. As anguished screams of "Pray, massa!" pierced the still rural air, the bellowing reply of "Dam you, pray yo'self!" left few in doubt as to the planter's singular lack of self-discipline. Whether dragged onto the levee or whipped before one's fellow slaves, the economic and social effects of physical intimidation left powerful memories that time seldom healed. When interviewed in the 1940s, the son of Mary Harris angrily remembered how his mother was brutally beaten, fuming, "Yes, I'm bitter—I have a right to be. My mother tells me about the brutality of those days, how they whipped unmercifully their slaves." The ex-slave Henrietta Butler had equally graphic memories: "You see dis finger here?—dare is where she bit it de day us was set free. Never will forgit how she said 'Come here, you little black bitch you!' and grabbed my finger—almos' bit it off." Understandably, Henrietta never forgot her mistress or her incisors. Bitter that he "could pass for white," Mary Harris's son likewise vilified the sugar masters for adding sexual intimidation to their arsenal of techniques for human control. Incensed by his questioner's defense of the slaveholding class, Harris raged, "The man who owned and sold my mother was her father . . . a brute like that who could sell his own child into unprincipled hands is a beast—The power, just because he had the power, and thirst for money." [32]

The thirst for money and power extended to social and residential organization throughout the antebellum cane world. By extending their sphere of control into slave housing and slave health, planters consciously sought to maximize the productive potential of the bondspeople and to reveal unambiguously the extent of their mastery. Standing atop the hurricane deck of a Mississippi steamboat as it churned its way downstream from Baton Rouge to New Orleans, visitors to the sugar country looked out over hundreds of planned communities. To port and starboard lay long, strip-like estates with a narrow river frontage, extending back from the levee crest into the distance. Passing over the lush greensward of the cane fields, the traveler's eye would quickly focus on the plantation mansion, the chimneys rising above the sugarhouse, and the regimented layout of the slave quarters. Touring south Louisiana just a few weeks before civil war seized the United States,

32. Interviews with Julia Woodrich, 13 May 1940 (first quote), Mary Harris, 28 October 1940 (second and fourth quotes), and Henrietta Butler, 28 May 1940 (third quote), WPA Ex-Slave Narrative Collection, LSU.

William Howard Russell described the sugar country as demarcated by lines drawn at right angles to the banks of the Mississippi River, in which rectilinear rows of whitewashed slave cabins seemed to stretch into the distance. "The sugar-house is the capitol of the negro quarters," Russell observed, "and to each of them is attached an enclosure, in which there is a double row of single storied wooden cottages divided into two or four rooms. An avenue of trees runs down the centre of the negro street, and behind each hut are rude poultry hutches." On this plantation, as on many others, the identical wooden cabins that lined the rectilinear street to the sugarhouse matched the planter's image of an ordered universe, in which power led in only one direction.[33]

As architectural historians have observed, buildings often serve as reflections of self-identity. No less representational than the big house or sugarhouse, the slave cabins reflected the cultural assumptions and racial convictions of the antebellum architects and builders. For Louisiana artists like Marie Adrien Persac, the mental and painted image of the plantation was akin to that of the planters, who idealized a model and stable community in which discipline, gentility, and order appeared to reign. Persac's panoramic canvases focused upon neatly built residences and conveyed an overwhelmingly positive impression of the planters' world. The plantation mansion, sugar mill, and outhouses dominated Persac's visual portraits, as did the immensity of the cane fields. African Americans (if pictured at all) appeared as tiny specks, and their diminutive cabins stood in the shadow of the master's palatial abode. The brutality of slavery faded from Persac's landscapes, much as it was exorcized from the planters' collective consciousness. Persac's late antebellum watercolors captured the planter's self-image as the lord of his demesne, where serenity, stability, and sugar flourished as one.[34]

Like Persac's landscapes, the nomenclature of plantation outhouses—or in slaveholding discourse, dependencies—reflected the sugar masters' firm sense of

33. Russell, *My Diary North and South,* 273 (quote); *Southern Cultivator* 5 (April 1847): 55; Bacot, "The Plantation," 90, 111–12. On Creole architectural antecedents, see Samuel Wilson Jr., "Architecture of Early Sugar Plantations," in *Greenfields: Two Hundred Years of Louisiana Sugar* (Lafayette, La., 1980), 51–82.

34. Kevin Lynch, *The Image of the City* (Boston, 1960), 1–13; Olsen, *City as a Work of Art,* 281–87; Mike Parker Pearson and Colin Richards, eds., *Architecture and Order* (London, 1994), 5–9. On Persac, see John Michael Vlach, *The Planter's Prospect: Privilege and Slavery in Plantation Paintings* (Chapel Hill, N.C., 2002), 91–111; H. Parrott Bacot et al., *Marie Adrien Persac: Louisiana Artist* (Baton Rouge, La., 2000).

social hierarchy; whether referring to their slaves or their buildings, slaveholders inscribed their sense of hegemony onto wooden sheds and human backs. The architecture of the plantation quarters assumed ideological proportions, reflecting the planters' monumental sway and their self-image as masters and lords. The massive columns of Nottoway Plantation reproduced in brick and mortar the grandeur of John Hampden Randolph, though on smaller plantations the big house could only be made to look really big by surrounding it with little cabins. The ornate classicism of Nottoway, with its stark white walls and columns, elevated whiteness as a symbol of civility and drew a color line across the landscape. White mansions, elegant loggias, and stucco facades stood in sharp relief to the pale wooden cabins, shorn of decoration, that stood diminutively beneath them. Moreover, by building cabins to a disciplined plan, slaveholders sought to engrave their dominance onto the landscape and upon their slaves. Throughout the sugar country, the rectilinear plantation streets reflected the slaveholders' idealized notions of community and their mastery of all beneath them. The hierarchical order of southern society thus appeared on the landscape in the shape of roads, grid patterns, perfect symmetry, and precisely measured distances between the identical slave cabins. The quarters themselves extended such rationalism, dotting the landscape as "bare geometric expressions," as one scholar observes, of the planters' sway. Usually dark inside and unpainted on the outside, these simple roofed boxes depersonalized their inhabitants and symbolized in graphic terms the dependency of those within them. In striking contrast to the elegant balustrades, ornate colonnades, and elliptical arches of the big house, the slave cabin had nothing but a stoop; it was not designed as a home, but rather as a simple shelter for those denied the right to independence and home ownership. Whether neat or dilapidated, these cabins served to propagate feelings of dependency and symbolized the total subordination of the bondsman to his master.[35]

The riparian topography of the cane world defined plantation geography. Estates followed three fundamental designs: the bayou-block pattern, the nodal-block pattern, and the linear pattern. On linear plantations, the slave quarters

35. On slave cabins, see Berlin, *Many Thousands Gone,* 97; Vlach, *Back of the Big House,* 162 (quote); Vlach, "Not Mansions . . . But Good Enough: Slave Quarters as Bi-Cultural Expression," in *Black and White Cultural Interaction in the Antebellum South,* ed. Ownby (Jackson, Miss., 1993), 89–114. On the physical dimensions of the slave community, see Laurie A. Wilkie, *Creating Freedom: Material Culture and African American Identity at Oakley Plantation, Louisiana, 1840–1950* (Baton Rouge, La., 2000), 85–94.

stood at right angles to the river and along a single road that led back from the levee to the master's house and the sugar mill. On block plantations, by contrast, the slave quarters or village constituted part of a rectangular grid pattern, where the sugarhouse stood at the heart of the settlement. These plantations often featured several slave streets that crossed each other. In both settlement patterns, however, the slave quarters remained fully within the planter's orbit. The frontier-like estates of southern Louisiana seldom matched the awe that estates like Nottoway were designed to induce, though they too aspired to a plantation ideal in which the slave cabins stood as appendages of planters' power and as symbols of the slaves' abject dependence.[36]

Ultimately, the sugar lords crafted a microcosm of their idealized universe, one that reflected their total exploitation of labor, gang work, and the slave village. Built around the mill house, slave quarters not infrequently resembled small industrial villages, where planters and overseers could speedily regulate their slaves' activities. Such regulatory control undoubtedly appealed to planters not only as a means of surveillance against slave insurgency, but also as a logical step toward disciplining and directing their laborers. Geographical proximity to the sugar mill proved particularly valuable during the grinding season, when centralized food preparation and rotating shift or watch labor prevailed. There was, however, something grimly familiar in the sugar masters' residential plans. Back-to-back housing at a Manchester textile mill produced scenes of urban horror; yet for industrialists, factory towns provided additional discipline and an opportunity to isolate their workers from all outside influences. In Cuba, *barracones* or barracks similarly stood as symbols of agro-industrial labor regulation, in which the roving eye of the planter could inspect all and minimize loafing. Slave compounds in Louisiana exploited managerial advantages similar to their Cuban counterparts, not infrequently resembling the rectilinear factory towns of the industrializing North. As one correspondent observed after touring the plantation belt, the "direct lines [of] uniform huts . . . exhibited the neatness of a clean New England Village." The reporter noted that the slave quarters stood adjacent to the sugarhouse and at a convenient distance for receiving the cane crop. Whether built to ceremonialize the planters' power over every plantation household or as a business decision to

36. Rehder, *Delta Sugar*, 90–100. For a detailed analysis of Ashland–Belle Helene, a linear plantation, see David W. Babson, *Pillars on the Levee: Archaeological Investigations at Ashland–Belle Helene Plantation, Geismar, Ascension Parish, Louisiana* (Normal, Ill., 1989), 16–18.

maximize the availability of labor, slave cabins in the sugar country symbolized the planters' world view.[37]

Not all sugar masters built or maintained decent accommodation for their bondspeople, but raw economic considerations led many planters to construct modest, though comparatively good, cabins. Business-conscious sugar planters realized that cold and poorly housed slaves stood a significantly higher risk of infection and that ill-treatment diminished profit returns. Asked by Theresa and Francis Pulszky why he built new quarters for his slaves, one sugar planter responded that "it was a good investment to have the slaves well lodged, as their health was then generally better." Sickness and ague chipped away at potential income and left few doubting the good sense of those who favored improved cabin construction. Profit-conscious capitalists, however, could not divorce themselves from their idealized notion of mastery or their perceived commitment to the slaves' welfare. As one planter observed, "Interest as well as duty ought to prompt every master to the erection of comfortable cabins for their slaves." Another "eminently distinguished planter" counseled readers in the local press, "You must . . . make him as comfortable at home as possible, affording him what is essentially necessary for his happiness—you must provide for him yourself, and by that means create in him a habit of perfect dependence on you."[38]

To orchestrate such dependency, planters built regimented lines of cabins, ordered regular cleaning, and coordinated health checks on a recurring basis. Along the Mississippi, slave compounds varied little, as identically built and indistinguishably painted wooden or brick houses dotted the landscape. At Houmas Plantation, Solon Robinson noted, the slaves resided in thirty double cabins. Describing the scene in characteristic detail, he observed that the cabins were "all neatly whitewashed frame houses, with brick chimneys, built in regular order upon both sides of a wide street, and which is the law, must be kept in a perfect state of cleanliness." Traveling a little further downstream, Robinson described the slave cabins on Myrtle Grove Plantation, where the bondspeople's living

37. *Southern Cultivator* 5 (April 1847): 55 (quote); Lanman, *Adventures in the Wilds of the United States and British American Provinces*, 1:209; Olmsted, *Seaboard Slave States*, 2:317. On *barracones* and industrial towns, see Scott, *Slave Emancipation in Cuba*, 19. David L. Carlton, *Mill and Town in South Carolina, 1880–1920* (Baton Rouge, La., 1982), 89–90.

38. Pulszky, *White, Red, Black*, 2:105 (first quote); *Southern Cultivator* 10 (1852): 41, quoted in Postell, *Health of Slaves*, 144 (second quote); *Franklin Planters' Banner*, 8 September 1842 (third quote).

quarters included thirty-two-feet square brick houses with elevated floors and chimneys in the center. Twelve of these structures comprised the total housing for the 139 slaves who toiled in the draining heat of the sugar bowl and no doubt sweated through the night in these packed huts. Most residences were built of wood, with either one or two rooms and a simple single-batten shutter as the door. At Ashland Plantation, archaeological evidence suggests that Duncan Kenner maintained double cabins with a central fireplace that served both rooms. Cheaply and uniformly constructed, these whitewashed cabins contained two rooms, each measuring twenty by twenty feet. They had plate-glass windows and a small vegetable garden in the back. The central fireplace saved money, since two families used the same flue. Located on a long street close to the sugarhouse, the cabins at Ashland were comparatively commodious. Archaeological research on other plantations suggests that rooms commonly measured just sixteen by twenty feet or even sixteen by sixteen feet. As the primary sleeping and eating locale for an entire family, even Kenner's cabins were minuscule, designed to enhance the planter's bigness by their relatively Lilliputian size.[39]

Irrespective of their design, cabins were frequently built by skilled slave workers on the plantation or by local white carpenters specially hired for the task. Initially built of available timber, these log cabins were subsequently replaced by a variety of wooden shacks that planters built as they mastered the landscape and turned frontier sugar land into a productive plantation. Former bondswoman Elizabeth Hines remembered that on S. G. Laycock's farm about four miles from Baton Rouge, some of her compatriots resided in log houses while others lived in "big old boxed houses," usually with two rooms. If separated at all, field hands and house slaves lived in cabins divided by a simple wicker fence that attempted (unsuccessfully) to draw a line through the slave community. Whether segregated by the presumed distinction of housework or not, these clapboard cabins were divided into two parts, each of which lodged a family or grouping of three or four people—a modest reduction from the mean of 5.2 slaves living in many slave cabins throughout the slave South. Still, former bondswoman Ceceil George remembered the cabins were "packed wid people."[40]

39. Robinson, *Solon Robinson*, 167, 172, 180 (quote); Yakubik and Méndez, *Beyond the Great House*, 20; Craig A. Bauer, *A Leader among Peers: The Life and Times of Duncan Farrar Kenner* (Lafayette, La., 1993), 54–55; Babson, *Pillars on the Levee*, 21; McDonald, *Economy and Material Culture*, 130.

40. Interviews with Elizabeth Hines (first quote) and Caroline Wright, WPA Slave Narrative Project, Arkansas Narratives, vol. 2, pt. 3, Federal Writers' Project, U.S. Work Projects Administration

Planters who sought to compel their slaves into a dependent relationship through housing also knew that the reciprocity of the master-slave relationship required them to supply—or at least appear to offer—basic, though adequate, lodging. The Reverend H. B. Price echoed contemporaries in noting that the slave cabins "are very comfortable houses, supplied with every necessary of life, arranged in proper method [and] presenting in a high degree an aspect of comfort." Others described the slave quarters as "perfect little villages," with brick-built quarters that "are most comfortable, clean, commodious, and desirable residences, such as we poor city folks would be glad to rent at $25 per month." Although the sugar masters provided their crews with marginally better housing than slaveowners provided for their bondspeople in the cotton country, they were hardly pleasant. The humid, sickly climate of south Louisiana made these cabins grim by anyone's standard, as ambient temperatures and contagious disease turned these brick and wooden shacks into forbidding billets where infection was rampant. Ventilation was poor, and while some cabins contained glass windows, others possessed only a primitive wooden slide or grating, which, William Howard Russell caustically noted, "admits all the air a Negro desires." Pigs, poultry, and dogs all lolled around these dismal huts, adding their feces to the human waste and stagnant pools that encircled the cabins. Animal and livestock excrement merely intensified the hazards of pollution, creating an environment in which fungi and bacteria proliferated, along with body lice, ringworm, and bedbugs. Mercifully, not all estates proved as decrepit as the ones described by Russell. Whether built of brick or wood, possessing a plate-glass window or a wooden grate, packed or not, most slave housing offered only rudimentary protection against extreme heat, bacterial growth, or Louisiana's meteorological fluctuations.[41]

Few of the antebellum sugar masters would have doubted the sagacity of maintaining basic, though clean, housing. Like other slaveholders, Louisiana planters

---

(USWPA), Manuscript Division, Library of Congress. See "Born in Slavery: Slave Narratives from the Federal Writers' Project, 1936–1938," http://memory.loc.gov/ammem/snhtml/snhome.html (accessed 30 April 2004); McDonald, *Economy and Material Culture of Slaves*, 132–33; Interview with Ceceil George, 15 February 1940, WPA Ex-Slave Narrative Collection, LSU (second quote). The literature on the relationship between slaves in the big house and field hands is extensive. See Genovese, *Roll, Jordan, Roll*, 327–65; Kolchin, *Unfree Labor*, 352–57. On residency per cabin, see Fogel and Engerman, *Time on the Cross*, 115.

41. *De Bow's Review* 8 (February 1850): 149 (first quote); *New Orleans Weekly Delta*, 18 October 1847 (second quote); Russell, *My Diary North and South*, 256 (third quote).

sought to reaffirm their power and authority by intervening in both the public and private lives of the slaves. As they did with housing, planters wrapped their concern for their slaves' health in apparent benevolence, which served further to validate their status as masters. With slave prices escalating rapidly in the 1850s and the unremitting risk of endemic disease stalking the plantation belt, planters ranked health care alongside clean and comparatively adequate housing as a necessary course of action. Only the most foolhardy completely ignored their slaves' health. Although slaves died in hundreds, most planters sought to introduce basic medical care, such as it was, on their estates. Economic factors aside, most Louisiana slaveholders resided on their plantations, and infection never discriminated across the color line. An outbreak of disease in the slave quarters ensured both black and white victims, and contagion spread swiftly from one estate to the next and from one bayou to another. Despite the relative isolation of the plantation compounds, they were not hermetically sealed units lacking contact with the outside world. The black community interacted with slaves on neighboring farms, and most estates received fairly regular visits from tradesmen, engineers, and diverse laborers. Maintaining the good health of all and preventing infection was imperative, though in the damp and humid environment of south Louisiana, disease control proved a continuous uphill battle. Valuable bondspeople swiftly fell prey to infection, and due to their mineral-deficient diets, their immune systems recovered slowly.

Solon Robinson grasped the essence of planters' medical care, writing in *De Bow's Review* in 1849 that "slaves are better treated now than formerly . . . partly from their masters becoming more temperate and better men, but mainly from the greatest of all moving causes in human action—self interest." Robinson's conclusion seemingly underscored the slaveholders' singleminded pursuit of profit, but he went on to present the master-slave relationship in a different light. Masters have discovered, he affirmed, that their true interests were inescapably bound to the "humane treatment, and comfort, and happiness of their slaves." Solomon Northup concurred, noting that "those who treated their slaves most leniently, were rewarded by the greatest amount of labor." Accordingly, medical care derived not solely from the values of the counting house, but also from the slaveholders' shared "interest" with the slaves. In his article, Robinson rationalized slavery as an organic relationship forged by benevolent masters and their wayward, childlike slaves. Slaveholders from the Carolinas to Louisiana espoused these blinkered values and introduced medical care partly out of self-interest but also because they ardently believed their own rhetoric. Even as their slaves died in appalling numbers, southern planters could build hospitals, provide medical care, and purchase the latest drugs confident in their acts of Christian stewardship. As the grim population records from the sugar country attest, the slaveholders' charade proved empty, yet they remained adamant that no one brook their attempt to orchestrate slave health.[42]

The sugar lords possessed one further reason for considering themselves suitably equipped to master the medical sciences. In health as in slave housing, planters sought to impress upon the slaves the power and authority of the master class. Total subordination required control over the heart, mind, and movements of each individual and the regulation of his or her medical condition. Planters sought to stamp their authority upon the slave's body and institutionalize their curative powers in the estate hospital. Frequently located at one end of the slave quarters—though not too close to risk contagion—the sick house was where African American nurses and visiting white doctors addressed the myriad medical complaints that emerged on every plantation. Such facilities enabled planters to demonstrate to themselves and others that they stood as paternalistic labor lords who built hospitals for their "people." Moreover, hospitals concretely displayed the planters' relative urbanity by reflecting the common nineteenth-century belief that hospital care derived from the charity and benign stewardship of the upper classes. While a sense of duty compelled many early Victorians to favor institutionalized health care, an overriding sense of reciprocity between social classes underpinned the emergence of the hospital system in the American North.[43]

Southern planters, who utilized a similar discourse of paternalism in health care, found themselves aptly positioned to present themselves as modern individuals. Building a hospital in antebellum America was a singular act of charity to one's social inferiors and one that the sugar masters manipulated to bind those in the sick house to the munificence of the master. Slave hospitals were plainly self-serving, but they also served as representational sites where planters could perform their patriarchal script before slaves and visitors. Perhaps embellishing her narrative, Matilda Houstoun—who portrayed slavery in bucolic

---

42. *De Bow's Review* 7 (September 1849): 220 (first quote); Northup, *Twelve Years a Slave*, 69–70 (second quote).

43. On antebellum hospitals, paternalistic values, and society, see Charles E. Rosenberg, *The Care of Strangers: The Rise of America's Hospital System* (New York, 1987), 19–21, 48–58. The key empirical work on slave medical care in the Louisiana sugar belt is Bankole, *Slavery and Medicine*; also see Whitten, "Medical Care of Slaves," 153–80.

terms—watched one sugar planter execute his role when she observed that the hospital featured interior arrangements that proved "as clean and comfortable . . . as one could wish to see." Displaying his modern sensibilities and impressing Houstoun with his stewardship, this planter stood firmly within a transatlantic community of paternalistic benefactors whose social responsibilities extended to asylum and hospital construction. Sick rooms and purpose-built infirmaries, however, had additional clear-cut purposes. They isolated the infectious from the rest of the plantation community and ensured that they would be cared for by retired and comparatively valueless slaves. Elizabeth Ross Hite remembered from her childhood on Bayou Lafourche that Grandma Delaite took charge of the hospital—a pattern that repeated itself on many estates, as estate managers utilized in the sick house older slaves who had passed their productive prime. At Oaklands Plantation, fifty-five-year-old Julia Ann served as the plantation nurse, where she perhaps brought a host of folk remedies to bear on her patients. Julia Ann's valuation of $400 was literally a drop in the ocean of the slave property on this estate, which was valued in tens of thousands of dollars. From the master's cold perspective, it was far better for the nurse to die than to have infection spread among the principal field gangs.[44]

Elderly slaves who toiled in the plantation hospital unquestionably relied on their own African American herbal remedies. Interviews conducted in the 1940s suggest that slaves in the sugar country frequently cured themselves with roots, medicinal herbs, and additional homemade remedies. Ex-slave Verice Brown of St. James Parish recalled that her mother used leaves to make a tea that controlled fever and utilized jackvine for purifying blood. Lindy Johnson likewise prescribed coal oil and salt for rheumatism and Jacob bush for fever, while Ellen Broomfield recalled that she ate sulphur and molasses to purify her blood and that candy made of Jimson weed and sugar was good for worms. Planters, however, were often unwilling to brook this challenge to their authority. Eschewing the medical expertise of "ruptured" or elderly slaves, they employed local physicians to attend to their slaves' health needs. Catherine Cornelius underscored the duality of black and white medical care, observing that a black nurse and midwife worked alongside

44. Houstoun, *Hesperos,* 157 (quote); Interview with Elizabeth Ross Hite, date unknown, WPA Ex-Slave Narrative Collection, LSU; Inventory and Valuation of Slaves, Stock, and Farming Utensils of Oaklands Plantation, 1859, Samuel D. McCutchon Papers, LSU.

the local white physicians who visited periodically. Planter J. P. Bowman, for instance, engaged the services of two doctors, who visited his estate thirty-nine times in the space of eight months—a frequency of over one visit per week. A physician examined the slaves for fevers, chronics, pleurisy, bowel complaints, flux, spasms, diseased legs, and abdominal pain, and his presence at Frogmoor proved a common occurrence. Clearly, the slaves took full advantage of Bowman's careful attention to their health. From January to mid-May 1857, overseer George Woodruff reported hundreds of incidents of slaves reporting sick, with over seven slaves citing a medical reason for not working on a daily basis. Probably feigning illness in a collective go-slow through the planting and cultivating seasons, these slaves utilized the planter's predilection for medical care to stretch the limits of his authority. Regular visits by Drs. Garret and Bell, however, ensured that feigning illness could only last so long before professional advice exposed the lie.[45]

While not facing Woodruff's predicament, planter David Weeks similarly arranged for a physician to visit his Grand Cote estate every six months. Vigilant in slave medical care on the family's Attakapas plantations and fearful of the impending risk of a cholera outbreak, David's son Alfred remained "very particular in regard to the diet and conditions of the negroes, and have told them all that the moment any symptoms of disease arise, they must inform me." Linking himself and his slaves in a personal bond in which the slave reported directly to the master, Alfred Weeks not only reaffirmed his paternalistic guise as the slaves' protector, but he also underlined the dependency of those on the estate on the bounty of the planter. Weeks had literally killed three birds with one stone—he protected his valuable property, he appeared before the slaves and his community as a benevolent steward, and he reminded those in the quarters that he alone was the master and the source of all things, be it housing or medicinal care. Given the overlapping social and economic benefits of medical coverage and the collective value of the slave crews, many other planters contracted with one or two doctors to conduct regular check-ups and supply medical services for a calendar year. Such care was not cheap—up to $300 per annum—and it was rarely effective, but it was an

45. Bankole, *Slavery and Medicine,* 143, 144, 145. Interviews with Ellen Broomfield, 20 February 1941, and Catherine Cornelius, date unknown, WPA Ex-Slave Narrative Collection, LSU; Frogmoor Plantation Diary, Plantation Management Papers, Turnbull-Bowman-Lyons Family Papers, LSU. On herbal medication, see Fett, *Working Cures,* 60–83, 111–41.

essential investment for those who wished to maintain slave health and check episodes of feigned illness.[46]

Given the relatively poor quality of rural physicians and their inability to stem the spread of infection, many sugar planters distrusted professional doctors and preferred to treat medical cases themselves. Most believed that disease and illness derived from the disordered state of an individual and that medical care served at best to support the body's recovery to fitness. Since sickness was deemed to be as much social as physical, the patients' character was frequently imputed as the source of illness; this in turn, most antebellum Americans believed, reflected God's wrath or displeasure. Myopic conclusions of this type conveniently obscured the physiological impact of the sugar order and enabled planters to release themselves from culpability after meddling in the medical sciences. Amateur care proved so widespread in southwest Louisiana that the Attakapas medical lobby (under the moniker "Medicus") published a two-part essay in the local press criticizing the planters for their homespun medical treatments and decrying the "total lack of confidence . . . toward their physicians, both as respect their skill and honesty." Urging the planters to employ professional medics, the Attakapas physicians cogently argued that since no rational planter would accept the advice of an amateur agrarian, they should not rely on laypersons in matters medical. To ignore professionals, Medicus fumed, contradicted the planters' normally rational managerial strategy. Perhaps more significantly, it also undercut medical business and left local physicians short on trade.[47]

Despite bleating from physicians, several planters clearly possessed medical skills or followed readily available guidelines in curing or preventing common illnesses. "Intelligent planters," *De Bow's Review* pronounced, should swiftly adopt "judicious precautions against cholera, as soon as they heard about its arrival on

---

46. Medical Charges, 24–27 October 1826, and Alfred W. Weeks to John Moore, 23 March 1849 (quote), both in David Weeks and Family Papers, LSU; Cash Book, 1851–1854 (vol. 5), Bruce, Seddon, and Wilkins Plantation Records, LSU; Expense Book, 1847–1854 (vol. 5), 22 February 1850, John H. Randolph Papers, LSU.

47. *Franklin Planters' Banner,* 26 April (quote) and 24 May 1849. On the commonly held belief among nineteenth-century Americans that indolent patients were responsible for bringing disease and illness upon themselves, see Charles E. Rosenberg, *The Cholera Years: The United States in 1832, 1849, and 1866* (Chicago, 1962), 56. God and divine intervention were similarly invoked as definitive forces in explaining illness and infant death. See Sylvia D. Hoffert, *Private Matters: American Attitudes toward Childbearing and Infant Nurture in the Urban North, 1800–1860* (Urbana, Ill., 1989), 182; Rosenberg, *Care of Strangers,* 24, 72.

the continent." After removing all refuse from the slave quarters, the correspondent continued, planters should whitewash all cabins, place lime beneath each slave hut, and pay particular attention to the quality of food. Former bondswoman Elizabeth Ross Hite similarly recalled of her childhood that slaves slept on wooden beds with fresh moss mattresses and that master Pierre Landreaux required them to scrub their cabins once a week. This directive, Hite recounted, ensured that "our bed was kep clean. Much cleaner den de beds of today." Landreaux supervised all operations, inspecting visiting slaves for fleas and bugs, and "sure fuss[ing] about it" when he saw a "chinch on a bed." Checking cabins for bugs and drilling visitors about their personal hygiene, Landreaux inscribed his authority over the quarters and the health of those within them. Such preventive measures attempted—albeit with limited success—to stem contagious outbreaks of disease. Other recommendations, however, were even less constructive. As soon as an outbreak of cholera was confirmed, one large sugar planter ordered a cessation of work and "permitted [the slaves] to go into a regular frolic" for two to three days. Fiddlers were called up to entertain the slaves, and the planter distributed sufficient whiskey "to produce a pleasant exhilaration," if not inebriation. This cure, he concluded, proved remarkably effective, as hardly a new case occurred after the commencement of the frolic.[48]

While not doubting the medicinal and psychological qualities of whiskey, planters believed that whitewashing the slave cabins was a significantly more effective precaution than the whiskey barrel against cholera. Adhering to customary practices, planter Elu Landry whitewashed the sugarhouse and the interiors of all the slave cabins following a cholera outbreak in June 1849. He then cleaned all the housing on the estate. When Big Henry fell ill nonetheless, Landry drew upon his own medical experience and administered laudanum, brandy, and "frictions of cayenne" to the bondsman. After consuming this heady and spicy brew, Henry, perhaps unsurprisingly, died within six hours. Having suffered the "terrible ravages of cholera upon his place," Bishop Leonidas Polk similarly experimented with cholera remedies, joining pharmacist C. Rabe in advertising an invaluable cure for the early stages of the disease. The efficacy of Polk's medication, however, remains debatable. As local medics observed, slaves' suffering and death often occurred due to a planter's incompetence in diagnosing illness and administering the correct cures. Satirizing the planting interest, Medicus recounted the story of one

48. Interview with Elizabeth Ross Hite, date unknown, WPA Ex-Slave Narrative Collection, LSU (second quote); *De Bow's Review* 11 (November 1851): 476 (first and third quotes).

slaveholder who administered ipecac (to provoke vomiting) and calomel (to purge the bowels) to a bondsman. When the slave failed to recover, his master bled him as a last resort. The unfortunate bondsman died the same day. "This man was astonished that the negro should die under such a treatment," Medicus observed, "for he insisted that he had seen 'many a Pleurisy' get well under it." The planter's ill-judgment and poor diagnosis, however, cost the slave his life. Intrusive medical management of this stripe enabled planters like Landry and Polk to exercise their authority over every aspect of the bondsmen's lives.[49]

Whereas planters like Elu Landry failed in their health management, others proved somewhat more successful. Former slave Rebecca Fletcher remembered that "ole missus useter give us blue mass pills when we needed medicine. It sho did make us sick. We had to get sick to get well, ole Missus said." These pills contained powdered mercury, which offered primitive assistance in easing fever and dysentery. Henrietta Butler underscored the juxtaposition of the slaveholding personality when she reported that her cruel mistress made a "fly-blister" for one of her bondsmen suffering from lockjaw and called out a physician to administer pills whenever any of her slaves fell ill. As masters and mistresses of medicine and men, slaveholders sought to oversee medicinal complaints and simultaneously to emphasize their authority over their patients. In the case of Rebecca Fletcher, "ole Missus's" treatment was covered with a paternalistic sheen that fostered a dependent relationship between slave and mistress, even though she refused to internalize her dependence as deeply as "ole Missus" sought. By orchestrating medical care, the slave mistress was carefully choreographing her power, her benevolence, and medical modernity. Above all, dispensing medical care enabled planters to orchestrate their authority over the slaves' bodies and appear as the pater- or materfamilias. As one slave mistress observed of her slaves, "You can't get them to understand anything about taking care of the sick." Constantly on the "look-out," this particularly attentive slaveholder doled out charcoal and roasted prickly pear to Ike, congratulating herself that such treatment "seemed to do him more good than all the medicine the Dr. gave him." Like many of his neighbors, planter John Lobdell similarly sought to intervene personally in the slaves' health, filling his medicine cabinet with the latest cures and panaceas that antebellum medicine afforded. Wisely preparing for the risk of malaria with a purchase of quinine, Lobdell also possessed the bark extract cinchona for marsh miasmata and

49. Elu Landry Estate Plantation Diary and Ledger, 11 July 1849, LSU (first quote); Robinson, *Solon Robinson*, 201; *Franklin Planters' Banner*, 26 April and 24 May 1849 (second quote).

fevers, sulphur for skin infections, calomel for fevers and ague, and cream of tar-tar for dropsy. Including camphor in his medical chest, Lobdell equipped himself to counter the mortal effects of dysentery and gonorrhea. He also stocked two ounces of ipecacuanha, a common cure for yellow fever when mixed with calomel. Finally, Lobdell bought two hundred cholera pills to protect his slaves against the ravages of that disease. Suitably equipped with a range of reputedly reliable med-ications, Lobdell could exploit the opportunity to enhance his mastery by attend-ing to patients himself and cultivating a dependency that seemed, in his myopic eyes, to symbolize the ideal master-slave relationship.[50]

At the root of planters' concerns with health care, however, lay raw self-interest and the economic desire to maintain the value of their investments. For-mer slave Elizabeth Ross Hite assessed medical care with telling accuracy. "De slaves got de best of attention," she noted in regard to the two white doctors who treated them. "No sir," she concluded, "Master Landro was makin' too much money off of his darkies to let 'em die lak mules. Dey was gud workers." As one plantation mistress similarly observed, slaves were often "too valuable . . . to tamper with." Yet to close the story here would portray the plantocracy as mono-chrome capitalists devoid of an ideology and shorn of the referents of southern slaveholding. In medical care as with many other aspects of their slave manage-ment, the sugar masters were unquestionably profit-minded, rational, and ex-ploitative. They utilized incentives and rewards with a clear eye for optimizing productivity, and few questioned the veracity of their economic claims to mastery. Like their brethren in the North, they recognized the value of time and labor management; but beyond that, sugar planters shared what we might refer to as an economy of desire, which encapsulated their vision of mastery. Above all, they sought to master sugar and men and compel all to bow to them in total sub-

50. Interview with Rebecca Fletcher, 20 July 1940 (first quote), and Henrietta Butler, May 28 1940 (second quote), WPA Ex-Slave Narrative Collection, LSU; Lydia Murphy McKerall to Emma Caffery Thomson, 13 September 1857, Caffery Family Papers, UNC (third quote); Anonymous Plan-tation Ledger, Entry 276, to James Shaw, December 1848, LSU. On the curative properties of the above drugs, see James Ewell, *The Planter's and Mariner's Medical Companion* (Philadelphia, 1807), 25, 46, 104, 110, 141, 145, 153; Jabez W. Heustis, *Physical Observations and Medical Tracts and Remarks, on the Topography and Diseases of Louisiana* (New York, 1817), 117, 137; Todd L. Savitt, *Medicine and Slav-ery: The Diseases and Health Care of Blacks in Antebellum Virginia* (Urbana, Ill., 1978), 155–56; John Duffy, ed., *The Rudolph Matas History of Medicine in Louisiana*, 2 vols. (Baton Rouge, La., 1958), 1: 272–76. For a similar analysis of slaveholders seeking to dictate the medical care of their slaves, see Whitten, *Andrew Durnford*, 94–101; Faust, *James Henry Hammond*, 77–82.

ordination. Imagining themselves as guardians and paternalists, the sugar masters—like slaveholders in older and more established sections of the South—held resolutely to a self-image of stewardship, benevolence, and an idealized master-slave relationship. Devoid of true authenticity, paternalism cloaked the brutal reality of their actions, proferring comforting images of stable relations and domestic reciprocity. Rather than focusing on the frequency of infant deaths, for example, Robert Ruffin Barrow recorded that the death of driver Andrew was met with "deep sorro . . . [by] every one who knew him both white and Black." Perhaps the bondsman's particular skills touched Barrow's emotions; in any event, he grieved for this "valuable man," who had an "uncommon good mind" and "good judgment" about plantation work. The tears shed by slaveholders on such occasions convinced them, if few others, that the idealized master-slave relationship flourished on every estate. Reality suggested otherwise; but fiction, not fact, shaped the slaveholders' identity. Still, the paternalistic vision that sugar masters cherished in their hearts proved priceless in their pockets. Paternalism provided slaveholders with a vocabulary to describe self-serving and highly exploitative estate management. They could parade economic incentives as acts of reciprocity and mutualism, coordinate the Christmas feast to exalt their paternal power, and build economic and social ties that bound master and slave in a web of mutual dependence. Financial incentives and paternalistic compromises ultimately aided plantation production, enabling the sugar masters systematically to exploit their workers and still hold to an idealized vision of mastery. African Americans, with good reason, had little faith in this charade. Rather than accepting their owners' discourse of kindly mutual benevolence, they struggled to secure rights, money, and autonomy from Louisiana bondage.[51]

51. Interview with Elizabeth Ross Hite, date unknown, WPA Ex-Slave Narrative Collection, LSU (first quote); Margaret Marsh Henshaw to Sarah Craig Marsh, 29 November 1852, Avery Family Papers, UNC (second quote); Residence Journal of R. R. Barrow, 21 April 1858, Robert Ruffin Barrow Papers, UNC (third quote).

six

"INCHIN' ALONG"

$\mathcal{T}$HE INSIDIOUS NATURE of Louisiana slaveholding was played out over the countertop of the plantation commissary. This small emporium was a building— or more often just a room—stocked with the produce of the market economy, yet one that denied its customers the full implications of the commercial ethos. Whether in the Caribbean, in Louisiana, or elsewhere in the Americas, bondspeople utilized their corn patches, poultry hutches, and free time to develop a flourishing internal market economy. Receiving payment for the produce of their labors, slaves derived both material and cultural capital from producing and trading their own commodities. In controlling their land, planting their crops, marketing their produce, and spending their profits, slaves in the West Indies and some parts of the American South resembled landed peasants who organized their labor and used their earnings for self-improvement. Throughout Louisiana's cane world, African Americans actively engaged in independent production or overtime labor—a process historians call "overwork." By working beyond the confines of a normal day, slaves found an erratic and conflicted space in which they could grasp the essence of liberty and temporarily invert the planters' claims to all their labor.

Irrespective of the social and economic importance African Americans attached to independent production, the slaves' internal economy ultimately reinforced chattel bondage. Not only did overwork promote labor stability, but it ensured that planters could maximize the working day and exploit the slaves' desire for pecuniary gain to optimize sugar production. More so than their neighbors on the cotton frontier, planters in the cane world were faced with a relative shortage of time. They had myriad tasks to complete, ranging from basic maintenance, ditching, draining, levee work, and cultivating the canes to the final process of manufacturing sugar. Added to this, planters raised cattle, sheep and poultry, and farmed small grains for home consumption. Finally, every planter required hundreds of cords of wood to be burned as fuel for the steam engine and beneath the sugar kettles. For most of the year, planters possessed enough hands to do all these tasks. But as the amount of time and effort increased for tasks such as timber collection or corn farming, the volume of cane sowed and harvested necessarily declined. Planters could evade this resource- and time-allocation problem by purchasing supplies and focusing only on sugar. Those on the extreme northern and western margins of the sugar country chose to do just that, maximizing their cane acreage at the expense of other commodities. Most planters, however, still opted to produce their own food and supplies. Self-sufficiency freed them from the volatility of the market and dependence on others for the provision of food and clothing. There was a way, however, to maximize cane land and still produce one's own food and provisions—overwork. By paying their slaves to conduct time-consuming tasks on Sundays and in their own time, time-conscious managers could save literally thousands of hours over the course of the agricultural year. The slaves gained money from this accommodation with which they could improve their modest lives, but the practice of overtime work did exactly what it said—it overworked the slaves.[1]

Throughout south Louisiana, Sabbath observance broke the working week and defined a single day for rest. During the harvest, planters often reappropriated Sundays, though on most estates the French tradition of permitting slaves to work for themselves on the Sabbath continued up to the Civil War. African Americans converted that colonial right into a customary practice, and on most weekends they continued to raise and sell the produce of their small garden plots. Whether collecting timber for the steam engine, raising cattle for the master, or cultivating corn for subsistence, slaves worked even more hours for their owners. Crucially, by transferring these tasks to the slave community and to evening and Sunday work, planters could focus additional energy, resources, and regular workday time on the cash crop. Moreover, by paying the slaves to be industrious workers in the

---

1. Roderick McDonald contends that while Louisiana sugar planters gained from the slaves' "economic enterprise, . . . the internal economy plainly cannot be viewed as being in the main a scheme by planters to manipulate the slave labor force." This chapter suggests otherwise. See McDonald, *Economy and Material Culture,* 168. On the way other planters similarly sought to establish self-sufficiency via provision grounds and the slaves' evening and Sunday labor, see Woodville K. Marshall, "Provision Ground and Plantation Labor in Four Windward Islands: Competition for Resources in Slavery," in *Cultivation and Culture,* ed. Berlin and Morgan (Charlottesville, Va., 1993), 205; Sheridan, *Doctors and Slaves,* 166–69; Mullin, *Africa in America,* 141–46. On overwork as literally overworking the slave, also see Bolland, "Proto-Proletarians?" 133–35.

evenings, planters hoped that they could inculcate habits of commitment and diligence that would extend to regular working hours. In addition, if a slave ran into debt at the estate commissary or store, planters could utilize overtime work to force him or her to clear the account. In short, overwork gave slaveholders another tool with which to discipline their workers. Furthermore, planters could also broadly direct their slaves' economic activities into work they deemed profitable, more for themselves than their slaves. Above all, this led to timber collection and the raising of corn and livestock—products that otherwise would have to be either purchased, cultivated, or collected on the planters' land and in his time. By paying for the bondspeoples' corn, planters additionally ensured that slaves were raising their own food in their own time, freeing hundreds of hours for other duties and saving their masters a small fortune in factors' and farmers' charges for commercially raised grains and pork. In paying the slaves to conduct extra work in the sugarhouse, planters also sealed an economic link between the slave and the use of the steam-powered machinery. Corn, wood, livestock, and overtime payments ultimately tied the slaves to the plantation order and forged an iron bond between the slaves' material well-being and agro-industrial sugar production. As Carolina rice lord Charles Manigault observed, the extra "comforts" slaves gained from the sale of their commodities "tends to attach them to their home." To be sure, African Americans derived many more benefits from the internal economy than planters assumed, but the entire structure was both subtly and explicitly oppressive.[2]

2. Louisiana slave law protected independent production but denied slaves' formal ownership of sums earned. Stating that it was illegal for slaves to sell "corn, rice, greens, fowls, or any other provision" without the written permission of their master, the state's slave code also stipulated that slaves must receive a plot of ground "to cultivate on their own account." Article 175 of the 1824 slave code similarly stated, "All that a slave possesses, belongs to his master; he possesses nothing of his own, except his *peculium*, that is to say, the sum of money, or movable estate, which his master chooses he should possess." As late as 1838, the Louisiana Supreme Court upheld the right of the slaves "to the produce of their labor on Sunday; even the master is bound to remunerate them, if he employs them." See Schafer, *Slavery, the Civil Law, and the Supreme Court of Louisiana,* 1, 8, 21–27; Willie Lee Rose, ed., *A Documentary History of Slavery in North America* (New York, 1976), 176. On the importance slaves in other southern economies attached to Sunday as "their" time for family, trading, and religion, see Betty Wood, "'Never on a Sunday?': Slaves and the Sabbath in Low-Country Georgia, 1750–1830," in *From Chattel Slaves to Wage Slaves,* ed. Turner (London, 1995), 97–122. On the way independent garden production was used to inculcate industriousness and give slaves a perception that they had a stake in the estate, see Dale Tomich, "*Une Petite Guinée*: Provision Ground and Plantation in Martinique, 1830–1848," in *Cultivation and Culture,* ed. Berlin and Morgan (Charlottesville, Va., 1993), 222; Dusinberre, *Them Dark Days,* 182 (quote).

Just as quickly as the planter opened his door to his slaves' commodities, he could slam it shut and halt the slaves' market economy. Such acts, or the threat of them, aimed to promote compliance with the planter's agenda. Moreover, by paying their slaves for their wood and corn just prior to the grinding season, planters could reaffirm their paternalistic identities and appear to balance their own and their slaves' expectations from chattel bondage. As the Jamaican planter-historian Bryan Edwards observed, the slaves' provision-ground culture created "a happy coalition of interests between master and the slave." Such a coalition proved ruthlessly profitable, Edwards remarked, because the "Negro who has acquired by his own labour a property in his master's land, has much to lose, and is therefore less inclined to desert his work . . . and the proprietor is eased . . . of the expense of feeding him." Those in chains might not have agreed with Edwards's depiction of a happy coalition. As ex-slave Octavia George observed, independent production only flourished at the planters' behest:

> We were never given any money, but were able to get a little money this way: our Master would let us have two or three acres of land each year to plant for ourselves, and we could have what we raised on it. We could not allow our work . . . to interfere with Master's work, but we had to work our little crops on Sundays. Now remind you, all the Negroes didn't get these two or three acres, only good masters allowed their slaves to have a little crop of their own. We would take the money from our little crops and buy a few clothes and something for Christmas. The men would save enough money out of their crops to buy their Christmas whiskey. . . . We were allowed to have a garden and from this we gathered vegetables to eat; on Sundays we could have duck, fish, and pork.

George's insistence that only "good masters" allowed their slaves to farm underlined both the apparent benevolence of her master and his capacity to strip away the slaves' independent production. Ever astute to the dynamics of plantation management, William F. Weeks similarly manipulated the paternalistic ties of master and slave to enhance his own status and to underscore the constrained limits to African American trading. Musing on the request of his slave Amos to spend a portion of his crop in neighboring New Town, Weeks granted permission "in consideration of his faithful services . . . and his really conscientious scruples about trading on Sunday." Leaving Amos with no doubt over whose consent he required and the benefits he would gain from faithful and conscientious work, Weeks—like Octavia George—revealed that overtime work occupied a transient space in which the slaves' internal economy could be regulated at the planter's whim.[3]

By the 1840s, overwork seldom hindered regular plantation duties; rather, it operated in tandem with the various tasks incumbent in sugar production. But even as it underscored the reciprocity of the idealized master-slave relationship, overtime work in no sense led to transactions between equals. In exchanging their slaves' commodities for credit at the plantation store, slaveholders circumscribed the availability of specie, minimized the potential for interaction with free labor, and ensured that the fruit of their slaves' overwork seldom exited the confines of the plantation world. Even as they valued African American domestic production, "good" and "bad" planters knew that the prospect of its curtailment could prove equally profitable. Having offered their slaves the taste of money and due recompense for their labor, slaveholders could reasonably expect their bonded workers to be more tractable if threatened with the prohibition of overwork. Such a course of action proved hazardous, sometimes inciting the very workplace activism and resistance that the planters sought to avoid; but it could be an effective weapon in their campaign to regulate slave labor. Few sugar masters ultimately risked challenging the slaves' internal economy as the harvest season approached. They knew the potential economic and social gains of independent production and realized that the internal economy worked to their advantage only if the slaves gained too.

African Americans perceived their market activities from a wholly different perspective. The sale of their labor power enabled all members of the slave community to deny temporarily the inequalities of southern society and to face their masters as equals in a market shorn of racial, social, and class hierarchies. The slaves' ability to sell their labor and own property, moreover, inverted plantocratic notions of property. As the bearers of commodities, slaves accordingly bartered and sold their goods and invested in a cash economy that fostered socially and communally significant property ownership. Building upon colonial antecedents and a flourishing frontier exchange economy, they profitably discovered that the diffusion of tasks with the onset of agro-industrial sugar production in the 1840s

3. Sheridan, "Strategies of Slave Subsistence," 53 (first quote); Interview with Octavia George, Oklahoma Writers' Project, WPA Slave Narrative Project, Oklahoma Narratives, vol. 13, Federal Writers' Project, U.S. Work Projects Administration (USWPA), Manuscript Division, Library of Congress. See "Born in Slavery: Slave Narratives from the Federal Writers' Project, 1936–1938," http://memory .loc.gov/ammem/snhtml/snhome.html (second quote; accessed 30 April 2004); William F. Weeks to Mary C. Moore, 31 January 1853, David Weeks and Family Papers, LSU (third quote).

proffered new opportunities for economic gain. Steam engines required timber, the greater complexity of sugar production provided new opportunities for paid harvest work, and the drive toward self-sufficiency created market opportunities for slaves to produce corn, small grains, and livestock for the larger plantation economy.[4]

The swift rise of steam power in the sugar country created a pressing demand for timber, since early steam engines burned between three and four cords of wood for every one hogshead of steam-milled sugar. Collecting wood proved a time-consuming and burdensome task that every planter had to oversee during harvest preparations. Given the pressing demand for cordage, overwork flourished in the mechanized era, as the slave community appropriated paid timber collection as a customary right and lucrative market. This type of paid overwork reached its apogee in the 1850s, due to the ever-growing number of steam engines and the increased power and capacity of the sugar mills. As the mean size of the sugar crop grew, planters made further demands on the timber supplies of the Louisiana backswamps. Astute to the risks of poor harvest preparation, they turned to their slaves to supply the steam engine with adequate fuel for the harvest. Either collecting driftwood along the rivers or standing knee-deep in swamp water among cypress trees, African Americans took over this unhealthy and at times dangerous task for weekend and night work. The heat, humidity, mosquitoes, and snakes made lumber collection a grim occupation and one that the sugar masters readily transferred to the slave community. Planters might not have explicitly selected this kind of scut work for payment, but in doing so, they demeaned wage work, racialized woodcutting as a black occupation, and ensured that every cent earned by a slave would be paid twice over by his sweat.

The enormous demand for cordage burdened estate managers in the summer and early autumn. During the lay-by, the maturing sugar crop required little attention, though timber collection added to the seasonal work requirements as slaves concurrently began attending and harvesting other crops on the estate and

4. On the social significance of property, see Hudson, "All that Cash," 77–94; Philip D. Morgan, "The Ownership of Property by Slaves in the Mid-Nineteenth-Century Low Country," *JSH* 49 (August 1983): 399–426; Lawrence T. McDonnell, "Money Knows No Master: Market Relations and the American Slave Community," in *Developing Dixie,* ed. Moore, Tripp, and Tyler (Westport, Conn., 1988), 31–44. On the family as the primary economic unit, see esp. Dylan C. Penningroth, *The Claims of Kinfolk: African American Property and Community in the Nineteenth-Century South* (Chapel Hill, N.C., 2003), 79–109.

conducting essential plantation maintenance before the coming harvest. On Robert Ruffin Barrow's Residence Plantation, annual woodcutting necessitated a considerable labor input. In preparation for the 1857 crop, overseer Ephraim Knowlton estimated that 1,580 cords of wood would be necessary for grinding, though to assure a comfortable surplus, he ultimately prepared 2,050 cords. As on every other sugar estate, the enormous demand for timber on Residence Plantation directly pressed on the plantation timetable, necessitating hundreds of hours of manpower that could be profitably employed elsewhere. By paying slaves fifty to fifty-five cents a cord to conduct this laborious work after the cessation of the working day, planters and overseers assuaged the slaves' desire for cash while simultaneously saving hundreds of hours for alternative duties.[5]

From aboard a Mississippi steamboat that passed through the heart of the sugar country in 1838, Harriet Martineau unconsciously observed the prevalence of overtime work, noting that groups of slaves continued to chop wood in moonlight and work along the shoreline well after dusk had turned to nightfall. As she was borne downstream, she passed Samuel Fagot's Constancia Plantation, an estate where both master and slave took full advantage of overwork. In preparation for the 1858 rolling season, slaves collected and chopped over two thousand cords of wood in their own time, for which they earned $1,047.50 "on their terms." Presumably, their terms stipulated a pay rate of between fifty-one and fifty-two cents per cord. Fagot either accepted the slaves' terms at face value, or a period of negotiation transpired, during which the rate perhaps oscillated between the employer and his temporarily waged employees. To all intents and purposes, the slaves at Constancia were selling their labor power, even though their experience of the market economy was fractured by the planters' countervailing force. The master provided the axes, saws, pirogues, and flatboats for timber collection, he owned the trees in the backswamp, and he ultimately possessed the authority to check the slaves' market-related activities. Like his slaves, Fagot nonetheless found slave overwork profitable. In 1859, he turned again to his slave woodsmen, who cut and hewed 2,018 cords of wood, for which he paid them $1,077. While this exhausting night work certainly proved taxing for the laborers, almost every slave dwelling profited by over $20, which in turn forged an economic link be-

5. Residence Journal of R. R. Barrow, 15 September and 18 October 1857, Robert Ruffin Barrow Papers, UNC; Memorandum Book 10 (1846–1848), and Accounts with Slaves, 1859, Landry Family Papers, LSU; May 1849, Anonymous Plantation Ledger, LSU; William Whitmell Pugh, Accounts with the Negro Men, Pugh Family Papers, TEX.

tween the slaves' material well-being and the rapacious demand of the mechanized sugar mill.[6]

Fagot's 130 slaves proved themselves to be highly capable woodsmen who produced not only firewood but also over 1,300 hewn wooden boards, which Fagot hoped to use as sheeting material. Having disbursed an additional $660 for the boards, Fagot paid his slaves over $4,600 for timber products alone between 1853 and 1859. In 1860, drought followed by sporadic though extraordinarily intensive storms delimited cane production. Even as Fagot once again turned to his own slaves and those of a Mr. Lewis for timber and firewood, he was gradually disassociating himself from almost complete dependence on his workers. Crediting his bondspeople $506 on October 30, Fagot also paid $1,700 to local whites for wood produce. Whether the slaves willingly reduced their share of the annual timber requirement remains unclear. Following the escape of slaves Sam and George Washington in the late spring and summer of 1860, it seems probable that Fagot decided to break his alliance with the slaves, reinscribe his power over independent production, and buy wood from other suppliers. For the African Americans on Constancia, Fagot's decision swiftly eroded household income and harshly indicated the limits to their internal economy.[7]

The same calculating regard for time and profit shaped overwork payments for feed and livestock, as planters paid their bondspeople to cultivate enough grain to meet the plantation's annual dietary requirements. The pervasiveness of the slaves' household economy forcibly struck visitors to the sugar country. As one traveler observed after his visit to Wade Hampton's estate, the slave cabins resembled the "neatness of a clean . . . village," where each hut possessed a garden plot "surrounded by fruit trees and shrubbery." While eschewing this theatrical Victorian romanticism, other travelers to the sugar plantations also observed the small gardens attached to slave cabins or glimpsed the separate fields where bondspeople could cultivate their own crops either for home consumption or for sale. In most cases, however, African Americans traded and sold corn that they had cultivated on Sundays and in the evenings. On one estate in the middle of the Mississippi Valley sugar belt, planter Benjamin Tureaud paid his bondspeople to raise enough

6. Harriet Martineau, *Retrospect of Western Travel*, 2 vols. (London, 1838), 2:166.

7. Plantation Journal, 1859–1872 (vol. 28), 10 October 1858, 25 October 1859, 30 April, 31 July, 30 October, and 28 December 1860, Uncle Sam Plantation Papers, LSU; P. A. Champomier, *Statement of the Sugar Crop, 1860–61*, 2. Also see Octave Colomb Plantation Journal (1849–1866), 10 February 1854, TUL.

corn to provide for everyone on the plantation. According to records of his transactions with 109 individuals, Tureaud expended over $1,500—a relatively mod-
est sum—remunerating his slaves for their cereal crop. Since corn sold in New
Orleans between 50¢ and $1 a barrel, Tureaud's slaves produced more than enough
grain to meet their own needs and supply additional quantities for livestock or
poultry feed. In orchestrating this scheme, Tureaud both defrayed his expenses
and saved hundreds of hours of slave labor by ensuring that the bondspeople raised
the crop in their free time and not in his. Whether cognizant of this fact or not,
Tureaud's slaves were actually subsidizing the plantation economy by reducing annual food costs and by saving him time for more profitable occupations. Their actions additionally freed Tureaud from the vagaries of the marketplace and sporadic
price hikes for corn and other commodities.[8]

Plantation stability lay in self-sufficiency and in the rational management of
available resources. The sugar masters were hardly unique in coming to this conclusion; like business executives who integrated their industries, antebellum
planters eagerly sought to control their raw materials, be they wood and corn rather
than coke and iron. Planter self-sufficiency, however, carried a uniquely southern
twist: although chopping wood and harvesting corn generated a small profit for the
bondspeople, it was the fear of black indolence that tormented slaveholders like
Benjamin Tureaud. Inactive slaves could challenge their condition by escape, sabotage, or collective action, dangerously threatening the entire edifice of bondage.
Slaves with the leisure to reflect on their fragmentary experience of wage labor
might discover the tortured contradictions of slave labor and extend the logic of
wage and contract to themselves. As Baron de Carondelet warned Louisiana sugar
planters in 1795, overtime work and independent production "advantageously employ[s] the time [slaves] might otherwise spend in riot and debauchery." Forty years
later, the antebellum sugar masters still held to this advice, actively favoring self-
sufficiency not solely on economic grounds, but also as a means of maintaining
racial stability. Benjamin Tureaud astutely gauged the risks incumbent in slave
underemployment when he complained, "Dr. Chauvin wants to obtain authority

8. *Southern Cultivator* 5 (April 1847): 55 (quote); Bayard Taylor, *Eldorado, or Adventures in the Path
of Empire* (New York, 1860), 6; Thomas Low Nichols, *Forty Years of American Life, 1821–1861* (1864;
reprint, New York, 1937), 123; Plantation Ledger, 1858–1872 (vol. 46), Benjamin Tureaud Family Papers, LSU; Menn, *Large Slaveholders of Louisiana*, 121. On annual dietary requirements (52 pecks of
corn per annum), see Sam B. Hilliard, *Hog Meat and Hoe Cake: Food Supply in the Old South, 1840–1860*
(Carbondale, Ill., 1972), 157.

to purchase some two or three hundred cords of wood-for the next rolling; Damnation, how does he intend to employ the hands?" Fuming to factor Martin Gordon over Chauvin's managerial incompetence, Tureaud snarled, "A short crop planted and yet he wishes to buy the wood necessary for the grinding season—Well my friend—I wish I may be damned—if I am not more and more convinced that the Doctor is not competent to manage a sugar plantation." Chauvin's critical errors lay in his mismanagement of the labor force and his disregard for the economic and social significance of self-sufficiency. For Tureaud, by contrast, overtime work resolved fully six overlapping concerns. It enabled self-sufficiency in two plantation consumables (wood and corn), ensured employment for the slave force, allayed fears of racial unrest, and resolved the reciprocal imperatives of the slave regime. It also transferred a significant proportion of annual maintenance costs to the slave community and made those slaves involved in the domestic economy indispensable to effective management and plantation-wide profits. Above all, woodcutting and corn farming bound the slaves' interests to those of their masters, while providing a veneer of independence and financial autonomy. For African Americans, by contrast, cash payments for plantation necessities and the sale of their labor power temporarily cut against the planter's hegemonic sway, however insidious they might ultimately prove to be.[9]

By trading their commodities, bondspeople secured some cash income, tantalizingly grasped the essence of free labor, and carved out a meaningful orbit of self- and communal identity. Drawing principally upon the social relationships of family, kin, and community, enslaved African Americans in the cane world (as elsewhere) fashioned their vision of property and waged income through the lens of overwork. Black Louisianans responded vigorously to the fragmentary promise of independence and the experience of wage relations by engaging in the market for slave-produced commodities and by expanding the boundaries of their economic interactions. Formally a male occupation—its transactions predominantly involved payment to men—independent production touched all members of the slave community; women and children not only cultivated and harvested corn on their plots but also raised poultry, livestock, and marketable produce, which frequently generated significant income. Most bondsmen who hauled their corn to Benjamin Tureaud's office on November 29, 1858, earned approximately $10 to

9. James A. Padgett, "A Decree for Louisiana Issued by the Baron de Carondelet, June 1, 1795," *LHQ* 20 (July 1937): 601, quoted in Morris, *Becoming Southern*, 76 (first quote); Martin Gordon Jr. to Benjamin Tureaud, 13 February 1850, Benjamin Tureaud Family Papers, LSU (second quote).

$15 for their year's crop, though others—like Aaron Butcher and Mitchell, who earned $120 and $110 respectively for their maize—clearly benefitted substantially from their extra labors. But neither Aaron nor Mitchell matched the productivity of Bill Siddon, who found time to harvest corn valued at $130 and chop thirty cords of wood as well. By crediting his slaves just days prior to the rolling season, Tureaud seemingly placated his bondspeople as they entered the most exacting time of the year, while expropriating hundreds of hours for other profitable work. Yet in strictly controlling the potential avenues for marketing the crop, Tureaud calculatingly ensured that neither corn nor timber would slip into the open market without his authority and that the value of his slaves' overwork would profit other planters. Not only did he exploit his laborers to maximize individual productivity, but, through the commissary, he could master all aspects of the plantation economy and even manipulate the slaves' free time almost as assuredly as he drilled them during the working day. Further, and perhaps most injurious of all, overwork ensured that slaves had a direct economic stake in the plantation and tied their material well-being to that of the master class.[10]

While the bondspeople gained some material wealth from overtime labor, the cost was steep, as they ceded their time off to the planter's economic interests. Still, many hundreds of individuals decided that the price of a modestly improved lifestyle was ultimately worth their sweat and labor, and overwork flourished from the mid-1840s. Corn and timber underpinned the slaves' domestic economy, but Louisiana slaves also earned extra revenue from livestock husbandry, moss trading, basket weaving, and additional shop-floor duties in the mill house or cane shed. Passing through the sugar country, travelers and itinerant journalists observed that slaves possessed their own small plots, where they raised poultry and other commodities for sale to the slavemaster. Inquiring of the plantation mistress as to the rationale behind paying their bondspeople for fowl, Theresa and Francis Pulszky remarked, "The planters think it mean to rear their own poultry, and not to leave the profits to the slaves." Touched by this pastiche of apparent benevolence, the Pulszkys underscored the reciprocity incumbent in paternalistic slaveholding. Not all planters, however, were so generous or left the profits to the slaves.[11]

Archaeological research of slaves' garden plots at Duncan Kenner's Ashland Plantation indicates that, like their brethren on other plantations, slaves fenced in

---

10. Plantation Ledger, 1858–1872 (vol. 46), Benjamin Tureaud Family Papers, LSU.
11. Pulszky, *White, Red, Black*, 2 : 104.

pens for raising livestock or tending poultry near their cabins. These small fenced areas evidently were of significant relevance within the slave community. As former slave Alexander Kenner recalled, bondspeople "are so anxious to make money that they work upon their little patches at night." The use of fences defined the economic context of independent production along strictly individualistic lines, though the collective output could be considerable. As Alexander remembered, slaves at Ashland raised up to a "thousand dozen" chickens per annum upon their master's estate, in effect running a broiler farm in the midst of Duncan Kenner's sugar operations. Alexander well remembered his master's chicanery and duplicity in the poultry trade. According to him, Duncan Kenner obliged the slaves at Ashland "to sell the chickens to him, instead of selling them to the hucksters, because he wanted to know how much money they had, and didn't want them to have too much." Kenner eagerly exploited his bondspeople by purchasing the chickens for twenty cents apiece and selling them to the hucksters at the "advanced price" of thirty cents per bird. Reaping a clear profit, the sugar master "didn't wish the slaves to accumulate any property," but rather "to spend whatever money they got." On occasion, Ashland slaves nonetheless were able to accumulate significant sums; old Cudjo (perhaps legendarily) was said to have hoarded over five hundred silver dollars. Alexander explained this success story as testament to the intelligence of those "negroes on the river," who "certainly take care of themselves." He categorically declared that slaves in the interior "are stupid," adding that "they see nothing, know nothing, and are very like cattle." When pushed to elaborate upon the differences between sugar slaves and slaves in neighboring Mississippi, the former bondsman tangentially brushed upon the disparate forms of slave management and the widespread practice of overwork that made slave fortunes, however modest, possible by shrewd and intelligent bargaining.[12]

Ellen McCollam, a slave mistress in rural Ascension Parish, also took full advantage of the slaves' internal economy. Unlike her Mississippi neighbor Duncan Kenner, however, Ellen's initiative was undone by her slaves' superior intelligence in poultry trading. Ellen initiated her poultry negotiations in the summer of 1848, and over the subsequent eighteen months she purchased dozens of chickens, roosters, and pullets from the slave community. Not satisfied with the extent of her

12. Alexander Kenner, "American Freedmen's Inquiry Commission Interviews, 1863," in *Slave Testimony,* ed. Blassingame (Baton Rouge, La., 1977), 393 (quote). On Kenner, see Bauer, *Leader among Peers;* Yakubik and Méndez, *Beyond the Great House,* 21–22. Also see Thorpe, "Sugar and the Sugar Region," 753.

purchases, McCollam later returned to the quarters to purchase additional hens from Little Jack and Molly—an action she repeated a year later, when she offered Little Isaac a dollar for five hens, two roosters, and a small chicken. Yet black agency imposed an indelible mark on McCollam's livestock operations; she lost eight hens to slave theft. Old Armistead's larceny converted Ellen McCollam's newly acquired property into extra slave capital that could be traded to hucksters, eaten, or even sold back to McCollam. Theft, in short, cemented ties of African American solidarity and proffered additional opportunities for pecuniary gain. Since children played a substantive role in raising livestock and tending poultry, the effects of Armistead's theft transcended generations, assuring community cohesion and mutual gain for children and adults alike. On Edward Gay's estates, overtime work functioned along broadly similar parameters; slaves gained significant revenue from husbandry and raising livestock or poultry. The latter required fairly little attention and had the additional attraction of yielding eggs that could be sold for profit or used to diversify the monotonous slave diet. Livestock husbandry was significantly more labor-intensive. On Gay's Iberville Parish estate, Harry Tunley and Mack raised swine in 1854 for sale to the master. Two years later, Tunley and other members of the slave community had expanded their operations and were raising cattle as well. Gay's decision to rely partially on the slaves' economy for meat and poultry internalized production within the estate, aided plantation self-sufficiency, and greatly lessened the planter's reliance on New Orleans factors and peripatetic traders who visited the plantation belt to truck their wares and trade their livestock. Self-sufficiency, moreover, checked the small but not insignificant trickle of capital out of the estate, kept expenditure within the plantation's credit network, and limited contact between black slaves, river hucksters, and poor whites.[13]

Beyond the occasional shoat, cord of firewood, ear of corn, or basket of slave-grown vegetables, African Americans on Gay's plantation established a complex and highly remunerative trade with Missouri furniture makers, who required Spanish moss for stuffing bed mattresses and lounge cushions. Like former bondsman Hunton Love, Gay's slaves swiftly realized that money "grows on trees" and that they could "git down that moss an' convert it" into revenue. After collecting and drying the moss, the slaves turned to their master to ship their material to

---

13. Diary of E. E. McCollam of Ellendale Plantation (1847–1851), 15 and 29 August 1847, 18 November 1848, Andrew McCollam Papers, UNC; Plantation Record Book (vol. 36), Edward J. Gay and Family Papers, LSU. Additionally see Plantation Journal, 1859–1872 (vol. 28), Uncle Sam Plantation Papers, LSU.

St. Louis, where Gay's factor sold over one thousand bales of moss between 1849 and 1861. The slaves earned a few extra dollars a year in the moss trade, but Gay maintained a firm grip over their market activities, managing the shipments from his estate and subsequently crediting their earnings to their accounts at the plantation commissary. Like bondspeople elsewhere, Gay's slaves found their entrance into the market fruitful, but they also gained a shadowy lesson in free labor and stark introduction to commerce and trade. Not only did the price of moss fluctuate between 1844 and the 1850s, but slaves paid all the shipping and marketing costs, which skimmed off some 20 percent of the value ascribed to their produce. Despite the fact that 160 slaves were actively engaged in the moss trade—most adults on Gay's Iberville Parish plantation collected and traded moss at some point during the 1850s—the moss market proved highly unstable for those involved in it. Good years preceded poorer spells, during which moss collection collapsed as a serious economic venture within the slave community. Furthermore, while some slaves traded moss throughout the 1850s, many individuals entered the market sporadically to trade their produce. The slave Ben, for instance, collected and marketed almost fifty bales of moss, for which Gay's estate managers credited him approximately $200. Ben's output peaked in 1856 and 1859, when he collected ten bales per year. In other years, he proved to be a competent moss collector who seldom earned less than $10 to $12 per annum. Despite Ben's relative success, he joined the majority of fellow bondsmen in producing just one bale of moss in three separate years between 1849 and 1861. Like Ben, Jim Shallowhorn entered and exited the moss trade after an initial burst of activity in which he sold five bales in three years. Shallowhorn's income proved as sporadic as his experience of the marketplace and the wage economy. Not all slaves proved as fitful in their trading operations as Jim Shallowhorn; indeed, the records of Home Plantation indicate that the slaves shared in a vibrant market economy that Edward Gay openly condoned. Income gained from the backswamp linked Ben and Jim to the estate and provided them with a modest—though not insignificant—source of revenue. Gay's contribution to the slaves' market activities extended to the use of his trees, his estate livestock for haulage, and his wharf for final transportation upriver. In return, Gay benefitted by being able to strengthen his commercial relationship with St. Louis merchants and cement trade agreements with steamboats plying the Mississippi.[14]

14. Interview with Hunton Love, date unknown, WPA Ex-Slave Narrative Collection, LSU (quote); Moss Record Book (vol. 35), Edward J. Gay and Family Papers, LSU. Alternatively, see McDonald, *Economy and Material Culture of Slaves,* 66–67.

While some of Gay's slaves profited by supplying moss to meet the northern white demand for comfortable furniture, other skilled bondsmen on Home Plantation discovered that the pressure to grind cane and the advent of the industrial age proved particularly lucrative for those who took on extra work during the planting and rolling seasons. Skilled slaves were uniquely positioned to tailor the demands of the industrial sugar mill to their internal economy and to derive income from the repercussions of the machine age. The slave Bill Garner, for instance, received $75 in 1854 for his services as an engineer during the rolling season, while Harry Tunley profited not only from raising livestock but by cleaning the steam engine boilers. Aleck, likewise, earned $3.50 by fixing and setting the sugar kettles, while Jacob Lennox and App both received $40 as recompense for their work as sugar makers during the grinding season. In a similar vein, Moses received $5 for his mastery of the kettle furnaces, while Patrick cleared $31 in 1854 for making five sugar coolers and completing several tasks in the carpenter's shed. Thornton also turned his carpentry skills to profit, constructing a cart in his own time, for which Gay credited him $25. Two years later, Joe Penny also benefitted from Gay's need for sturdy plantation wagons, receiving $5 for encasing five cartwheels with iron. While he credited his skilled slaves for extra tasks during the grinding season, Gay also paid his slaves over $80 for work completed during the post-harvest vacation. Finding Gay's pay rate of $1 a day attractive, thirty-three slaves accepted his offer to work overtime in the sugarhouse, where they potted and drained the sugar. Although he allowed his slaves to profit from these opportunities to earn additional money, Edward Gay was self-interestedly able to ensure that his labor force viewed the sugar season and the machinery not as their enemies, but as a source of additional income. Through payments to his slaves for firewood, cultivating corn, raising livestock, ditching canals, sharpening plow points, constructing hogsheads, building carts, weaving baskets, and potting sugar, Gay, like many of his contemporaries, discovered that slave overwork rested at the axis of estate productivity.[15]

Through overtime payments, monetary rewards, and primitive forms of wage work, the sugar lords fashioned a plantation order in which both master and slave

15. Plantation Record Book (vol. 36), Edward J. Gay and Family Papers, LSU. For other examples of slave overwork, see Plantation Diary, 1842–1863, Ransdell Papers, LSU; Cash Books, 1847–1851 (vol. 2), 1848–1851 (vol. 4), and 1851–1854 (vol. 5), Bruce, Seddon, and Wilkins Plantation Records, LSU; Valerin Ledoux and J. C. Van Wickle Account Book, 1849–1883, John G. Devereux Papers, UNC; Accounts of Wood Cut by Slaves and Moss Accounts, Journal of St. Rosalie Plantation, 1840–1868, Andrew Durnford Plantation Journal, TUL.

shared a direct economic stake in the plantation. Mutualism within this realm of estate management, however, in no sense signified the sharing of collective values. As Ellen McCollam discovered, slaves could challenge white power, seizing back the produce of their labors and inverting the slaveholder's definition of property. Domestic production complicated definitions of property rights, as planters recognized—however grudgingly—the slaves' ownership of their labor and their commodities. The inevitable clash between master and slave over appropriate property rights left both parties profoundly at odds about notions of one's rights to goods, services, and, most fundamentally, labor. Yet in the end, both black and white Louisianans nudged the slave regime toward recognizing the contract and reciprocity of wage work, in which there was no obvious breach between a slave mode of production and a free-labor model of paid work in the cane fields. For the two decades preceding the onset of the Civil War, masters and slaves had effectively combined waged and coerced labor for profit. The varied structures of labor extraction lay at the core of the sugar masters' business success, revealing the flexibility and hybridity of plantation capitalism in Louisiana's cane world.

While Gay and Kenner gained from the slaves' domestic production, African Americans utilized the internal economy to buy a few simple items to which they alone laid claim. In a world that legally denied the ownership of self or property, the slaves' ability to acquire a few objects slightly improved the desperate boredom, penury, and bleakness of life in the sugar country. Moreover, by controlling their own free time, crops, land, and goods, slaves could dissociate themselves from their masters, redefine power and authority, and reaffirm their vision of semi-autonomous wage work and domestic production. Time devoted to money-making activities cut down on slaves' hours for rest and relaxation; yet as historian Roderick McDonald concludes, "The *process* of the internal economy doubtless proved cathartic to slaves . . . [and] their motivation for involvement undoubtedly rested largely rested with the internal economy's end product, the purchase and acquisition of goods."[16]

The slaves' zeal for ownership derived from a sophisticated sense of their own property rights. Twenty years of negotiating their terms of payment left them with

16. McDonald, *Economy and Material Culture of Slaves*, 79 (quote). On slaves' vision of wage work and peasant production, see Eric Foner, *Nothing But Freedom: Emancipation and Its Legacy* (Baton Rouge, La., 1983), 85–110; Rebecca J. Scott, "Fault Lines, Color Lines, and Party Lines: Race, Labor, and Collective Action in Louisiana and Cuba, 1862–1912," in *Beyond Slavery,* ed. Cooper, Holt and Scott (Chapel Hill, N.C., 2000), 70.

a rich reservoir of experience in selling their labor power. Their very familiarity with waged work, cash incentives, and payment were important concessions wrung from their masters. Yet the more reliant slaves became on domestic production for revenue and payment, the stronger grew their dependence on the market and their masters' consent. Market values encroached still further upon the slave quarters, exposing slaves to the cut and thrust of commercial relations, tying their interests to those of the planter, and linking them to the wider business community. On Benjamin Tureaud's plantation, for instance, Little Jesse was so fully ensnared by the market culture that he invested his overwork payment in wire, twine, and a lock to protect his corn patches and personal possessions (which included $42 in cash). Jesse not only worried for the security of his possessions, but his relative prosperity surely made him an object of jealousy among the less successful. Perhaps inevitably, the prevailing logic of paid work led toward individualism and occasional resentment among the haves and the have-nots. The regimented credit and debit lines in the planters' accounts had little room for the communal ethos of the plantation quarters; rather, they subdivided the slaves by income, wealth, and expenditure, and transformed the social ethos of African American labor from communalism to independent production, wealth accumulation, and individualistic thought and action.[17]

Although self-interest and jealousy crept into the slave quarters, most slaves balanced communal and individualistic values. Their work culture stressed gang collectivity, their effective resistance hinged on teamwork, and their social arrangements favored kinship, faith, and leisure networks whenever possible. Shared behavioral values—a sense of place, solidarity, and collective responsibilities—shaped the slave community as a physical and social space that defined the slaves' lives. Overwork occasionally chipped away at those values, but it could also reaffirm community ties. Former slave Elizabeth Ross Hite viewed her family's production in both independent and collective terms. Recalling life before the Civil War, Hite affirmed that her parents maintained a garden in which they planted

17. Plantation Ledger, 1858–1872 (vol. 46), Benjamin Tureaud Family Papers, LSU. On the intrusion of market values in the slave community, see Hudson, "All that Cash," 89. On the way that African Americans were increasingly forced to embrace an individualistic ethos of independent production, see McDonnell, "Work, Culture, and Society," 139. Norrece Jones similarly argues that planters sought to orchestrate garden production to encourage individualism over communalism. See Norrece T. Jones, *Born a Child of Freedom, Yet a Slave: Mechanisms of Control and Strategies of Resistance in Antebellum South Carolina* (Middletown, Conn., 1990), chap. 4.

corn, watermelon, mushmelon, and flowers. While these products surely bettered the family diet and added color to their home, the corn that Hite's mother sold for fifty cents a barrel also provided income, with which her mother "bought good clothes . . . nothin' but silk dresses." These exotic items were passed down through generations, as Hite added that her children "went to school in dem dresses." Although her mother in all probability bought more than silk dresses, Hite's memories nonetheless address the signal importance of self-improvement and the significance that bondspeople attached to the purchase of luxury items.[18]

Forced to wear the dull garb of slave clothing, Hite's mother's purchase of silk reflected a comprehensive rejection of the inexpensive linens and rough cloth that comprised slave garments. Hite's fellow slaves utilized their savings to purchase material from the plantation commissary that they could cut and sew into designs reflecting their own conception of clothing rather than that of the plantation owner. On Benjamin Tureaud's plantation, Jack Locket similarly utilized the extralegal internal economy to forge a distinctively African American economic space that facilitated individual and collective uplift through acquisition and measured monetary independence. Spending in the plantation commissary his credit earned by woodcutting, Locket purchased twenty yards of calico, fourteen yards of cotton, three yards of checked cloth, four handkerchiefs, two pairs of shoes, one barrel of flour, ten pounds of chewing tobacco, and two iron hoes for agricultural work. Locket's acquisitions enabled his family to fashion a distinctive appearance for themselves with calico, checked, and high-quality fabrics. As cultural historians Shane and Graham White indicate, the slaves' ability to define their own clothing meshed within a broader African American aesthetic that asserted their own autonomous identities and subtly challenged white authority. By rejecting the "badge of slavery"—as Harriet Jacobs poignantly called her slave garments—African Americans negated white racial models and wrought novel constructions of independence and self-assertiveness. Through woven calico and cotton,

---

18. Interview with Elizabeth Ross Hite, date unknown, WPA Ex-Slave Narrative Collection, LSU (quote). On the concept of slave community, see Anthony E. Kaye, "Neighbourhoods and Solidarity in the Natchez District of Mississippi: Rethinking the Antebellum Slave Community," *Slavery and Abolition* 23 (April 2002): 1–24; Kolchin, *Unfree Labor*, 199–200; Margaret Washington Creel, *A Peculiar People: Slave Religion and Community Culture among the Gullahs* (New York, 1988), 2. On community formation, also consult Blassingame, *Slave Community*, 147–48, 315–17; Lawrence W. Levine, *Black Culture and Black Consciousness: Afro-American Thought from Slavery to Freedom* (New York, 1977), 29–39; Stevenson, *Life in Black and White*, 254–55, 323–24; Schwalm, *Hard Fight*, 64–70.

bondspeople could mask the degradation of bondage, craft their own identities, confront white stereotypes of black sexuality, and lay claim to their own bodies.[19]

In these efforts to redefine their appearances and identities through dress, slave men and women vigorously forged a vibrant cultural aesthetic that rejected subservience in favor of dignity and personal sovereignty. Nor did they limit their purchases at the estate commissary to fabric. As Thomas B. Thorpe recalled on visiting the sugar country immediately after the harvest, African Americans expended their accumulated wealth on "cheap crockery, bales of gayly-colored handkerchiefs . . . [and] ribbons and nick-nacks, that have no other recommendation than the possession of staring colors in the most glaring contrasts." In a slightly less patronizing—though equally oblivious—manner, William Howard Russell remarked upon the young women rushing to the weekly sugarhouse dance in "snow-white dresses, crinolines, pink sashes, and gaudily coloured handkerchiefs on their heads." Blind to the complex symbolism of color or the significance of African American fashion, these travelers misconstrued the slaves' intricate use of diverse dyes and shades that wove vital African traditions through their plain, "ruflined," and "couse" American dress. Ex-slave Solomon Northup provided a gloss upon the cultural importance of the slaves' purchasing behavior when he noted that the "females . . . are apt to expend their little revenue in the purchase of gaudy ribbons, wherewithal to deck their hair in the merry season of the holidays." Freed from the domineering influence of white hegemony, black women seized the initiative during these brief post-harvest celebrations to invert established white constructions of African American beauty. Colorful ribbons added vibrancy to their braided hair and clashed with the utilitarian bandana-wearing stereotype that dominated the Caucasian mind. By distinguishing themselves through their newly acquired apparel, slaves embraced the consumer market; in the process, they also gained a new type of currency, since clothing could be exchanged and swapped.[20]

19. Plantation Ledger, 1858–1872 (vol. 46), Benjamin Tureaud Family Papers, LSU. On clothing within the African American aesthetic, see White and White, *Stylin'*, 37–62; Prude, "To Look upon the Lower Sort," 148–149. On the social significance of clothing and style as oppositional culture, also see Helen Bradley Foster, *New Raiments of Self: African American Clothing in the Antebellum South* (Oxford, 1997), 134–223; Patricia K. Hunt, "The Struggle to Achieve Individual Expression through Clothing and Adornment: African American Women under and after Slavery," in *Discovering the Women in Slavery*, ed. Morton (Athens, Ga., 1996), 227–40.

20. Thorpe, "Sugar and the Sugar Region," 737 (first quote); Russell, *My Diary North and South*, 256 (second quote); Interviews with Elizabeth Ross Hite, date unknown (third quote), and Mandy

Moreover, the purchase of silk dresses, pink sashes, and elegant handkerchiefs challenged white-only claims to elite attire and confused the racial lines that clothing allegedly revealed. Slave women wearing sashes and slave men donning imported Russian hats muddied racial categories and complicated the demarcation line between the enslaved and the enslavers. Worse still from the perspective of the big house, gaudy handkerchiefs and crinoline dresses parodied the plantation order. African American attire typically exaggerated normative Anglo-American fashion, and by combining vibrant colors with European jackets, slaves could forge a distinct appearance that satirized white clothing still further. Within the slave community, rich reds and bright yellows symbolized wealth and hierarchy, and they arrogated the slaves' claim to independence and personal mastery. Fashioned from the bolts of cloth in the commissary and the peddler's ribbons, African American clothing stamped its own cultural aesthetic onto the visual landscape. Whether "dressing out," as Alexander Kenner observed, in their Sunday go-to-meeting suits or in russet shifts banded with orange, black Louisianans reordered the world according to their own values and priorities.[21]

While apparel served as a primary channel for safeguarding one's self-respect, other commodities also functioned to ameliorate the drudgery of slave life. Indeed, sometimes overtime pay was the only means to secure basic utensils. As Solomon Northup observed in the only nineteenth-century slave narrative to describe life in the cane world, the bondsman frequently received neither cutlery, dishes, kettles, or crockery, nor furniture of any type. Instead, he was furnished solely with a blanket and was expected to find a gourd to keep his meal. As Northup described, to ask the master for a knife, skillet, or any other basic commodity "would be answered with a kick, or laughed at as a joke." Every item for the cabin, he added, had to be purchased with Sunday money. While such an arrangement might have offended his church-going readers, the former bondsman underlined the fact that "it is certainly a blessing to the physical condition of the slave to be permitted to break the Sabbath." Former bondswoman Ceceil George

---

Rollins, 14 June 1937 (fourth quote), WPA Ex-Slave Narrative Collection, LSU; Northup, *Twelve Years a Slave*, 149 (fifth quote); Shane White and Graham White, "Slave Hair and African-American Culture in the Eighteenth and Nineteenth Centuries," *JSH* 61 (February 1995): 45–76.

21. McDonald, *Slavery and Material Culture*, 81; White and White, *Stylin'*, 18–36; Kenner, "American Freedmen's Inquiry Commission Interviews, 1863," in *Slave Testimony*, ed. Blassingame, 393 (quote).

similarly recalled the stingy management that prevailed on Dick Proctor's farm, where one's only recourse lay in overwork: "In dis country, dey give yo' de ole clothes, one pair shoes a year, no stockin's an' in de winter, sometimes yo' so cold— Lawd have mercy." Others went "bare-footed and half-naked." As Carlyle Stewart remembered, "We didn't know what 'draws' wuz." Annie Flowers evoked the poverty of many slaves when she added, "I didn't work for a salary, it was something to eat and a few rags to wear." Mercifully, not all planters were so brutally minimalist in supplying their slaves with provisions, particularly planters who subscribed to market paternalism; most provided basic housing, though any additional commodities usually could be gained only from overwork payment. On George Lanaux's Bellevue Plantation, the slave Frederick utilized long-term credit to purchase and trade an array of goods that augmented both his diet and his sense of self-assurance and community status. Like many bondspeople, Frederick systematically employed the overwork system to his benefit, raising poultry in his own time to pay his rolling debts at the plantation commissary. Purchasing a barrel of flour for $5.50 in 1851, Frederick attempted to repay his debt quickly by bringing $2.50 in cash and eight chickens to the plantation store, where Lanaux credited him $4.50. Somewhat later, Frederick, who had perhaps adopted the time-conscious values and consumerism of the industrial age, purchased a silver watch for $5. With his debts growing, the bondsman decided to expand his chicken sales—not only to pay for his newly acquired timepiece, but also for the half-barrel of rice that he purchased in December 1853. Selling Lanaux a total of seventeen chickens and twelve dozen eggs, Frederick methodically reduced his debt to just 55¢ by January 1854. Although Lanaux never ceded control over the internal economy of his plantation, his acceptance of overwork presented individuals like Frederick with a loosely defined path to autonomy.[22]

Frederick's poultry operations partially satiated his own wants and perhaps whetted his appetite for more, but his chicken sales did not transform him into a nascent peasant worker living upon the fruits of his land and labor. Like many others, Frederick's experience of the wage market proved limited and his humble trade offered at best a scrappy and no doubt torturous vision of independence. In Frederick's every transaction, George Lanaux possessed the authority and physical

22. Northup, *Twelve Years a Slave*, 149 (first quote). Interviews with Ceceil George, 15 February 1940 (second quote), Annie Flowers, 1 May 1940 (fourth quote), and Carlyle Stewart, 3 May 1940 (third quote), WPA Ex-Slave Narrative Collection, LSU; Journal, 1851–1860 (vol. 14), George Lanaux and Family Papers, LSU.

resources to check his poultry business, and his credit and debt register at the commissary plainly recorded the extent of his dependency upon his master. Not only could Lanaux compel Frederick to sell to him and clear his debts, but he could largely dictate the type of items available for purchase, and he could grant or deny the bondsman credit at his discretion. To be sure, the commissary's packed shelves confirmed the slave's right to the value of his labor, yet simultaneously it denied him the liberty to spend his income freely or outside the plantation compound. Insulating the slave from the freedom attendant in wage work, the commissary allowed planters to enjoy the advantages of free labor while denying its laissez-faire implications. Ever anxious to minimize the slaves' contact with the world beyond the estate boundaries, planters resolutely guarded the slaves against potentially hazardous symbols of freedom, be they in the shape of river boatmen, white traders, or townsfolk. Henrietta Butler accurately fathomed mistress Emily Haidee's stringent approach to imprisoning her slaves. "You know none of the white folks didn't want the niggers to get out," she added. "They was afraid they would learn somethin'." The plantocracy was never wholly successful in stemming the slaves' illicit trading in neighboring towns or their subversive chats on the levee. Nevertheless, Haidee and others like her sought to segregate the bondspeople from local peddlers and seal the plantation gates firmly shut. The estate commissary thus served the dual function of delimiting the slaves' experience of the market economy and channeling their economic activities into plantation profits. Yet above all, commissary trading enabled planters to stamp their authority upon the slaves' trading experiences and to intervene in their lives in yet another way. Local towns like Baton Rouge, Franklin, and Donaldsonville consequently never experienced the dynamism of their Caribbean counterparts, where slave-run Sunday trading flourished. New Orleans remained an exception to this rule, but for most African Americans in Louisiana's cane world, New Orleans was simply too far from the main plantation belts to attend, save for their own sale. Louisiana's enslaved workforce accordingly remained locked within their prison-like estates, strangers to the outside world—if not the market philosophy—of the nineteenth-century sugar country.[23]

23. Interview with Henrietta Butler, 28 May 1940, WPA Ex-Slave Narrative Collection, LSU (quote). On white alarm over African American town trading, see Robert Olwell, "'Loose, Idle, and Disorderly': Slave Women in the Eighteenth-Century Charleston Marketplace," in *More Than Chattel*, ed. Gaspar and Hine (Bloomington, Ind., 1996), 97–110; Timothy James Lockley, "Trading Encounters between Non-Elite Whites and African Americans in Savannah, 1790–1860," *JSH* 66 (February 2000): 25–48.

If envy occasionally crept into the plantation quarters, overwork and wealth accumulation often fostered community pride. Solomon Northup revealed the significance accorded to even comparatively modest sums of money, noting that after earning $17 for his expert fiddle play at a party in nearby Centreville, "Up through all rose the triumphant contemplation, that I was the wealthiest nigger on Bayou Boeuf." Northup's earnings were clearly a source of satisfaction to him— and to the larger community as well. He admitted that "visions of cabin furniture, of water pails, of pocket knives, new shoes and coats and hats, floated through my fancy," and that he counted his money "over and over again, day after day." Since income was so highly valued as a means of material improvement and a source of self-esteem, African Americans embraced overwork and elementary consumerism as a conduit for material and cultural enhancement. Little Jesse, Frederick, Jack Locket, and Elizabeth Hite all manipulated the overwork system to their own pecuniary advantage. Through modest purchasing they achieved a degree of autonomy that collapsed the crude inequalities of slavery and provided welcome degrees of economic, cultural, and—in the case of watch-bearing Frederick—temporal autonomy from white rule. Partially a safety valve to curb black resistance, overtime work simultaneously affirmed the slaves' right to their own labor and exposed the contradiction of wage-earning slaves in a slaveholders' economy.[24]

Across the countertop of the plantation commissary, master and slave came to a grudging compromise. Overwork and domestic production established a remunerative system in which planters gained wood, corn, and services, while bondspeople modestly improved the quality of their lives by purchasing a range of consumer items. Structured to accommodate planters and slaves, internal economies furthered black autonomy yet also benefitted the master by alleviating potential slave hostility to the industrial order and by gaining workfloor stability during the grinding season. Moreover, overwork enabled planters to achieve self-sufficiency in corn and wood; to a lesser extent, it served as a tool to divide the slave community. As lord and master, the planter utilized his power as the adjudicator and recipient of the slaves' produce to forge an economic and psychological link that bound slaves like Jack Locket and Elizabeth Hite to their proprietor or his agent. By dramatizing this link in an orchestrated display of their power, slaveholders who received and credited their bondspeople's produce just prior to the harvest trumpeted their paternalist credentials in a public charade that exalted the

24. Northup, *Twelve Years a Slave,* 149.

master's prestige and disclosed his compassion to white and black audiences alike. Slaves rarely accepted the master's swagger at face value, yet the apparent reciprocity of overwork ultimately dulled collective resistance in favor of individual and sporadic acts such as running away. In accommodating the slaves' desires, overwork proved to be the master stroke of the plantocracy, as the sugar masters defined a free-market sector that enhanced their prestige and promised stability during the grinding season. For them, overwork further reinforced class rule by offering apparent concessions that masked the crudest form of labor exploitation and slave management. From the slaves' perspective, payment through rewards, extra work, bonuses, and improved material conditions commodified their labor power. African Americans exploited their position as both slave and waged workers to improve their modest lives and rework the terms of wage payment into dues and expectations within bondage. To this extent, the slaves were pressing on the inherent contradictions within slavery and claiming money, rights, and power for themselves. Yet by accommodating the machines and the planters' compensation systems, Louisiana slaves were incorporated into—even if they did not internalize—the exploitative economy of the sugar regime.

The era of the sugar masters stands as a depressing episode in the history of humankind. Their management combined elements of nineteenth-century modernity with primeval thuggery; for all its hybridity, however, it proved coldly rational and efficient in its structured and total exploitation. Given this, it seems hard to believe that African Americans survived in the sugar country—and many thousands did not. Yet despite suffering and death, Louisiana's sugar-making slaves made the best of their lives. They adapted to mechanization because it held out the promise of compensation, and they rebuilt community ties in the relative privacy of the plantation quarters. For every key-bearing slave who jealously guarded his commodities, there were dozens who embraced a communalistic vision that included domestic production as one aspect of a wider communitarian ethos. Indeed, gang and team work probably strengthened the slaves' collective identity: individuals worked directly with their neighbors, were awarded incentives as part of a squad, shared a work culture, and collectively exercised their power over the production process by slowing down or speeding up their labors. Likewise, overwork—while resting on individual effort—functioned within the context of a larger family, kin, and group effort. However encumbered Louisiana slaves were, the horrors—and occasional joys—of their lives ultimately brought them together more than it thrust them apart, even as they struggled to find an element of stability in what was surely an alienating and deeply hostile environment. Encapsulating these feelings, former slave Ceceil George sang the simple refrain: "I'm inchin' along, inchin' along . . . like de pore lowly worm, I'm inchin' along." Like George, generations of African Americans inched along, availing themselves of every opportunity to restore their dignity through faith and by action.[25]

Whether following the lead of Edward Gay's slaves, earning extra money during the vacation, or dancing to the beat of an African American drum, slaves in the sugar country greeted the close of the harvest and onset of the vacation with considerable delight. On her visit to the sugar country, Matilda Houstoun observed that the slaves "seemed to look forward with intense delight to the harvest home festivities," where they could "dance and sing, and drink incessantly, never pausing in their merriment for an hour." In this stylized and highly fanciful description, Houstoun underscored the slaves' ardent interest in preserving—and vigorously enjoying—the week-long Christmas holiday. But even as planters self-applaudingly ascribed the slaves' enjoyment of the post-harvest ritual to their benevolent paternalism, African Americans reconstituted the meaning of the vacation and transgressed white authority. As late as the 1920s, African Americans attached significant value to working hard at one's leisure time. Indeed, as historians Tera Hunter and Robin Kelley observe, dancing hard in a black juke joint inverted traditional notions of comportment and sexuality and challenged white authority. Sweating hard on the dance floor enabled African Americans to reclaim physical labor and exertion for themselves and to reassert their autonomy over their own bodies and leisure time. In the slave quarters of antebellum Louisiana, working hard and playing hard functioned in parallel. Both in the sugarhouse and plantation quarters, slaves eagerly embraced the opportunity to relax and share a fleeting moment of independence from the punishing order of the sugar country. Whether sharing a holiday meal, expressing themselves through dance, or displaying their purchases, they forged a communal sense of themselves out of their daily struggles and collective exhaustion. Hewn from experience and a rich vein of West African and Afro-Haitian traitions, the slaves' value system provided a moral code distinct

25. Interview with Ceceil George, 15 February 1940, WPA Ex-Slave Narrative Collection, LSU (quote). Although wealth and occupational stratification could erode community values, slaves shared the same values and aspirations; the harsh reality of bondage "unified slaves" more than "rudimentary stratification divided them." Kolchin, *Unfree Labor,* 352. Also see John W. Blassingame, "Status and Social Structure in the Slave Community: Evidence from New Sources," in *Perspectives and Irony in American Slavery,* ed. Owens (Jackson, Miss., 1976), 137–51.

from that of the big house by which they could order their lives. Conflict and jealousy no doubt crept in, but arguments and their resolution are an essential part of community formation. Insults and gossip simultaneously divide and cohere cultural groups and are testimony to the vibrancy of a community, not to its demise.[26]

Whether in the sugarhouse or among the slave cabins, post-harvest festivities presented an opportunity to cultivate friendships and reaffirm community ties through dance. Calling upon the "pleasure-seeking sons and daughters of idleness who move with measured step, listless and snail-like, through the slow-winding cotillion" to look upon "genuine happiness, rampant and unrestrained," Solomon Northup urged his readers to "go down to Louisiana and see the slaves dancing in the starlight of a Christmas night." With fiddle or by patting juba—a musical accompaniment in which individuals kept time by rhythmically slapping their hands, knees, and shoulders—slaves fashioned an energetic dance that subverted white morality, challenged Victorian notions of sexuality, and extolled a black construction of leisure and dance. Often with little more than a simple drum, Louisiana slaves adapted French dance and transformed it into a pulsating and strenuous step that brought musicians and dancers to a frenzy of excitement and perspiration. Frances Doby remembered that as soon as one slave quickly tacked skin over a barrel to construct a drum, "We was a dancin', a beatin de flooh wid de foots." Dancing until midnight, Frances and her compatriots ended the ball chanting "*Balancez, Calinda,*" before twisting, turning, and repeating the command. In one undated antebellum account, Alcée Fortier vividly described the fusion of European and African influences in slave dance:

> In dancing the '*carabiné*' the man took his '*danseuse*' by the hand and whirled her rapidly and almost interminably while she waved her red handkerchief on high and sang over and over again some strange French couplet. In dancing the less graceful but more strange '*pilé Chactas*' the order was

26. Houstoun, *Hesperos,* 156 (quote); Robin D. G. Kelley, "Rethinking Black Working-Class Opposition in the Jim Crow South," *JAH* 80 (June 1993): 84–86; Tera W. Hunter, *To 'Joy My Freedom: Southern Black Women's Lives and Labors after the Civil War* (Cambridge, Mass., 1997), 179–83. On the impact of West African cultural traditions, particularly from Senegambia, on Afro-Creole culture in south Louisiana, see Hall, *Africans in Colonial Louisiana,* 175–200, 288–94; Thomas M. Fiehrer, "The African Presence in Colonial Louisiana: An Essay on the Continuity of Caribbean Culture," in *Louisiana's Black Heritage,* ed. McDonald, Kemp, and Haas (New Orleans, 1979), 25–26. On conflict as a central force for the constitution of social relations, institutions, and ideologies, see Charles L. Briggs, "Introduction," in *Disorderly Discourse,* ed. Briggs (New York, 1996), 4–13.

reversed. The woman scarcely moved her feet as she danced while the man was turning around her, kneeling, making grotesque faces and writhing and twisting like a serpent. When the object of his affections had become sufficiently enchanted she became excited, and sang and waved her gaudy handkerchief with her jubilant companion. She concluded by obligingly drying the perspiration from the face of her *danseur* and finally, from the faces of the musicians.[27]

On this occasion, communalism, sweat, hard work, and gaudy handkerchiefs combined to celebrate a distinct black cultural aesthetic. Like Elizabeth Ross Hite's conscious rejection of slave garments in favor of silk dresses, the African Americans who twisted, turned, and perspired on the dance floor purposely defied the conventions of white dance with unpredictability and improvisation. Recalling the subversive context of black dance, Hite observed, "De slaves had balls in de sugar house. Dey would start late an' was way out in de field whar de master could not heah dem. Not a bit of noise could be heahed." Elaborating on her point, the former bondswoman concluded, "No sir, de slaves had some swell times." Hite captured the dissident element of the sugarhouse dance, noting that not only did the slaves dance by candlelight, but that the following day they had to conceal their exhaustion or else a "whippin would follow." On the dance floor, Hite recalled that her bonded compatriots "did de buck dance an de shimmie," in which women took steps alternatively to the right and left, "shake[ing] dere skirts" as the men danced around them. Former slave Ellen Betts similarly remembered "all the dancing and capering you ever seen" in the aftermath of the harvest, reminiscing, "We'd cut the pigeon wing and cut the buck and every other kind of dance." When her father grew tired of providing the musical accompaniment, Ellen and her friends refused to stop the festivities; to entice the fiddler to play more, "us gals git busy" and bribed him with popped corn and candy.[28]

27. Northup, *Twelve Years a Slave*, 166 (first quote); Interview with Frances Doby, 6 December 1938, WPA Ex-Slave Narrative Collection, LSU (second quote); Alcée Fortier, *Louisiana Studies: Literature, Customs, and Dialects, History and Education* (New Orleans, 1894), 127 (third quote). On patting juba, see Dena J. Epstein, *Sinful Tunes and Spirituals: Black Folk Music to the Civil War* (Urbana, Ill., 1977), 141–44.

28. Interview with Elizabeth Ross Hite, date unknown, WPA Ex-Slave Narrative Collection, LSU (first quote); Botkin, ed., *Lay My Burden Down*, 133 (second quote). On the legal restrictions imposed on dancing and excessive noise, particularly drum play, see Epstein, *Sinful Tunes and Spirituals*, 52, 93, 95.

Repossessing one's body through dress or dance—preferably both—proved especially meaningful in a society that was so committed to the evaluation of physiognomy and reproductive potential. The planter's prodding fingers and calculating mind placed a valuation on skin, muscle, and teeth, while his searching eye assessed the meaning attached to scars. Threatened and maimed by the master's whip, the slave's body was a uniquely contested site, where black women and men strove to maintain their dignity and freedom from the planter's roving sexuality, stinging lash, and mean-spirited diet. In their little free time, African Americans seized the initiative in dance and dress, forging their own communal values in a world that assessed them solely as "sugar machines." [29]

As Ellen Betts and Elizabeth Ross Hite testified, the dance culture of the antebellum sugar country encompassed a number of styles that combined European reels and schottisches with dances that drew upon West African traditions. Indeed, by the 1830s European and African dances were fusing, with slaves performing jigs and fandangos in New Orleans's Place Congo. Such incidents of white-black interaction across the color line enhanced the emergence of syncretic dance forms in south Louisiana that combined polyrhythmic meter and heterogeneous styles. The buck dance that both described evolved as a flexible fluid dance in which the slaves, bending low to the ground, pounded the earth with rhythmic intensity. With hunched bodies, flexed joints, and shuffling feet, these dances improvised upon the polyrhythmic meter of African American music and served as a means of collective cultural expression. In "cuttin' the pigeon wing," the dancers held their necks rigid and their heads still while beating their arms. This dance placed a premium on balance and matching the rhythm and meter of the music with composed precision. Musicologists argue that in contrast to occidental dance, traditional forms of West African dance not only respond to the rhythm of the music but also add another rhythm, one that is not readily apparent. The dancer "tunes his ear to hidden rhythms, and he dances to the gaps in the music." Slave dance reflected the multiplicity of direct and refracted voices in African American music; like black autobiography, it adopted a double voice that enabled the performer to converse in a participatory relationship with his or her audiences. Like the modern poet Maya Angelou, who retooled English "to fit a rhythm which is

29. Stirling, *Letters from the Slave States*, 124 (first quote); Nutall, *Journey of Travels into the Arkansas Territory*, 239 (second quote). On the slave's body as a contested site, also see Stephanie M. H. Camp, "The Pleasures of Resistance: Enslaved Women and Body Politics in the Plantation South, 1830–1861," *JSH* 68 (August 2002): 533–72.

really African and of Creole descension," slave dance triangulated between European, American, and African cultures. It reflected the transference of sub-Saharan identities and consciousness to the lives of slaves and embodied the syncretic and hybrid pattern of African American culture. Most significantly, black dance evolved as a means of personal and group communication. Antiphonal dance, in which the leading performer poses rhythmic and bodily questions to his or her partner, mirrors the call and response of both religious and secular music. Above all, harmonious communication between individual dance partners, rhythmic accompaniment, and fellow participants—be they dancers or onlookers—choreographed a cultural aesthetic of strength, community bonds, compassion, and stability. In this way, African Americans rendered their ethical values and everyday experiences into melodies and gave physical shape to their sentiments.[30]

Deaf to the overlapping rhythms and metric mosaic of West African and African American music, white observers—like Thorpe and Alcée Fortier— missed the elisions and cultural complexity of the dances they described. The slaves' shuffling feet, gliding moves, upper body shimmying, and pelvic thrusts appeared both alarming and incomprehensible to white observers. The *carabiné*, a dance with rapid and raging choreography, and its inverse, the *pilé Chactas,* both choreographed the African dancing ring. The West African ring encircled the community, encompassing the renewal of life and embodying a profound and innate cultural memory of movement and sound upon which African Americans drew. Contemporary descriptions of slave dance in New Orleans reveal a vibrant community of black dancers, "fancifully dressed with fringes, [and] ribbons," who

30. On double-voicedness, see M. M. Bakhtin, *The Dialogic Imagination: Four Essays,* ed. Michael Holquist, trans. Caryl Emerson and Michael Holquist (Austin, Tex., 1981), 324–28. Angelou quoted in Helen Taylor, *Circling Dixie: Contemporary Southern Culture through a Transatlantic Lens* (New Brunswick, N.J., 2001), 169. On rhythm and dance aesthetics, see Paul Gilroy, *Small Acts: Thoughts on the Politics of Black Cultures* (London, 1993), 246; Thomas F. DeFrantz, "Black Bodies Dancing Black Culture—Black Atlantic Transformations," in *EmBODYing Liberation,* ed. Fischer-Hornung and Goeller (Hamburg, 2001), 11–16. On sub-Saharan antiphonal dance, see John Miller Chernoff, *African Rhythm and African Sensibility: Aesthetics and Social Action in African Musical Idioms* (Chicago, 1977), 144–51; Robert Farris Thompson, "An Aesthetic of the Cool: West African Dance," *African Forum* 2 (fall 1966): 85–102, esp. 91–95. Also see Thompson, *Flash of the Spirit: African and Afro-American Art and Philosophy* (New York, 1984); Shane White and Graham White, "'Us Likes a Mixtery': Listening to African American Slave Music," *Slavery and Abolition* 20 (December 1999): 31, 35, 38; Gary A. Donaldson, "A Window on Slave Culture: Dances at Congo Square in New Orleans, 1800–1862," *Journal of Negro History* 69 (spring 1984): 63–72.

kept the "most perfect time . . . making the beats with the feet, heads, or hands, or all, as correctly as a well-regulated metronome." The dancers that James Creecy described in 1834 had been performing in Congo Square for generations. As early as 1819, Benjamin Latrobe described five or six hundred people dancing collectively in circular rings, ten feet in diameter. Within several circles, two women danced as they held handkerchiefs by the corners, each "hardly moving their feet or bodies" and responding to the rapid beat of the drum by repetitively singing two notes. Latrobe found the noise unspeakably loud, but he was singularly deaf to the musical and danced conversation that was unfolding before his eyes. The waving handkerchiefs added to the communicative effect, as the dancers conversed in a rhythmic discourse that emphasized self-control, composure of mind, and both collective and individual involvement.[31]

In dancing the *carabiné*, the danseur and danseuse performed in a manner consistent with other West African ring dances. The man circled his partner (probably counter-clockwise) and whirled her to one meter, and she responded to his rhythmic call by repeatedly calling out "some strange French couplet." The red handkerchief that she waved above her head accentuated the communicative effect of the dance, adding an improvised meter to the call and response of the danseur. The partners' rhythmic steps articulated the black cultural aesthetic and reaffirmed communal solidarity and individual control. In the *pilé Chactas*, the female dancer—like those in the ring dance—moved slowly to one rhythm while her partner darted around her, moving to the gaps in the music and responding to the mixed and overlapping tones of the instruments. Good dancers not only responded to the complex multiple-meter structure of African American music but critically balanced it. As Robert Farris Thompson argues, balance is the aesthetic acid test of West African dance: "The weak dancer soon loses his metric bearings

---

31. On discrete styles in slave dance, see Blassingame, *Slave Community*, 36–40, 109; Epstein, *Sinful Tunes and Spirituals*, 44, 98 (second quote), 134 (first quote), 159; Levine, *Black Culture and Black Consciousness*, 16–17; Roger D. Abrahams, *Singing the Master: The Emergence of African American Culture in the Plantation South* (New York, 1993), 98–100. On ring dances, see Samuel A. Floyd Jr., *The Power of Black Music: Interpreting Its History from Africa to the United States* (New York, 1995), 21, quoted in White and White, "'Us Likes a Mixtery,'" 27, 29; Sterling Stuckey, *Slave Culture: Nationalist Theory and the Foundations of Black America* (New York, 1987), 12–16, 25, 57–58, 86–94. On dance as mediated involvement, see Chernoff, *African Rhythm and African Sensibility*, 140. Carabiné, *adj. fièvre*, raging, violent. Pilé, *past participle*, to pound; *Dictionnaire de la langue française du XVI siècle* (Paris, 1961); *Dictionnaire étymologique de la langue française* (Paris, 1960).

in the welter of competing countermeters and is, so to speak, knocked off balance." The male dancer thus not only matched his rhythmic accompaniment with composure, but his bodily movements asked questions of his partner and the musicians. Such personal and group communication required guile, evasion, deception, balance, and, above all, a coolness of mind. His partner harmoniously responded to the darting discourse of the male danseur, and, having reached its crescendo, the vibrant communicative thrust of the music and dance was rewarded when the bondswoman dried the perspiration from the faces of the participants. If she used her handkerchief to do so, her act carried yet more symbolism. Colored and checked handkerchiefs served as a means of self-adornment and aesthetic display, so that in mopping her partner's brow, the bondswoman consummated the ritual, faced her musicians, and completed the circle of reciprocity displayed in the entire dance.[32]

In a society where time was brutally regimented, harvest and weekend dances ritualized the slaves' periods of leisure, strictly demarcating their time from that of the masters. Dances, furthermore, provided opportunities to satirize the plantocracy. By waving gaudy handkerchiefs and dressing in snow-white dresses, crinolines, and pink sashes, slave women restyled occidental dress to meet an African American agenda. On these occasions emulation often masked derision, as black Louisianans mocked their owners' dress and lampooned their inflexible and clumsy dances. Thus when Elizabeth Ross Hite specifically remembered the dance skills of Jolly, she might well have described a wholly satirical scene. "Dis darkey sur could dance," she added. "Boy, when he started twirling his legs and stickin out his back old master would holler." Jolly's eccentric dances appeared all the more amusing to Pierre Landreaux because they imitated and exaggerated— with telling accuracy—the stultifying reels and cotillions of the white world. "Old master" might have hollered, but he was ultimately laughing at himself. Moreover, slave dances transformed buildings symbolic of the planters' plower— like the sugarhouse—into vibrant cultural spaces where African Americans could engrave their own cultural pattern onto the land- and soundscape. Packed with the technical symbols of the planter's commercial power, the sugarhouse was the symbolic and economic hub of the estate and the primary production zone come harvest. For those six to eight weeks during which the mills turned, the building

32. Thompson, "Aesthetic of the Cool," 91 (quote). On head and handkerchiefs, see White and White, *Stylin'*, 59; White and White, "Slave Hair and African-American Culture," 45–76. On parallels to Surinam dances, see Stuckey, *Slave Culture*, 20.

symbolized the commercial thrust of plantation slavery; yet once the grinding season came to a close, African Americans reconstituted the sugarhouse as their own space. Whether mocking whites or communing with each other through dance, African Americans in the sugar country grasped a sliver of personal and group autonomy and redefined a dissident space in the midst of the planter's regime.[33]

The slaves' nocturnal culture flourished on the boundaries of the plantation world. Whether out in the fields "whar de master could not heah dem" or in the sugarhouse where the planter could not understand them, African Americans transformed these physical and emotional spaces. Although rectilinear streets led symbolically from the big house to the sugarhouse, the slaves who stole away from the quarters to moonlit vigils followed their own tracks through the plantation. Runaways and fugitives had secret passageways from the woods to the quarters, reappearing—literally under the nose of the master—to receive a nightly meal from their loved ones. And those slaves who illicitly traded with river peddlers followed the ruts and trampled-down paths that led to the levee crest and a world beyond the confines of the plantation kingdom.[34]

By following their own tracks, slaves contested the disciplined geography of the sugar country and forged a definably distinct black landscape, in which Christianity also marched to an African American beat. Elizabeth Ross Hite vividly described her own church as an invisible institution flourishing in the dark of the night on the margins of the plantation, far from the prying eyes of Master Pierre Landreaux. With just the flickering flame of a little grease burning in a can through the early morning hours, slaves on Trinity Plantation converted Christianity into a vibrant faith that met their own emotional and religious desires. Like many south Louisiana slaveholders, Pierre Landreaux sought to institutionalize the Catholic faith on his plantation. Hite recalled that he introduced a black preacher from France to "teach us how to pray, read, and write," but, as she added, "lots didn't like that 'ligion . . . caise ya could not do lak ya wanted ta do." Above all, African Americans sought to control their own faith "an shout and pray lak ya

33. Interview with Elizabeth Ross Hite, date unknown, WPA Ex-Slave Narrative Collection, LSU (quote). On satire in black performative arts, see Stuckey, *Slave Culture*, 71–72; Levine, *Black Culture and Black Consciousness*, 17.

34. Interview with Elizabeth Ross Hite, date unknown, WPA Ex-Slave Narrative Collection, LSU (quote). On how slaves created a black landscape out of the rigidities of plantation topography, see Vlach, *Back of the Big House*, 12–17.

wanted to." Strict Catholicism provided little opportunity for participation, and liturgical recitation denied the parishioner the prospect of individual biblical interpretation. The austerity of Catholicism won few converts on Trinity Plantation; instead, the slaves embraced a syncretic Afro-Christian devotion that served as a "weapon of personal and community survival." Mixing with Afro-Protestant slaves who were sold from Virginia into the sugar country, the slave community on Landreaux's estate forged a dissident church that linked fiery evangelicalism with their own African socioreligious and cultural systems. As Elizabeth Ross Hite asserted, true and cathartic faith derived from "prayin an be[ing] free to shout . . . Dats what Ise call religion."[35]

Stealing away from the plantation quarters in the dead of night, African Americans at Trinity reconvened at their own church, located far out in the fields. Hiding behind a mound of bricks, Hite and others assembled nightly to listen to old man Mingo preach and conduct Bible lessons. These slaves were putting themselves at no small risk, because, as Hite declared, master Landreaux "didn't want us to hav too much religion caise de darkies would giv' him all religion an' no work." Landreaux recognized that he was powerless to stop the slaves' illicit activities, though he energetically sought to cap any activities that ate into his time, ordering his drivers to whip those individuals "who stayed up too late singing and so could not do a gud day's work de next day." Landreaux's drivers might prowl the slave quarters and whip miscreants, but, as Hite remembered, "We hid and prayed just de same. No sir, nothing could stop us from prayin to Gawd." With just enough light to read the scripture, Mingo led his flock in singing spirituals "all night and way into de mornin'." "Yeah, we shouted all night" too, Hite admitted. "Gawd says

---

35. Interview with Elizabeth Ross Hite, date unknown, WPA Ex-Slave Narrative Collection, LSU (first, second, fourth quotes). Albert Raboteau first coined the term "invisible institution" to describe slave religion beyond the confines of the institutional church and thus "hidden from the eyes of the master." Albert J. Raboteau, *Slave Religion: The "Invisible Institution" in the Antebellum South* (New York, 1978), 212. On slave religion in Louisiana and the acquisition of religious items from traders, see Taylor, *Negro Slavery in Louisiana*, 133–52; Malone, *Sweet Chariot*, 241–50. On the centrality of religion as the primary means for self and communal identity among slaves, see Genovese, *Roll, Jordan, Roll*, 161–284 (third quote, 212). On the limitations of Catholicism as a faith-culture among slaves and the relative success of Baptism and Methodism as vehicles for Afro-Christian faith and the retention of African religious practices and values, see Randall M. Miller, "Slaves and Southern Catholicism," in *Masters and Slaves in the House of the Lord*, ed. Boles (Lexington, Ky., 1988), 127–52; Sylvia R. Frey and Betty Wood, *Come Shouting to Zion: African American Protestantism in the American South and British Caribbean to 1830* (Chapel Hill, N.C., 1998), 178; Creel, *Peculiar People*, 184–85.

dat ya must shout if ya want to be saved," she added. "Dats in de Bible. Don't ya know it?" Regaling interviewer Robert McKinney with the details of the slaves' midnight services, the eighty-year-old declared that "de old preacher would stomp his foot and all de people would pray an shout . . . ya ought er heahed dem." Whether leading the spiritual, calling the shout, or poring over his Bible by just a flickering flame, Mingo must have been a true source of inspiration within the slave quarters. Like most other black preachers, he "worked on de plantation lak ev'rybody else," and in addition to his nightly ministrations, he conducted baptisms three times a year. Holding these sacred rites in the pond "whar dey drew water for the sugar house," Mingo reappropriated the planter's water supply and consecrated it for the holy sacrament of baptism. Inverting the planter's control over their spiritual lives and seizing the outhouses and ponds as their own, the slaves contributed to the ceremony and the shared African American cultural experience by singing "I baptize yo in de ribber Jordan." But the scene was packed with yet more significance. As one cultural historian observes, even as they took part in Christian ceremonies (such as baptism), African American slaves were often expressing profound and longstanding African religious concerns. In this context, Mingo and the slaves on Trinity Plantation were involved in a communal and communicative act, in which the bondspeople sang as Mingo baptized his parishioner. Such displays of communal and individual reciprocity were fundamental to African American religious ceremonies. Much like the mutualism and reciprocity of dance, baptism served as a culturally significant ritual among those of African descent.[36]

Mingo must have been a busy man. "Ev'rybody," Hite recalled, "was anxious to git baptized . . . to sav dere soul," and although the slaves were supposed to have church only on Sundays, "We wanted ta pray all de time." John Adams, the "cullered man . . . from France," who came to Trinity to preach the Holy Trinity undoubtedly found Mingo to be an irritating meddler. The slaves, however, thought otherwise. They judged their preachers by the emotive power each wielded, determining a spiritual hierarchy based on the charisma of each minister and the admiration in which he was held. Mingo climbed high on this spiritual ladder as he guided his nightly congregation in reaffirming their individual and group identities. Like other slave preachers who recast white evangelical faith along a new and

36. Interview with Elizabeth Ross Hite, date unknown, WPA Ex-Slave Narrative Collection, LSU (quotes); Stuckey, *Slave Culture,* 36.

meaningful trajectory for African Americans, Mingo transformed the gospel into a powerful weapon with which to challenge the ideological hegemony of the master-slave relationship. Mingo wielded this weapon with precision. He used the planter's water for baptism and the planter's bricks to muffle the sound of the spirituals. In doing so, he created midnight institutions to protect and nurture black faith.[37]

In following Mingo, the slaves at Trinity consciously rejected their master's faith for a more convincing gospel that confirmed their humanity and dignity. Their faith was hewn from their daily struggle with the sugar masters and evolved as a socio-religious system which allowed them to reconceptualize their world and resist the brutal suffering of slavery. Promising liberty and equality in the afterlife, it served as a way to vent one's anger and resentment in the present. Afro-Christianity reaffirmed a sense of community, strengthened the ties of kinship and family, and gave all involved a constitutive role in shaping a faith that mixed African customs and practices with the conventions of orthodox Christianity. While outwardly recognizable to white observers, the slaves' faith masked a dissident agenda of black assertiveness.[38]

Nowhere was the communicative impact of African American religion seen more forcefully than in the expressive culture of ring shouts and spirituals. Shouting in the Afro-diasporic tradition was a deeply religious and cathartic exercise that enabled slaves to express their faith in an overtly African manner. Joined in a circle that moved counter-clockwise at ever increasing speed, Mingo stamped his foot to the rhythm as the slaves collectively chanted their devotion to their heavenly Lord. In shouting, Elizabeth Ross Hite found unity within her community and an emotive source of strength in the liturgical dance of which she was a part. Deriving from western Africa—and Dahomey in particular—the ring shout reaffirmed the slaves' emotional and spiritual needs. The creative and impulsive spontaneity of the shout flourished in the large Louisiana slave communities, where Afro-Haitian traditions fused with those of Virginia and Carolina slaves to

37. On black and white competition for spiritual hierarchy within the community, see Christine Leigh Heyrman, *Southern Cross: The Beginnings of the Bible Belt* (Chapel Hill, N.C., 1997), 221. On the power and influence of black ministers, see Raboteau, *Slave Religion*, 231–39; Genovese, *Roll, Jordan, Roll*, 255–79; Blassingame, *Slave Community*, 130–32.

38. On how Afro-Christian religion served as way to structure community values, see Frey and Wood, *Come Shouting to Zion*, 182–208; Albert J. Raboteau, "The Black Experience in American Evangelicalism: The Meaning of Slavery," in *African-American Religion*, ed. Fulop and Raboteau (London, 1997), 90–106; Creel, *Peculiar People*, 188–210.

create a rich and hybrid culture. Shuffling along in a slightly stooped pose, slaves in the ring shout moved to the increasing beat, vocalizing their faith and shouting fragments of gospel in rhythm with the dance. White observers were completely confounded. The slaves' flexed position reminded them of the supplicant on bended knee, and the jarring contrasts of faith and sinful dance left them at a loss to explain the scene. Slaves, however, had grasped the upper hand. By hunching and bending, black Louisianans expressed the West African belief that flexed and bended joints were indicative of energy and life, whereas locked knees and elbows implied death and rigidity. Slaves who bent and flexed in dance thus reaffirmed their collective and individual humanity and communicated their inner desires to their fellow slaves and the Deity.[39]

Elizabeth Ross Hite was not alone in subverting Landreaux's religion and landscape. As former slave Carlyle Stewart recalled, "I knowed my prayers." Using candles that she made from beef tallow, Carlyle's mother "teached me behind the Marse's back in her old dark house." The creole J. B. Roudanez similarly remembered how secret religious meetings were held in the woods, far from the prying eyes of the planter class. Not all slaves, however, had to be so circumspect. Many planters ascribed to the belief that religion made African Americans more tractable, more readily controlled, and less likely to escape or plot against their secular masters. Moreover, they assumed that it was their Christian duty to foster religion among their slaves. Leonidas Polk underscored the divine responsibilities of slaveholding when he counseled his fellow planters, "You may not save him [the slave], but you will save yourself." Paternalists like Polk recognized that the reciprocal and mutual obligations of the master-slave relationship compelled them to evangelize their chattel, and that in teaching the gospel they fulfilled their divine obligation to safeguard their slaves' spiritual lives. But to orchestrate their power and demonstrate their stewardship, slaveholders actively intervened and arranged the slaves' religious services, either by hiring a preacher, leading the slaves to Sunday service, or even conducting the ministry themselves. To ensure that he appeared before his slaves and the community as a Christian master, Polk encouraged Episcopalian services on his Lafourche Parish estate and employed a full-time chaplain to minister to the slaves' spiritual needs. Every newborn received

39. On ring shouts, see Stuckey, *Slave Culture*, 12, 24, 85–87; Raboteau, *Slave Religion*, 68–71. On the slaves' predilection for flexed bodies, see Peter H. Wood, "'Gimme de Kneebone Bent': African Body Language and the Evolution of American Dance Forms," in *The Black Tradition in American Dance*, ed. Myers (Durham, N.C., 1988), 1–15, esp. 7–9.

baptism, children learned the catechism in Sunday School, and all slaves attended the weekly church service. Occasionally Polk took the pulpit himself, impressing upon the slaves the power of both their heavenly and earthly master. Preaching at a confirmation ceremony, Polk laid his hands upon each head and orchestrated a masterful display of paternalism, in which the plantation lord shared the body and word of Christ with the bondsman weeping before him.[40]

Other sugar masters, like Polk, occasionally used the established church to cement bonds of dependency. On Smithfield Plantation near Baton Rouge, for instance, bondswoman Catherine Cornelius was named after her young mistress and christened in the local Episcopalian church, while slavemaster William Hamilton exercised his authority by marrying slave couples on his estate. Proclaiming that the slaves "have had a great time of it," Hamilton celebrated his mastery over his "happy fellows," declaring that "they say I act the parson very well." Astute to the overlapping secular and spiritual significance of the sugarhouse, Hamilton's kinsman noted the common practice of converting the mill and plantation outhouses into a makeshift chapel: "The sugarhouse is a very fine place for preaching; there they can enjoy the sweets of various kinds . . . [and] be constantly nourished by the odouriferous smell of sugar pots and steam engine. May the Holyman succeed in instilling moral principles and actions into the whole community." While some slavemasters converted parts of the sugarhouse into an apse and performed services below the temporarily motionless cane carriers, others constructed small chapels on their plantations and left the religious duties to a designated slave, such as thirty-five-year-old Sam on Bernard Marigny's estate. Still others built churches on their plantations. John Dymond, established the Mt. Zion Baptist church on the grounds of his estate for his slaves, installing Nicholas Williams as the resident preacher. Such acts enabled planters to preen their paternalistic egos, but the sounds emanating from the sugarhouse expressed a wholly different cultural aesthetic at work. The spirituals that echoed throughout

40. Interview with Carlyle Stewart, 3 May 1940, WPA Ex-Slave Narrative Collection, LSU (first quote); "Testimony of a New Orleans Free Man of Color," in *Freedom,* ser. 1, vol. 3, *Wartime Genesis of Free Labor: The Lower South,* ed. Berlin et. al., 523 (second quote); Touchstone, "Planters and Slave Religion," 102, 113 (third quote). On Polk, see Mrs. Isaac H. Hilliard Diary, 3 February 1850, LSU; Polk, *Leonidas Polk,* 1:197; Genovese, *Roll, Jordan, Roll,* 204; Moody, *Slavery on Louisiana Sugar Plantations,* 93. On the slaveholders' Christian duty to evangelize, see ibid., 103–10; Young, *Domesticating Slavery,* 123–60; Anne C. Loveland, *Southern Evangelicals and the Social Order, 1800–1860* (Baton Rouge, La., 1980), 219–56.

the slave quarters resoundingly fused the slaves' secular desires, religious aspira-tions, and collective memory.[41]

The slaves' faith culture incorporated folk religion, syncretic Afro-Haitian voodoo, and conjuring. Like their African forebears, they believed in the power of good and evil spirits. Some, as Albert Patterson remarked, "don do no harm"; others scared Edward De Buiew so much that he ran two miles to escape them. Voodoo remained a predominantly New Orleans phenomenon, but former slaves occasionally referred to it. Odel Jackson, for instance, recalled a woman who "hoodooed me" and "put a snake in my leg. . . . I's still cripple yet." Snake cults and conjuring coexisted harmoniously with Christianity and added a further level of cultural complexity to African American faith. However emotionally and socially significant religion proved to be for Elizabeth Ross Hite and others, it was nonethe-less a relatively blunt weapon with which to contest the slaveholders. To be sure, slaves resisted the emotional bludgeoning of slavery by sharing in faith, but religion tended to channel anger and resentment into non-revolutionary acts. African American Christianity did not charge the slave community with a rebellious fer-ment, and although it reaffirmed life, it did not shatter Louisiana slavery or topple the sugar masters. Ultimately, religion transformed the struggle between master and slave into an ideological contest rather than a structural one. Like overwork and the slaves' domestic economy, religion proved to be a defensive weapon, as the slaves battled internally to wrest some rights and privileges from the horrors of Louisiana bondage.[42]

41. William B. Hamilton to William S. Hamilton, 1 January 1858, and John D. Hamilton to William S. Hamilton, 17 April 1851, William S. Hamilton and Family Papers, LSU (quotes); Copy of Act of Sale of Slaves and Plantation by Bernard Marigny and William and Haywood Stackhouse on 13 March 1852 (dated 26 January 1872), Ross/Stackhouse Records, Louisiana State Museum, New Orleans; Interviews with Catherine Cornelius, date unknown; St. Ann Johnson, 8 February 1940; and Rebecca Fletcher, 21 August 1940, WPA Ex-Slave Narrative Collection, LSU; Taylor, *Negro Slavery in Louisiana,* 149; Malone, *Sweet Chariot,* 244–245. On spirituals, see Levine, *Black Culture and Black Consciousness,* 17–55; Ronald Radano, "Denoting Difference: The Writing of Slave Spirituals," *Critical Inquiry* 22 (spring 1996): 506–44.

42. Interviews with Albert Patterson, 22 May 1940 (first quote), Edward De Buiew, 10 June 1940, and Odel Jackson, 31 May 1940 (second quote), WPA Ex-Slave Narrative Collection, LSU. On the slaves' folk religion and voodoo, see Gomez, *Exchanging Our Country Marks,* 54–58; Joyner, *Down by the Riverside,* 141–71; Bankole, *Slavery and Medicine,* 136–40; John Blassingame, *Black New Orleans, 1860–1880* (Chicago, 1973), 5–6. On the limits to religion as a force for structural change, see Genovese, *Roll, Jordan, Roll,* 273–74, 283–84.

African Americans never ceased to contest the slaveholders' goal of total subordination, and in the sugarhouse they secured spiritual autonomy, emotional
freedom, and a sphere in which they could satirize the planters. They adapted the
terms of industrial revolution to advance their own agenda, accepting payment for
extra work and grabbing, if only fleetingly, the freedom inherent in wage work.
Lured by incentives that they transferred into customary rights, bondspeople ac-
quiesced to the regimented and labor-intensive demands of the industrialized
grinding season. Over countertops, across dance floors, and on cabin stoops, slaves
in the cane world attempted to redefine the terms of the master-slave relationship
to advance their own vision of bondage. In faith and in leisure, they found temporary release from the planters' sway, and they engaged in socially and economically significant trade in the domestic economy. Black activism, however, carried
a darker and more prophetic ring. By accommodating the machine and their masters' economic agenda, the slaves descended into greater entrapment. The internal
economy buttressed that of the plantation, converting the agro-industrial machinery into sources of income. Planters and slaves viewed the internal economy from
different perspectives, but both ultimately found a way to articulate their own interests through the bargaining and contestation of the master-slave relationship.

*I*N THE FALL of 1887, the Knights of Labor struck for increased wages in Louisiana's sugar country. Timing the strike to the onset of the sugar harvest, cane workers in Lafourche, Terrebonne, and St. Mary Parishes downed their tools, walked off the job, and demanded cash payment instead of plantation scrip, higher pay for night work, and better wages throughout the agricultural year. The sugar masters responded with characteristic force to this episode of labor insurgency. Strikebreakers and state militiamen prowled the streets of Thibodaux, threatening workers and their families and turning them out of their homes. Intimidation swiftly gave way to open acts of violence, as white vigilantes joined with the militia and local citizens to crush the incipient labor movement. On the morning of Wednesday, November 23, the first shots rang out, as white Louisianans turned their firearms against their former bondspeople. In the space of a few brief hours, dozens of sugar workers lost their lives in the notorious Thibodaux Massacre. Union activists took flight, the strike ground to a halt, the workers returned to the gangs, and the harvest proceeded. Mary Pugh rejoiced over the sugar masters' victory, declaring that the massacre "will settle the question of who is to rule, the nigger or the White man? for the next 50 years." The sugar masters would brook no questioning of their authority. Like their predecessors in the slave era, they suppressed the slightest resistance to their racial and class primacy.[1]

1. Jeffrey Gould, "The Strike of 1887: Louisiana Sugar War," *Southern Exposure* 12 (November–December 1984): 45–55; Rebecca J. Scott, "'Stubborn and Disposed to Stand Their Ground': Black Militia, Sugar Workers, and the Dynamics of Collective Action in the Louisiana Sugar Bowl, 1863–87," in *From Slavery to Emancipation in the Atlantic World,* ed. Frey and Wood (London, 1999), 103–26 (quote, 119); Scott, "Fault Lines," 72–82; John C. Rodrigue, "'The Great Law of Demand and Supply': The

The black sugar workers who rallied to strike fully grasped the significance of wage income and the right to contract; from recent memory, they also recalled the bitter legacy of slavery. The lessons that their predecessors had learned under bondage remained pertinent a quarter of century after they had seized the promise of emancipation from the sugar masters and their slaveholding ilk. Resistance and negotiation defined Louisiana's cane world. And just as enslaved sugar workers had struggled to secure rights and liberties from Louisiana's slaveholding magnates, so too did their successors, who fought for higher wages from the unscrupulous heirs to the great antebellum estates. Withholding their labor as the harvest season neared, the freedmen contested the terms of postbellum agriculture, even though —like their predecessors who had negotiated over customary rights and over-work—they ultimately ceded their independence to the wage and commercial or-der that exploited them. As John Rodrigue recently observed, African Americans in the cane belt internalized the "capitalist market-place's values" and revealed a willingness to labor as long as planters offered a decent wage and adequate work-ing conditions and no longer exercised intrusive mastery over their workers. Tied to the planters by the iron links of wage labor, freedmen in the sugar country vig-orously challenged their former slave lords, only to find themselves—like their predecessors—entrapped once again within a vortex of dependence.[2]

Postbellum sugar masters bemoaned loafers and idlers and romanticized the past as a "grand and lordly life," but in reality neither planters nor freedmen were new to the principles of wage labor. To be sure, the sugar elite had every reason to believe that the social pyramid had inverted itself. In Reconstruction-era Louisiana, former slaves exercised political power and violently contested the terms of free la-bor, while legal codes overturned white-only privileges. As the most recent scholar of postbellum Louisiana has suggested, the war ushered a "seismic shift . . . in the way planters and workers related to one another." Emancipation, in short, divided postbellum Louisiana from its antecedents and ensured that freedom would impart landmark changes to the sugar country. But in fact no such rubicon existed. Although the scale of cash payments evidently changed with the introduction of free labor, neither planters nor former slaves were novices in the principles of wage

---

Contest over Wages in Louisiana's Sugar Region, 1870–1880," *AH* 72 (spring 1998): 159–82. On the Knights of Labor, see Melton A. McLaurin, *The Knights of Labor in the South* (Westport, Conn., 1978).

2. Rodrigue, *Reconstruction in the Cane Fields,* 41.

work, reward structures, or the art of compromising over the terms of labor. As practiced negotiators who firmly understood the contours of wage payment and who knew how to use their influence to wring benefits from the land and labor lords, freedmen—like the slaves before them—accommodated themselves to the machine and the agro-industrial age. The compensation systems of bondage offered a fraction of what landlords were forced to cede after emancipation, yet both masters and slaves understood the advantages and disadvantages of negotiation, and both were aware of the complex labor market emerging in the sugar country.[3]

Like their grandchildren, who took to the streets of Thibodaux in search of higher wages and cash payment, slaves in Louisiana's cane world utilized their position at the axis of production to gain both payment and rights. These achievements proved enormously significant to those in chains, but the compromises hammered out between masters and slaves ultimately strengthened the regime. Depressingly so, I might add. The sugar masters' power remained relatively unassailed as the plantation elite furthered their economic and political influence. But massive global competition and expansion of beet sugar as a crop in the 1890s eventually toppled the sugar masters from their lofty pinnacle, imposing challenges that Louisiana failed to meet. The apogee of the 1850s found no match a half-century later, but for thirty years before the Civil War, slaveholders in the lower Mississippi Valley had been combining wage and coerced labor for profit. The sugar masters often viewed their actions through a paternalistic lens, masking bare-faced incentives with a sheen of benevolence, yet their use of distinct forms of labor ensured profit, stability, and above all, a satisfactory way to induce slave to act in the planters' interest. As major slaveholders, Louisiana's sugar-planting elite held firmly to a social, racial, and economic agenda in which black labor stability was foundational to their business success and class position. Hardened by their experience on the southwestern frontier of American slaveholding and schooled in the vagaries of sugar production, Louisiana's cane lords mastered an industry

---

3. Ibid., 2 (second quote), 114 (first quote), 127. Also see Richard Follett and Rick Halpern, "From Slavery to Freedom in Louisiana's Sugar Country: Changing Labor Systems and Workers' Power, 1861–1913," in *Sugar, Slavery, and Society,* ed. Moitt (Gainesville, Fla., forthcoming 2004); Sitterson, "Minor Plantations," 216–24. For a discussion of the way planters perceived their loss of mastery, see James L. Roark, *Masters without Slaves: Southern Planters in the Civil War and Reconstruction* (New York, 1977), 68–108. On the way planters reestablished their power and control in the sugar country, see Rebecca J. Scott, "Defining the Boundaries of Freedom in the World of Cane: Cuba, Brazil, and Louisiana after Emancipation," *AHR* 99 (February 1994): 76–78.

where commercial success rested on the savage, though profitable, combination of slave labor and plantation capitalism. Frighteningly exploitative and ruthlessly conducted, agro-industry in the cane world hinged on manipulative management and upon the slaves' tolerance of the business regime. African Americans utilized the new expectations and duties within slavery to advance their own agenda, but by accommodating the planters, black Louisianans inadvertently and tragically fortified the sugar masters' power and sway.

PRIMARY SOURCES

MANUSCRIPT COLLECTIONS

BARKER TEXAS HISTORICAL CENTER, UNIVERSITY OF TEXAS LIBRARIES, AUSTIN

Microfilm: Kenneth M. Stampp, gen. ed., *Records of Ante-Bellum Southern Plantations: From the Revolution through the Civil War*, ser. G, pt. 1: *Texas and Louisiana Collections* (Frederick, Md., 1987).
Pugh Family Papers

HOWARD-TILTON MEMORIAL LIBRARY, TULANE UNIVERSITY, NEW ORLEANS

Microfilm: Kenneth M. Stampp, gen. ed., *Records of Ante-Bellum Southern Plantations: From the Revolution through the Civil War*, ser. H (Frederick, Md., 1987).
Robert Ruffin Barrow Papers
Citizens Bank of Louisiana Minute Books and Records
Octave Colomb Plantation Journal
De la Vergne Family Papers (Hughes Lavergne Letterbooks)
Andrew Durnford Plantation Journal
Eugene Forstall Letterbooks
John McDonogh Estate Inventory
John McDonogh Papers
Jean Baptiste Meullion Papers
John Minor Wisdom Collection (John McDonogh Series)

LOUISIANA AND LOWER MISSISSIPPI VALLEY COLLECTIONS, HILL MEMORIAL LIBRARY, LOUISIANA STATE UNIVERSITY, BATON ROUGE

William Acy Jr. Papers
Michel Thomassin Andry and Family Papers
Anonymous Plantation Ledger
Ashland Plantation Book
Ashton Plantation Auction Broadside

Charles Barbier Papers

Albert A. Batchelor Papers

Boucry Family Papers

J. P. Bowman Family Papers

Louis Bringier and Family Papers

Bruce, Seddon, and Wilkins Plantation Records

Thomas Butler and Family Papers

Thomas W. Butler Papers

Andrew E. Crane Papers

Eugenie Dardenne Document

Alexander E. DeClouet and Family Papers

Stephen Duncan and Stephen Duncan Jr. Papers

James E. Elam Letterbook

Isaac Erwin Diary

Edward J. Gay and Family Papers

Joseph Girod Papers

William S. Hamilton Papers

Daniel and Philip Hicky Papers

Mrs. Isaac H. Hilliard Diary

Patrick F. Keary Letters

Duncan Farrar Kenner Papers

Kenner Family Papers

Joseph Kleinpeter and Family Papers

George Lanaux and Family Papers

Elu Landry Estate Plantation Diary and Ledger

Severin Landry and Family Papers

Landry Family Papers

A. Ledoux and Company Record Book

Charles P. Leverich Correspondence

Moses and St. John Richardson Liddell Family Papers

George Mather Account Books

Charles L. Mathews Family Papers

Samuel McCutchon Papers

William J. Minor and Family Papers

Thomas O. Moore Papers

John Moore and Family Papers

John D. Murrell Papers

Palfrey Family Papers

Alexander Franklin Pugh Papers

William W. Pugh and Family Plantation Records
John H. Randolph Papers
John H. Ransdell Papers
Reggio Family Papers
Frederick D. Robertson Account Books
Jared Young Sanders Family Papers
H. M. Seale Diary
Lewis Stirling and Family Papers
Charles William Taussig Collection
Miles Taylor Family Papers
Benjamin Tureaud Family Papers
Turnbull-Allain Family Papers
Turnbull-Bowman Family Papers
Turnbull-Bowman-Lyons Family Papers
Uncle Sam Plantation Papers
Levin Wailes Letter
David Weeks and Family Papers
William P. Welham Plantation Records
William Webb Wilkins Papers
WPA Ex-Slave Narrative Collection

LOUISIANA STATE MUSEUM, NEW ORLEANS

Microfilm: Kenneth M. Stampp, gen. ed., Records of Ante-Bellum Southern Plantations: From the Revolution through the Civil War, ser. H (Frederick, Md., 1987).
Valcour Aime Slave Records
Nicholas Bauer Collection (John McDonogh)
John McDonogh Papers
Ross/Stackhouse Records
Ste. Sophie/Live Oak Plantation Records

SOUTHERN HISTORICAL COLLECTION, MANUSCRIPTS DEPARTMENT, WILSON LIBRARY, UNIVERSITY OF NORTH CAROLINA, CHAPEL HILL

Microfilm: Kenneth M. Stampp, gen. ed., Records of Ante-Bellum Southern Plantations: From the Revolution through the Civil War, ser. J, pt. 5: Louisiana (Frederick, Md., 1987).
Avery Family Papers
Robert Ruffin Barrow Papers
Bayside Plantation Records
John Boyd Diary
Brashear and Lawrence Family Papers

Caffery Family Papers
John G. Devereux Papers
Evan Hall Plantation Book
Franklin A. Hudson Diaries
Jackson, Riddle, and Company Papers
Joseph W. Lyman Letter
Andrew McCollam Papers
William A. Shaffer Papers
Simpson and Brumby Family Papers
Slack Family Papers
Maunsell White Papers
Trist Wood Papers

PRINTED SOURCES

CONTEMPORARY PERIODICALS AND OTHER PUBLICATIONS

American Agriculturist, 1842–1860
American Cotton Planter and Soil of the South, 1858–1860
American Farmer, 1849–1860
American Farmer's Magazine, 1858–59
American Journal of Agriculture and Science, 1847
American Railroad Journal, 1849–1860
Champomier, P. A. *Statement of the Sugar Crop Made in Louisiana*, 1844–1846, 1849–1861.
De Bow's Review, 1846–1860
Farmer's Cabinet and American Herd Book. 1840–1848
Farmer's Register, 1833–1842
Harper's New Monthly Magazine, 1850–1860
Harper's Weekly, 1857–1860
Hunt's Merchants' Magazine and Commercial Review, 1848–1860
Louisiana Planter and Sugar Manufacturer, 1888.
Monthly Journal of Agriculture, 1846–1848
The Plough, the Loom, and the Anvil, 1848–1857
Scientific American, 1848–1849, 1850–1860
Soil of the South, 1852–1856
Southern Cultivator, 1843–1855
Southern Planter, 1841–1860
Southern Agriculturist, 1853

LOCAL NEWSPAPERS

Alexandria Planter's Intelligencer and Rapides Advertiser, 1834–1836
Baton Rouge Democratic Advocate, 1845–1848, 1850–1852, 1854–1855

Baton Rouge Gazette, 1827–1835, 1837–1853

Baton Rouge Weekly Advocate, 1855–1860

Baton Rouge Weekly Comet, 1853–1855

Franklin Planters' Banner, 1849–1854

Franklin Planters' Banner and Louisiana Agriculturist, 1842–1848

Houma Ceres, 1855–1858, 1860

Lucy, St. John the Baptist L'Avant Coureur, 1854–1856, 1858–1860

Napoleonville Pioneer de L'Assomption, 1850–1855

New Orleans Bee, 1835–1860

New Orleans Weekly Delta, 1846–1853

New Orleans Weekly Picayune, 1838–1847, 1850, 1852–1856

New Roads Pointe Coupee Democrat, 1858–1860

New Roads Pointe Coupee Echo, 1848–1849

Opelousas St. Landry Whig, 1844–1846

Plaquemine Gazette and Sentinel, 1858–1860

Plaquemine Southern Sentinel, 1848–1858

Port Allen Capitolian Vis-à-Vis, 1852–1854

Port Allen Sugar Planter, 1856–1860

St. Francisville Louisiana Chronicle, 1838, 1841–1844

St. Martinsville Attakapas Gazette, 1824–1826

Thibodaux Lafourche Gazette, 1844–1845

Thibodaux Minerva, 1853–1856

TRAVELERS' ACCOUNTS, MEMOIRS, COMMENTARIES, AND PUBLISHED
CORRESPONDENCE

Abdy, Edward S. *Journal of a Residence and Tour in the United States of North America, from April 1833 to October 1834*. London: J. Murray, 1835.

Aime, Valcour. *Plantation Diary of the Late Valcour Aime*. New Orleans: Clark and Hofeline, 1878.

Ampere, J. J. *Promenade en Amerique; États-Unis—Cuba—Mexique*. Paris: M. Levy, 1855.

Ashe, Thomas. *Travels in America, Performed in the Year 1806, for the Purpose of Exploring the Rivers Allegheny, Monongahela, Ohio, and Mississippi, and Ascertaining the Produce and Condition of Their Banks and Vicinity*. London: R. Phillips, 1809.

Baird, Robert. *Impressions and Experiences of the West Indies and North America in 1849*. Philadelphia: Lea and Blanchard, 1850.

———. *View of the Valley of the Mississippi; or, The Emigrant's and Traveller's Guide to the West*. Philadelphia: H. S. Tanner, 1832.

Benwell, J. *An Englishman's Travels in America: His Observations of Life and Manners in the Free and Slave States*. London: Binns and Goodwin, 1853.

Berlin, Ira, et al., eds. *Freedom: A Documentary History of Emancipation, 1861–1867*. Ser. 1, vol. 3, *The Wartime Genesis of Free Labor: The Lower South.* Cambridge: Cambridge University Press, 1990.

Berquin-Duvallon, Pierre Louis. *Travels in Louisiana and the Floridas in the Year 1802, Giving a Correct Picture of Those Countries.* Translated by John Davis. New York: I. Riley, 1806.

Blassingame, John W., ed. *Slave Testimony: Two Centuries of Letters, Speeches, Interviews, and Autobiographies.* Baton Rouge: Louisiana State University Press, 1977.

Botkin, B. A., ed. *Lay My Burden Down: A Folk History of Slavery.* Chicago: University of Chicago Press, 1945.

Brackenridge, Henry Marie. *Views of Louisiana, Together with a Journal of a Voyage up the Mississippi River in 1811.* 1814. Reprint, Chicago: Quadrangle Books, 1962.

Breeden, James O., ed. *Advice among Masters: The Ideal in Slave Management in the Old South.* Westport, Conn.: Greenwood Press, 1980.

Bremer, Frederika. *The Homes of the New World; Impressions of America.* New York: Harper and Brothers, 1854.

Claiborne, W. C. C. *Official Letter Books.* Edited by Dunbar Rowland. Jackson, Miss.: State Department of Archives and History, 1917.

Clay, Henry. *The Private Correspondence of Henry Clay.* Edited by Calvin Colton. New York: A. S. Barnes, 1856.

Collot, Georges-Henri-Victor. *A Journey in North America.* 1826. Reprint, Florence: O. Lange, 1924.

Creecy, James R. *Scenes in the South and Other Late Miscellaneous Pieces.* Washington, D.C.: T. McGill, 1860.

Crete, Liliane. *Daily Life in Louisiana, 1815–1830.* Translated by Patrick Gregory. Baton Rouge: Louisiana State University Press, 1978.

Darby, William. *The Emigrant's Guide to the Western and Southwestern States and Territories: Comprising a Geographical and Statistical Description of the States.* New York: Kirk and Mercein, 1818.

———. *A Geographical Description of the State of Louisiana . . . with an Account of the Character and Manners of the Inhabitants.* Philadelphia: J. Bioren, 1816.

Davis, Edwin A., ed. *Plantation Life in the Florida Parishes of Louisiana, 1836–1846, as Reflected in the Diary of Bennet H. Barrow.* New York: Columbia University Press, 1943.

De Bow, James D. B. *The Industrial Resources, Etc., of the Southern and Western States.* 3 vols. New Orleans: Office of De Bow's Review, 1853.

De Laussat, Pierre Clément. *Memoirs of My Life to My Son during the Years 1803 and After.* Translated by Sister Agnes-Josephina Pastwa and edited by Robert D. Bush. Baton Rouge: Louisiana State University Press, 1978.

Dennett, Daniel. *Louisiana As It Is: Its Topography and Material Resources, Its Cotton, Sugar Cane, Rice and Tobacco Fields, Its Corn and Grain Lands, Climate and People of the State.* New Orleans: Eureka Press, 1876.

Douglass, Frederick. *Narrative of the Life of Frederick Douglass, An American Slave.* Edited by Houston A. Baker Jr. 1845. Reprint, New York: Penguin, 1982.

Dureau, B. *Notice sur la culture de la canne à sucre et sur la fabrication du sucre en Louisiane.* Paris: M. Mathias, 1852.

Evans, W. J. *The Sugar Planter's Manual, Being a Treatise on the Art of Obtaining Sugar from the Sugar Cane.* Philadelphia: Lea and Blanchard, 1848.

Everest, Robert. *A Journey through the United States and Part of Canada.* London: J. Chapman, 1855.

Ewell, James. *The Planter's and Mariner's Medical Companion.* Philadelphia: John Bioren, 1807.

Featherstonhaugh, G. W. *Excursion through the Slave States, from Washington on the Potomac to the Frontier of Mexico.* London: Murray, 1844.

Flint, Timothy. *The History and Geography of the Mississippi Valley. To Which Is Appended a Condensed Physical Geography of the Atlantic United States and the Whole American Continent.* 2 vols. Cincinnati: E. H. Flint, 1833.

———. *Recollections of the Last Ten Years in the Valley of the Mississippi, Passed in Occasional Residences and Journeyings in the Valley of the Mississippi from Pittsburg and the Missouri to the Gulf of Mexico, and from Florida to the Spanish Frontier.* Edited by George R. Brooks. 1826. Reprint, Carbondale: Southern Illinois University Press, 1968.

Forstall, Edward. *Agricultural Productions of Louisiana, Embracing Valuable Information Relative to the Cotton, Sugar, and Molasses Interests, and the Effects upon the Same of the Tariff of 1842.* New Orleans: Tropic Print, 1845.

Fortier, Alcée. *Louisiana Studies: Literature, Customs, and Dialects, History and Education.* New Orleans: F. F. Hansell and Brothers, 1894.

Foster, Lillian. *Way-Side Glimpses, North and South.* New York: Rudd and Carleton, 1860.

Gayarré, Charles. *History of Louisiana.* 4th ed. 4 vols. 1854. Reprint, New Orleans: F. F. Hansell and Brothers, 1903.

Hall, A. Oakey. *The Manhattaner in New Orleans; or, Phases of "Crescent City" Life.* New York: J. S. Redfield, 1851.

Hall, Basil. *Travels in North America in the Years 1827 and 1828.* Philadelphia: Carey, Lea, and Carey, 1829.

Heustis, Jabez W. *Physical Observations and Medical Tracts and Remarks, on the Topography and Diseases of Louisiana.* New York: T. and J. Swords, 1817.

Houstoun, Matilda C. *Hesperos, or Travels in the West.* London: J. W. Parker, 1850.

Ingraham, Joseph H. *The South-West, by a Yankee.* 2 vols. New York: Harper and Brothers, 1835.

Johnston, J. S. *Letter of Mr. Johnston of Louisiana, to the Secretary of the Treasury, in Reply to his Circular of the 1st July, 1830, Relative to the Culture of the Sugar Cane.* Washington, D.C.: Gales and Seaton, 1831.

Kingsford, William. *Impressions of the West and South during a Six Weeks' Holiday.* Toronto: A. H. Armour, 1858.

Lanman, Charles. *Adventures in the Wilds of the United States and British American Provinces.* 2 vols. Philadelphia: J. W. Moore, 1856.

Leon, J. A. *On Sugar Cultivation in Louisiana, Cuba, Etc., and the British Possessions.* London: J. Ollivier, 1848.

Lyell, Charles. *A Second Visit to the United States of North America.* 2 vols. New York: Harper and Bros., 1849.

Martineau, Harriet. *Retrospect of Western Travel.* 2 vols. London: Saunders and Otley, 1838.

Montgomery, Cora. *The Queen of Islands and the King of Rivers.* New York: C. Wood, 1850.

Murray, Amelia. *Letters from the United States, Cuba, and Canada.* New York: G. P. Putnam, 1856.

Nichols, Thomas Low. *Forty Years of American Life.* 1864. Reprint, New York: Stackpole, 1937.

Northup, Solomon. *Twelve Years a Slave.* Edited by Sue Eakin and Joseph Logsdon. 1853. Reprint, Baton Rouge: Louisiana State University Press, 1968.

Nutall, Thomas. *A Journey of Travels into the Arkansas Territory during the Year 1819. with Occasional Observations on the Manners of the Aborigines.* Philadelphia: Thomas A. Palmer, 1821.

Olmsted, Frederick Law. *A Journey in the Back Country in the Winter of 1853–54.* 1860. Reprint, New York: G. P. Putnam's Sons, 1907.

———. *A Journey in the Seaboard Slave States in the Years 1853–1854.* 2 vols. 1856. Reprint, New York: G. P. Putnam's Sons, 1904.

Parker, Amos A. *Trip to the West and Texas. Comprising a Journey of Eight Thousand Miles through New York, Michigan, Illinois, Missouri, Louisiana, and Texas, in the Autumn and Winter of 1834–35.* Concord, Mass.: White and Fisher, 1835.

Pierce, George. *Incidents of Western Travel, in a Series of Letters.* Nashville: E. Stevenson and F. A. Owen, 1857.

Pitot, James. *Observations on the Colony of Louisiana from 1796 to 1802.* Translated by Henry C. Pitot and edited by Robert D. Bush. Baton Rouge: Louisiana State University Press, 1979.

Porter, George Richardson. *The Nature and Properties of the Sugar Cane, with Practical Directions for the Improvement of Its Culture, and the Manufacture of Its Products.* London: Smith, Elder, 1830.

Power, Tyrone. *Impressions of America during the Years 1833, 1834, 1835.* Philadelphia: Carey, Lea, and Blanchard, 1836.

Pulszky, Francis and Theresa. *White, Red, Black: Sketches of American Society in the United States during the Visit of Their Guest.* 2 vols. New York: Redfield, 1853.

Reed, William, *The History of Sugar and Sugar-Yielding Plants, Together with an Epitome of*

Every Notable Process of Sugar Extraction from the Earliest Times to the Present. London: Longman, Green, 1866.

Richardson, F. D. "The Teche Country Fifty Years Ago." *Southern Bivouac* 4 (March 1886): 593–98.

Robertson, James. *A Few Months in America: Containing Remarks on Some of Its Industrial and Commercial Interests.* London: Longman, 1855.

Robin, Claude C. *Voyage to Louisiana, 1803–1805.* Translated by Stuart O. Landry Jr. 1807. Reprint, New Orleans: Pelican, 1966.

Robinson, Solon. *Solon Robinson: Pioneer and Agriculturist.* Edited by Herbert Anthony Kellar. Indianapolis: Indiana Historical Bureau, 1936.

Rose, Willie Lee, ed. *A Documentary History of Slavery in North America.* New York: Oxford University Press, 1976.

Russell, Robert. *North America, Its Agriculture and Climate; Containing Observations on the Agriculture and Climate of Canada, the United States, and the Island of Cuba.* Edinburgh: A. and C. Black, 1857.

Russell, William Howard. *My Diary North and South.* Boston: T.O.H.P. Burnham, 1863.

Silliman, Benjamin. *Manual on the Cultivation of the Sugar Cane, and the Fabrication and Refinement of Sugar.* Washington, D.C.: F. P. Blair, 1833.

Stirling, James. *Letters from the Slave States.* London: J. W. Parker, 1857.

Stroyer, Jacob. *My Life in the South.* Salem, Mass.: Salem Observer Book and Job Print, 1885.

Taylor, Bayard. *Eldorado, or Adventures in the Path of Empire: Comprising a Voyage to California, via Panama; Life in San Francisco and Monterey; Pictures of the Gold Region; and Experiences of Mexican Travel.* New York: G. P. Putnam, 1860.

Thomassy, Marie Joseph. *Géologie pratique de la Louisiane.* Paris: Lacrois et Baudry, 1860.

Tixier, Victor. *Travels on the Osage Prairies.* Translated by Albert J. Salvan and edited by John Francis McDermott. 1844. Reprint, Norman: University of Oklahoma Press, 1940.

Wailes, B. L. C. *Report on the Agriculture and Geology of Mississippi, Embracing a Sketch of the Social and Natural History of the State.* Jackson, Miss.: E. Barksdale, 1854.

Warden, D. B. *A Statistical, Political, and Historical Account of the United States of North America from the Period of the First Colonization to the Present Day.* Edinburgh: Archibald Constable, 1819.

Webb, Allie Bayne Windham, ed. *Mistress of Evergreen Plantation: Rachel O'Connor's Legacy of Letters, 1823–1845.* Albany: State University of New York Press, 1983.

Welby, Victoria. *A Young Traveller's Journal of a Tour in North and South America during the Year 1850.* London: T. Bosworth, 1852.

Weld, Theodore. *Slavery As It Is: Testimony of a Thousand Witnesses.* New York: American Anti-Slavery Society, 1839.

Wortley, Emmeline Stuart. *Travels in the United States during 1849 and 1850*. London: R. Bentley, 1850.

GOVERNMENT DOCUMENTS AND PUBLICATIONS

U.S. Bureau of the Census. *Agriculture in the United States, 1860: Compiled from the Original Returns of the Eighth Census*. Washington, D.C.: Government Printing Office, 1864.

————. *Compendium of the Enumeration of the Inhabitants and States of the United States Obtained at the Department of State from the Returns of the Sixth Census*. Washington, D.C.: Thomas Allen, 1841.

————. *Eighth Census of the United States (1860)*. New York: J. H. Colton, 1861.

————. *Population of the United States in 1860: Compiled from the Original Returns of the Eighth Census*. Washington, D.C.: Government Printing Office, 1864.

————. *Seventh Census of the United States, 1850*. Washington, D.C.: Robert Armstrong, 1853.

U.S. Bureau of the Census. Manuscript Returns. *Agricultural Schedule*, 1850, 1860. Ascension, St. Mary, and Terrebonne Parishes, Louisiana. (microfilm).

————. *Population Schedule*, 1850, 1860. Ascension, St. Mary, and Terrebonne Parishes, Louisiana. (microfilm).

————. *Slave Schedules*, 1850, 1860. Ascension, St. Mary, and Terrebonne Parishes, Louisiana. (microfilm).

U.S. Congress. Senate. *Report of the Secretary of the Treasury: Investigations in Relation to Cane Sugar: A Report of Scientific Investigations Relative to the Chemical Nature of Saccharine Substances, and the Art of Manufacturing Sugar. Made under the Direction of Professor A. D. Bache by Professor R. S. McCulloh*. 29th Cong., 2nd sess. Senate Doc. No. 209. Washington, D.C.: Ritchie and Heiss, 1847.

U.S. Patent Office. *Annual Report of the Commissioner of Patents for the Year 1848*. 30th Cong., 2nd sess., House of Representatives Doc. No. 59. Washington, D.C.: Wendell and Van Benthuysen, 1849.

————. *Annual Report of the Commissioner of Patents for the Year 1858*. 35th Cong., 2nd sess., House of Representatives Doc. No. 105. Washington, D.C.: James B. Steedman, 1859.

SECONDARY SOURCES

BOOKS

Abrahams, Roger D. *Singing the Master: The Emergence of African American Culture in the Plantation South*. New York: Penguin, 1993.

Alston, Lee J., and Joseph P. Ferrie. *Southern Paternalism and the American Welfare State: Economics, Politics, and Institutions in the South, 1865–1965*. Cambridge: Cambridge University Press, 1999.

Aptheker, Herbert. *American Negro Slave Revolts.* New York: International Publishers, 1943.

Babson, David W. *Pillars on the Levee: Archeological Investigations at Ashland–Belle Helene Plantation, Geismar, Ascension Parish, Louisiana.* Normal: Midwestern Archaeological Research Center, Illinois State University, 1989.

Bacot, H. Parrott, et al. *Marie Adrien Persac: Louisiana Artist.* Baton Rouge: Louisiana State University Press, 2000.

Bakhtin, M. M. *The Dialogic Imagination: Four Essays.* Translated by Caryl Emerson and Michael Holquist and edited by Michael Holquist. Austin: University of Texas Press, 1981.

Bankole, Katherine. *Slavery and Medicine: Enslavement and Medical Practices in Antebellum Louisiana.* New York: Garland, 1998.

Baptist, Edward E. *Creating an Old South: Middle Florida's Plantation Frontier before the Civil War.* Chapel Hill: University of North Carolina Press, 2002.

Bardaglio, Peter W. *Reconstructing the Household: Families, Sex, and the Law in the Nineteenth-Century South.* Chapel Hill: University of North Carolina Press, 1995.

Barnes, A. C. *The Sugar Cane.* New York: John Wiley, 1974.

Bateman, Fred, and Thomas Weiss. *A Deplorable Scarcity: The Failure of Industrialization in the Slave Economy.* Chapel Hill: University of North Carolina, 1981.

Bauer, Craig A. *A Leader among Peers: The Life and Times of Duncan Farrar Kenner.* Lafayette: University of Southwestern Louisiana Press, 1993.

Becnel, Thomas A. *The Barrow Family and the Barataria and Lafourche Canal: The Transportation Revolution in Louisiana, 1829–1925.* Baton Rouge: Louisiana State University Press, 1989.

Bergad, Laird W. *Cuban Rural Society in the Nineteenth Century: The Social and Economic History of Monoculture in Matanzas.* Princeton, N.J.: Princeton University Press, 1990.

Berlin, Ira. *Generations of Captivity: A History of African-American Slaves.* Cambridge, Mass.: Harvard University Press, 2003.

———. *Many Thousands Gone: The First Two Centuries of Slavery in North America.* New York: W. W. Norton, 1998.

Blackburn, Frank. *Sugar-Cane.* London: Longman, 1984.

Blassingame, John. *Black New Orleans, 1860–1880.* Chicago: University of Chicago Press, 1973.

———. *The Slave Community: Plantation Life in the Antebellum South.* Rev. ed. New York: Oxford University Press, 1979.

Blewett, Mary. *Men, Women, and Work: Class, Gender, and Protest in the New England Shoe Industry.* Urbana: University of Illinois Press, 1988.

Bodenhorn, Howard. *A History of Banking in Antebellum America: Financial Development in an Era of Nation-Building.* Cambridge: Cambridge University Press, 2000.

Bowman, Shearer Davis. *Masters and Lords: Mid-Nineteenth-Century U.S. Planters and Prussian Junkers.* New York: Oxford University Press, 1993.

Bush, Barbara. *Slave Women in Caribbean Society, 1650–1838.* Bloomington: Indiana University Press, 1990.

Butler, W. E. *Down among the Sugar Cane: The Story of Louisiana Sugar Plantations and Their Railroads.* Baton Rouge: Moran, 1981.

Bynum, Victoria E. *Unruly Women: The Politics of Social and Sexual Control in the Old South.* Chapel Hill: University of North Carolina Press, 1992.

Caldwell, Stephen A. *A Banking History of Louisiana.* Baton Rouge: Louisiana State University Press, 1935.

Carlton, David L. *Mill and Town in South Carolina, 1880–1920.* Baton Rouge: Louisiana State University Press, 1982.

Carney, Judith A. *Black Rice: The African Origins of Rice Cultivation in the Americas.* Cambridge, Mass.: Harvard University Press, 2001.

Cecelski, David S. *The Waterman's Song: Slavery and Freedom in Maritime North Carolina.* Chapel Hill: University of North Carolina Press, 2001.

Chaplin, Joyce E. *An Anxious Pursuit: Agricultural Innovation and Modernity in the Lower South, 1730–1815.* Chapel Hill: University of North Carolina Press, 1993.

Chernoff, John Miller. *African Rhythm and African Sensibility: Aesthetics and Social Action in African Musical Idioms.* Chicago: University of Chicago Press, 1977.

Clayton, Ronnie W. *Mother Wit: The Ex-Slave Narratives of the Louisiana Writers' Project.* New York: Peter Lang, 1990.

Coclanis, Peter A. *The Shadow of a Dream: Economic Life and Death in the South Carolina Low Country, 1670–1920.* New York: Oxford University Press, 1989.

Cole, Arthur Harrison. *Wholesale Commodity Prices in the United States, 1700–1861.* Cambridge, Mass.: Harvard University Press, 1938.

Conrad, Glenn R. *Green Fields: Two Hundred Years of Louisiana Sugar.* Lafayette: Center for Louisiana Studies, University of Southwestern Louisiana Press, 1980.

———, and Ray F. Lucas. *White Gold: A Brief History of the Louisiana Sugar Industry, 1795–1995.* Louisiana Life Series, No. 8. Lafayette: Center for Louisiana Studies, University of Southwestern Louisiana Press, 1995.

Cooper, Frederick, Thomas C. Holt, and Rebecca J. Scott. *Beyond Slavery: Explorations of Race, Labor, and Citizenship in Postemancipation Societies.* Chapel Hill: University of North Carolina Press, 2000.

Coulter, E. Merton. *Thomas Spalding of Sapelo.* Baton Rouge: Louisiana State University Press, 1940.

Craton, Michael. *Searching for the Invisible Man: Slaves and Plantation Life in Jamaica.* Cambridge, Mass.: Harvard University Press, 1978.

Craven, Avery O. *Rachel of Old Louisiana.* Baton Rouge: Louisiana State University Press, 1975.

Creel, Margaret Washington. *"A Peculiar People": Slave Religion and Community Culture among the Gullahs.* New York: New York University Press, 1988.

Davis, David Brion. *Slavery and Human Progress.* New York: Oxford University Press, 1984.

Deerr, Noël. *Cane Sugar: A Textbook on the Agriculture of Sugar Cane, the Manufacture of Cane Sugar, and the Analysis of Sugar House Products.* London: Norman Rodger, 1921.

———. *The History of Sugar.* 2 vols. London: Chapman and Hall, 1950.

Dew, Charles B. *Bond of Iron: Master and Slave at Buffalo Forge.* New York: W. W. Norton, 1994.

———. *Ironmaker to the Confederacy: Joseph R. Anderson and the Tredegar Iron Works.* New Haven, Conn.: Yale University Press, 1966.

Dublin, Thomas. *Women at Work: The Transformation of Work and Community in Lowell, Massachusetts, 1826–1860.* New York: Oxford University Press, 1979.

Duffy, John, ed. *The Rudolph Matas History of Medicine in Louisiana.* 2 vols. Baton Rouge: Louisiana State University Press, 1958.

Dunaway, Wilma A. *The African-American Family in Slavery and Emancipation.* Cambridge: Cambridge University Press, 2003.

———. *The First American Frontier: Transition to Capitalism in Southern Appalachia, 1700–1860.* Chapel Hill: University of North Carolina Press, 1996.

Dunn, Richard S. *Sugar and Slaves: The Rise of the Planter Class in the English West Indies, 1624–1713.* New York: W. W. Norton, 1972.

Dusinberre, William. *Slavemaster President: The Double Career of James Polk.* New York: Oxford University Press, 2003.

———. *Them Dark Days: Slavery in the American Rice Swamps.* New York: Oxford University Press, 1996.

Earle, Carville. *Geographical Inquiry and American Historical Problems.* Stanford, Calif.: Stanford University Press, 1992.

Epstein, Dena J. *Sinful Tunes and Spirituals: Black Folk Music to the Civil War.* Urbana: University of Illinois Press, 1977.

Evans, Curtis J. *The Conquest of Labor: Daniel Pratt and Southern Industrialization.* Baton Rouge: Louisiana State University Press, 2001.

Faust, Drew Gilpin. *James Henry Hammond and the Old South: A Design for Mastery.* Baton Rouge: Louisiana State University Press, 1982.

Fett, Sharla M. *Working Cures: Healing, Health, and Power on Southern Slave Plantations.* Chapel Hill: University of North Carolina Press, 2002.

Flamming, Douglas. *Creating the Modern South: Masters and Millhands in Dalton, Georgia, 1884–1894.* Chapel Hill: University of North Carolina Press, 1992.

Fleming, Walter L. *Louisiana State University, 1860–1986*. Baton Rouge: Louisiana State University Press, 1936.

Fogel, Robert William. *Without Consent or Contract: The Rise and Fall of American Slavery*. New York: W. W. Norton, 1989.

———, and Stanley L. Engerman. *Time on the Cross: The Economics of American Negro Slavery*. New York: W. W. Norton, 1974.

Foner, Eric. *Nothing But Freedom: Emancipation and Its Legacy*. Baton Rouge: Louisiana State University Press, 1983.

Foster, Helen Bradley. *"New Raiments of Self": African American Clothing in the Antebellum South*. Oxford: Berg, 1997.

Fox-Genovese, Elizabeth. *Within the Plantation Household: Black and White Women of the Old South*. Chapel Hill: University of North Carolina Press, 1988.

Franklin, John Hope, and Loren Schweninger. *Runaway Slaves: Rebels on the Plantation*. New York: Oxford University Press, 1999.

Frederickson, George M. *The Black Image in the White Mind: The Debate on Afro-American Character and Destiny, 1817–1914*. New York: Oxford University Press, 1971.

Freehling, William W. *Road to Disunion: Secessionists at Bay, 1776–1854*. New York: Oxford University Press, 1990.

Frey, Sylvia R., and Betty Wood. *Come Shouting to Zion: African American Protestantism in the American South and British Caribbean to 1830*. Chapel Hill: University of North Carolina Press, 1998.

Frisch, Rose E. *Female Fertility and the Body Fat Connection*. Chicago: University of Chicago Press, 2002.

Gartman, David. *Auto Slavery: The Labor Process in the American Automobile Industry, 1897–1950*. New Brunswick, N.J.: Rutgers University Press, 1989.

Genovese, Eugene D. *A Consuming Fire: The Fall of the Confederacy in the Mind of the White Christian South*. Athens: University of Georgia Press, 1998.

———. *The Political Economy of Slavery: Studies in the Economy and Society of the Slave South*. New York: Random House, 1965.

———. *Roll, Jordan, Roll: The World the Slaves Made*. New York: Vintage, 1972.

———. *The Slaveholders' Dilemma: Freedom and Progress in Southern Conservative Thought, 1820–1860*. Columbia: University of South Carolina Press, 1992.

———. *The World the Slaveholders Made: Two Essays in Interpretation*. New York: Vintage, 1969.

———, and Elizabeth Fox-Genovese. *Fruits of Merchant Capital: Slavery and Bourgeois Property in the Rise and Expansion of Capitalism*. New York: Oxford University Press, 1983.

Giedion, Siegfried. *Mechanization Takes Command: A Contribution to Anonymous History*. New York: Oxford University Press, 1948.

Gilroy, Paul. *Small Acts: Thoughts on the Politics of Black Cultures.* London: Serpent's Tail, 1993.

Gleeson, David T. *The Irish in the South, 1815–1877.* Chapel Hill: University of North Carolina Press, 2001.

Glickman, Lawrence B. *A Living Wage: American Workers and the Making of Consumer Society.* Ithaca, N.Y.: Cornell University Press, 1997.

Gomez, Michael A. *Exchanging Our Country Marks: The Transformation of African Identities in the Colonial and Antebellum South.* Chapel Hill: University of North Carolina Press, 1998.

Gray, Lewis Cecil. *History of Agriculture in the Southern United States to 1860.* 2 vols. Washington, D.C.: Carnegie Institution of Washington, 1933.

Green, George D. *Finance and Economic Development in the Old South: Louisiana Banking, 1804–1861.* Stanford, Calif.: Stanford University Press, 1972.

Greenberg, Kenneth S. *Masters and Statesmen: The Political Culture of American Slavery.* Baltimore: Johns Hopkins University Press, 1985.

Gross, Ariela J. *Double Character: Slavery and Mastery in the Antebellum Southern Courtroom.* Princeton, N.J.: Princeton University Press, 2000.

Guillaumin, Colette. *Racism, Sexism, Power, and Ideology.* London: Routledge, 1995.

Gutman, Herbert G. *The Black Family in Slavery and Freedom, 1750–1925.* New York: Vintage, 1976.

Hadden, Sally E. *Slave Patrols: Law and Violence in Virginia and the Carolinas.* Cambridge, Mass.: Harvard University Press, 2001.

Hall, Gwendolyn Midlo. *Africans in Colonial Louisiana: The Development of Afro-Creole Culture in the Eighteenth Century.* Baton Rouge: Louisiana State University Press, 1992.

Heitmann, John Alfred. *The Modernization of the Louisiana Sugar Industry, 1830–1910.* Baton Rouge: Louisiana State University Press, 1987.

Heyrman, Christine Leigh. *Southern Cross: The Beginnings of the Bible Belt.* Chapel Hill: University of North Carolina Press, 1997.

Higman, Barry. *Slave Population and Economy in Jamaica, 1807–1834.* 1976. Reprint, Kingston: University of West Indies Press, 1995.

———. *Slave Populations of the British Caribbean, 1807–1834.* Baltimore: Johns Hopkins University Press, 1984.

Hilliard, Sam B. *Hog Meat and Hoe Cake: Food Supply in the Old South, 1840–1860.* Carbondale: Southern Illinois University Press, 1972.

Hodes, Martha. *White Women, Black Men: Illicit Sex in the Nineteenth-Century South.* New Haven, Conn.: Yale University Press, 1997.

Hoffert, Sylvia D. *Private Matters: American Attitudes toward Childbearing and Infant Nurture in the Urban North, 1800–1860.* Urbana: University of Illinois Press, 1989.

Horton, Robin. *Patterns of Thought in Africa and the West: Essays on Magic, Religion, and Science.* Cambridge: Cambridge University Press, 1993.

Hounshell, David A. *From the American System to Mass Production, 1800–1932.* Baltimore: Johns Hopkins University Press, 1984.

Hudson, Larry E., Jr. *To Have and To Hold; Slave Work and Family Life in Antebellum South Carolina.* Athens: University of Georgia Press, 1997.

Hunter, Tera W. *To 'Joy My Freedom: Southern Black Women's Lives and Labors after the Civil War.* Cambridge, Mass.: Harvard University Press, 1997.

Hurt, R. Douglas. *American Farm Tools: From Hand-Power to Steam-Power.* Manhattan, Kans.: Sunflower University Press, 1982.

Ingersoll, Thomas N. *Mammon and Manon in Early New Orleans: The First Slave Society in the Deep South, 1718–1819.* Knoxville: University of Tennessee Press, 1999.

Isaac, Rhys. *The Transformation of Virginia, 1740–1790.* New York: W. W. Norton, 1982.

Jaynes, Gerald. *Branches without Roots: Genesis of the Black Working Class in the American South, 1862–1882.* New York: Oxford University Press, 1986.

Johnson, Walter. *Soul by Soul: Life inside the Antebellum Slave Market.* Cambridge, Mass.: Harvard University Press, 1999.

Jones, Norrece T. *Born a Child of Freedom, Yet a Slave: Mechanisms of Control and Strategies of Resistance in Antebellum South Carolina.* Middletown, Conn.: Wesleyan University Press, 1990.

Joyner, Charles. *Down by the Riverside: A South Carolina Slave Community.* Urbana: University of Illinois Press, 1984.

———. *Shared Traditions: Southern History and Folk Culture.* Urbana: University of Illinois Press, 1999.

Kale, Madhavi. *Fragments of Empire: Capital, Slavery, and Indian Indentured Labor Migration in the British Caribbean.* Philadelphia: University of Pennsylvania Press, 1998.

Kilbourne, Richard H., Jr. *Debt, Investment, Slaves: Credit Relations in East Feliciana Parish, Louisiana, 1825–1885.* Tuscaloosa: University of Alabama Press, 1995.

King, Norman J. *Manual of Cane-Growing.* New York: Elsevier, 1965.

King, Wilma. *Stolen Childhood: Slave Youth in Nineteenth-Century America.* Bloomington: Indiana University Press, 1995.

Kniffen, Fred B. *Louisiana: Its Land and People.* Baton Rouge: Louisiana State University Press, 1968.

Knight, Franklin W. *Slave Society in Cuba during the Nineteenth Century.* Madison: University of Wisconsin Press, 1970.

Kolchin, Peter. *American Slavery, 1619–1877.* New York: Hill and Wang, 1993.

———. *A Sphinx on the Land: The Nineteenth-Century South in Comparative Perspective.* Baton Rouge: Louisiana State University Press, 2003.

———. *Unfree Labor: American Slavery and Russian Serfdom.* Cambridge, Mass.: Harvard University Press, 1987.

Landes, David. *Revolution in Time: Clocks and the Making of the Modern World.* Cambridge, Mass.: Harvard University Press, 1983.

Lerner, Gerda. *The Creation of Patriarchy.* New York: Oxford University Press, 1986.

Levine, Lawrence W. *Black Culture and Black Consciousness: Afro-American Folk Thought from Slavery to Freedom.* New York: Oxford University Press, 1977.

Lewis, Ronald L. *Coal, Iron, and Slaves: Industrial Slavery in Maryland and Virginia, 1715–1865.* Westport, Conn.: Greenwood Press, 1979.

Lewis, Sinclair. *Cheap and Contented Labor: The Picture of a Southern Mill Town in 1929.* New York: United Textile Workers of America and Women's Trade Union League, 1929.

Licht, Walter. *Industrializing America: The Nineteenth Century.* Baltimore: Johns Hopkins University Press, 1995.

———. *Working for the Railroad: The Organization of Work in Nineteenth-Century America.* Princeton, N.J.: Princeton University Press, 1983.

Loveland, Anne C. *Southern Evangelicals and the Social Order, 1800–1860.* Baton Rouge: Louisiana State University Press, 1980.

Lynch, Kevin. *The Image of the City.* Boston: MIT Press, 1960.

Malone, Ann Patton. *Sweet Chariot: Slave Family and Household Structure in Nineteenth-Century Louisiana.* Chapel Hill: University of North Carolina Press, 1992.

Martinez-Alier, Verena. *Marriage, Class, and Colour in Nineteenth-Century Cuba: A Study of Racial Attitudes and Sexual Values in a Slave Society.* Cambridge: Cambridge University Press, 1974.

May, Robert E. *The Southern Dream of a Caribbean Empire, 1854–1861.* Baton Rouge: Louisiana State University Press, 1973.

Mbiti, John S. *African Religions and Philosophy.* New York: Frederick A. Praeger, 1969.

McDonald, Roderick A. *The Economy and Material Culture of Slaves: Goods and Chattels on the Sugar Plantations of Jamaica and Louisiana.* Baton Rouge: Louisiana State University Press, 1993.

Menn, Joseph Karl. *The Large Slaveholders of Louisiana—1860.* New Orleans: Pelican, 1964.

Miller, James David. *South by Southwest: Planter Emigration and Identity in the Slave South.* Charlottesville: University Press of Virginia, 2002.

Mintz, Sidney W. *Sweetness and Power: The Place of Sugar in Modern History.* New York: Viking, 1985.

Moody, V. Alton. *Slavery on Louisiana Sugar Plantations.* 1924. Reprint, New York: AMS Press, 1976.

Moore, John Hebron. *The Emergence of the Cotton Kingdom in the Old Southwest: Mississippi, 1770–1860.* Baton Rouge: Louisiana State University Press, 1988.

Moreno Fraginals, Manuel. *The Sugar Mill: The Socioeconomic Complex of Sugar in Cuba, 1760–1860.* New York: Monthly Review Press, 1976.

Morris, Christopher. *Becoming Southern: The Evolution of a Way of Life, Warren Country and Vicksburg, Mississippi, 1770–1860.* New York: Oxford University Press, 1995.

Morris, Thomas D. *Southern Slavery and the Law, 1619–1860.* Chapel Hill: University of North Carolina Press, 1996.

Mullin, Michael. *Africa in America: Slave Acculturation and Resistance in the American South and British Caribbean, 1736–1831.* Urbana: University of Illinois Press, 1992.

Niehaus, Earl F. *The Irish in New Orleans, 1800–1860.* Baton Rouge: Louisiana State University Press, 1956.

Oakes, James. *The Ruling Race: A History of American Slaveholders.* New York: Vintage Books, 1982.

———. *Slavery and Freedom: An Interpretation of the Old South.* New York: Alfred A. Knopf, 1990.

Olsen, D. J. *The City as a Work of Art: London, Paris, Vienna.* New Haven, Conn.: Yale University Press, 1986.

Olwell, Robert. *Masters, Slaves, and Subjects: The Culture of Power in the South Carolina Low Country, 1740–1790.* Ithaca, N.Y.: Cornell University Press, 1998.

Owens, Leslie Howard. *This Species of Property: Slave Life and Culture in the Old South.* New York: Oxford University Press, 1976.

Parks, Joseph H. *General Leonidas Polk, C.S.A.: The Fighting Bishop.* Baton Rouge: Louisiana State University Press, 1990.

Patterson, Orlando. *Rituals of Blood: Consequences of Slavery in Two American Centuries.* Washington, D.C.: Civitas Counterpoint, 1998.

Pearson, Mike Parker, and Colin Richards, eds. *Architecture and Order.* London: Routledge, 1994.

Penningroth, Dylan C. *The Claims of Kinfolk: African American Property and Community in the Nineteenth-Century South.* Chapel Hill: University of North Carolina Press, 2003.

Polk, William M. *Leonidas Polk: Bishop and General.* 2 vols. London: Longman, Green, 1915.

Postell, William Dosite. *The Health of Slaves on Southern Plantations.* Baton Rouge: Louisiana State University Press, 1951.

Prude, Jonathan. *The Coming of the Industrial Order: Town and Factory Life in Massachusetts, 1810–1860.* Cambridge: Cambridge University Press, 1983.

Raboteau, Albert J. *Slave Religion: The "Invisible Institution" in the Antebellum South.* New York: Oxford University Press, 1978.

Reed, Merl E. *New Orleans and the Railroads: The Struggle for Commercial Empire, 1830–1860.* Baton Rouge: Louisiana State University Press, 1966.

Rehder, John B. *Delta Sugar: Louisiana's Vanishing Plantation Landscape.* Baltimore: Johns Hopkins University Press, 1999.

Reidy, Joseph P. *From Slavery to Agrarian Capitalism in the Cotton Plantation South: Central Georgia, 1800–1880.* Chapel Hill: University of North Carolina Press, 1992.

Rivers, Larry Eugene. *Slavery in Florida: Territorial Days to Emancipation*. Gainesville: University Press of Florida, 2000.

Roark, James L. *Masters without Slaves: Southern Planters in the Civil War and Reconstruction*. New York: W. W. Norton, 1977.

Rodrigue, John C. *Reconstruction in the Cane Fields: From Slavery to Free Labor in Louisiana's Sugar Parishes, 1862–1880*. Baton Rouge: Louisiana State University Press, 2001.

Roediger, David R. *The Wages of Whiteness: Race and the Making of the American Working Class*. Rev. ed. London: Verso, 1999.

Rogin, Leo. *The Introduction of Farm Machinery in Its Relation to the Productivity of Labor in the Agriculture of the United States during the Nineteenth Century*. University of California Publications in Economics, vol. 9. Berkeley: University of California Press, 1931.

Roland, Charles P. *Louisiana Sugar Plantations during the American Civil War*. Leiden: E. J. Brill, 1957.

Rorabaugh, William J. *The Alcoholic Republic: An American Tradition*. New York: Oxford University Press, 1979.

Rose, Anne C. *Victorian America and the Civil War*. Cambridge: Cambridge University Press, 1992.

Rosenberg, Charles E. *The Care of Strangers: The Rise of America's Hospital System*. New York: Basic, 1987.

———. *The Cholera Years: The United States in 1832, 1849, and 1866*. Chicago: University of Chicago Press, 1962.

Rothman, Joshua D. *Notorious in the Neighborhood: Sex and Families across the Color Line in Virginia, 1787–1861*. Chapel Hill: University of North Carolina Press, 2003.

Rozenweig, Roy. *Eight Hours for What We Will: Workers and Leisure in an Industrial City*. Cambridge: Cambridge University Press, 1983.

Ryan, Mary P. *Cradle of the Middle Class: The Family in Oneida County, New York, 1790–1865*. Cambridge: Cambridge University Press, 1981.

Sacher, John. *A Perfect War of Politics: Parties, Politicians, and Democracy in Louisiana, 1824–1861*. Baton Rouge: Louisiana State University Press, 2002.

Savitt, Todd L. *Medicine and Slavery: The Diseases and Health Care of Blacks in Antebellum Virginia*. Urbana: University of Illinois Press, 1978.

Scarborough, William K. *Masters of the Big House: Elite Slaveholders of the Mid-Nineteenth-Century South*. Baton Rouge: Louisiana State University Press, 2003.

———. *The Overseer: Plantation Management in the Old South*. Baton Rouge: Louisiana State University Press, 1966.

Schafer, Judith K. *Becoming Free, Remaining Free: Manumission and Enslavement in New Orleans, 1846–1862*. Baton Rouge: Louisiana State University Press, 2003.

———. *Slavery, the Civil Law, and the Supreme Court of Louisiana*. Baton Rouge: Louisiana State University Press, 1994.

Schwalm, Leslie A. *A Hard Fight for We: Women's Transition from Slavery to Freedom in South Carolina.* Urbana: University of Illinois Press, 1997.

Schwartz, Marie Jenkins. *Born in Bondage: Growing Up Enslaved in the Antebellum South.* Cambridge, Mass.: Harvard University Press, 2000.

Schwartz, Stuart B. *Sugar Plantations in the Formation of Brazilian Society: Bahia, 1550–1835.* Cambridge: Cambridge University Press, 1985.

Schweikart, Larry. *Banking in the American South from the Age of Jackson to Reconstruction.* Baton Rouge: Louisiana State University Press, 1987.

Scott, James. *Weapons of the Weak: Everyday Forms of Peasant Resistance.* New Haven, Conn.: Yale University Press, 1985.

Scott, Joan Wallach. *Gender and the Politics of History.* New York: Columbia University Press, 1988.

Scott, Rebecca J. *Slave Emancipation in Cuba: The Transition to Free Labor, 1860–1899.* Princeton, N.J.: Princeton University Press, 1985.

Sheridan, Richard B. *Doctors and Slaves: A Medical and Demographic History of Slavery in the British West Indies, 1680–1834.* Cambridge: Cambridge University Press, 1985.

Shore, Laurence. *Southern Capitalists: Ideological Leadership of an Elite, 1832–1885.* Chapel Hill: University of North Carolina Press, 1986.

Shugg, Roger W. *Origins of Class Struggle in Louisiana: A Social History of White Farmers and Laborers during Slavery and After, 1840–1875.* Baton Rouge: Louisiana State University Press, 1939.

Siegel, Frederick E. *The Roots of Southern Distinctiveness: Tobacco and Society in Danville, Virginia, 1780–1865.* Chapel Hill: University of North Carolina Press, 1987.

Sitterson, J. Carlyle. *Sugar Country: The Cane Sugar Industry in the South, 1753–1950.* Lexington: University of Kentucky Press, 1953.

Smith, Mark M. *Debating Slavery: Economy and Society in the Antebellum American South.* Cambridge: Cambridge University Press, 1998.

———. *Listening to Nineteenth-Century America.* Chapel Hill: University of North Carolina Press, 2001.

———. *Mastered by the Clock: Time, Slavery, and Freedom in the American South.* Chapel Hill: University of North Carolina Press, 1997.

Smith-Rosenberg, Carroll. *Disorderly Conduct: Visions of Gender in Victorian America.* New York: Oxford University Press, 1985.

Stampp, Kenneth. *The Peculiar Institution: Slavery in the Ante-Bellum South.* New York: Vintage, 1956.

Stanley, Amy Dru. *From Bondage to Contract: Wage Labor, Marriage, and the Market in the Age of Slave Emancipation.* Cambridge: Cambridge University Press, 1998.

Starobin, Robert S. *Industrial Slavery in the Old South.* New York: Oxford University Press, 1970.

Stephenson, Wendell Holmes. *Alexander Porter: Whig Planter of the Old South*. Baton Rouge: Louisiana State University Press, 1934.

———. *Isaac Franklin: Slave Trader and Planter of the Old South*. Baton Rouge: Louisiana State University Press, 1938.

Stevenson, Brenda E. *Life in Black and White: Family and Community in the Slave South*. New York: Oxford University Press, 1996.

Stewart, Mart A. *"What Nature Suffers to Groe": Life, Labor, and Landscape on the Georgia Coast, 1680–1920*. Athens: University of Georgia Press, 2002.

Stowe, Steven M. *Intimacy and Power in the Old South: Ritual in the Lives of the Planters*. Baltimore: Johns Hopkins University Press, 1987.

Stuckey, Sterling, *Slave Culture: Nationalist Theory and the Foundations of Black America*. New York: Oxford University Press, 1987.

Tadman, Michael. *Speculators and Slaves: Masters, Traders, and Slaves in the Old South*. Madison: University of Wisconsin Press, 1989.

Tanner, J. M. *Foetus into Man: Physical Growth from Conception to Maturity*. Cambridge, Mass.: Harvard University Press, 1990.

Taylor, Helen. *Circling Dixie: Contemporary Southern Culture through a Transatlantic Lens*. New Brunswick, N.J.: Rutgers University Press, 2001.

Taylor, Joe Gray. *Negro Slavery in Louisiana*. Baton Rouge: Louisiana State University Press, 1963.

Thompson, Robert Farris. *Flash of the Spirit: African and Afro-American Art and Philosophy*. New York: Vintage, 1984.

Tise, Larry E. *Proslavery: A History of the Defense of Slavery in America, 1701–1840*. Athens: University of Georgia Press, 1988.

Usner, Daniel H., Jr. *Indians, Settlers, and Slaves in a Frontier Exchange Economy: The Lower Mississippi Valley before 1783*. Chapel Hill: University of North Carolina Press, 1992.

Varon, Elizabeth. *We Mean To Be Counted: White Women and Politics in Antebellum Virginia*. Chapel Hill: University of North Carolina Press, 1998.

Van Deburg, William L. *The Slave Drivers: Black Agricultural Labor Supervisors in the Antebellum South*. Westport, Conn.: Greenwood, 1979.

Veblen, Thorstein. *The Instinct of Workmanship and the State of the Industrial Arts*. New York: B. W. Huebsch, 1914.

Vlach, John M. *Back of the Big House: The Architecture of Plantation Slavery*. Chapel Hill: University of North Carolina Press, 1993.

———. *The Planter's Prospect: Privilege and Slavery in Plantation Paintings*. Chapel Hill: University of North Carolina Press, 2002.

Wade, Michael G. *Sugar Dynasty: M. A. Patout & Son, Ltd., 1791–1993*. Lafayette: University of Southwestern Louisiana Press, 1995.

Wahl, Jenny Bourne. *The Bondsman's Burden: An Economic Analysis of the Common Law of Southern Slavery.* Cambridge: Cambridge University Press, 1998.

Wallace, Anthony F. C. *Rockdale: The Growth of an American Village in the Early Industrial Revolution.* New York: Alfred A. Knopf, 1978.

Wallerstein, Immanuel. *The Capitalist World-Economy: Essays.* Cambridge: Cambridge University Press, 1979.

Way, Peter. *Common Labor: Workers and the Digging of North American Canals, 1780–1860.* 1993. Reprint, Baltimore: Johns Hopkins University Press, 1997.

Weiner, Marli F. *Mistresses and Slaves: Plantation Women in South Carolina, 1830–1860.* Urbana: University of Illinois Press, 1998.

White, Deborah Gray. *Ar'n't I a Woman? Female Slaves in the Plantation South.* New York: W. W. Norton, 1985.

White, Shane, and Graham White. *Stylin': African American Expressive Culture from Its Beginnings to the Zoot Suit.* Ithaca, N.Y.: Cornell University Press, 1998.

Whitman, T. Stephen. *The Price of Freedom: Slavery and Manumission in Baltimore and Early National Maryland.* London: Routledge, 2001.

Whitten, David O. *Andrew Durnford: A Black Sugar Planter in Antebellum Louisiana.* Natchitoches, La.: Northwestern State University Press, 1981.

Wilkie, Laurie A. *Creating Freedom: Material Culture and African American Identity at Oakley Plantation, Louisiana, 1840–1950.* Baton Rouge: Louisiana State University Press, 2000.

Woloson, Wendy A. *Refined Tastes: Sugar, Confectionary, and Consumers in Nineteenth-Century America.* Baltimore: Johns Hopkins University Press, 2002.

Wood, Betty. *Women's Work, Men's Work: The Informal Slave Economy of Lowcountry Georgia.* Athens: University of Georgia Press, 1995.

Wood, Marcus. *Blind Memory: The Visual Representation of Slavery in England and America, 1780–1865.* Manchester: Manchester University Press, 2000.

Wright, Gavin. *Old South, New South: Revolutions in the Southern Economy since the Civil War.* New York: Basic, 1986.

———. *The Political Economy of the Cotton South: Households, Markets, and Wealth in the Nineteenth Century.* New York: W. W. Norton, 1978.

Wyatt-Brown, Bertram. *The Shaping of Southern Culture: Honor, Grace, and War, 1760s–1880s.* Chapel Hill: University of North Carolina Press, 2001.

———. *Southern Honor: Ethics and Behavior in the Old South.* New York: Oxford University Press, 1982.

Yakubik, Jill-Karen, and Rosalinda Méndez. *Beyond the Great House: Archaeology at Ashland–Belle Helene Plantation.* Discovering Louisiana Archeology 1. Baton Rouge: Louisiana Department of Culture, Recreation, and Tourism, 1995.

Young, Jeffrey R. *Domesticating Slavery: The Master Class in Georgia and South Carolina, 1670–1837.* Chapel Hill: University of North Carolina Press, 1999.

ARTICLES

Anderson, Ralph, and Robert E. Gallman. "Slaves as Fixed Capital: Slave Labor and Southern Economic Development." *Journal of American History* 64 (June 1977): 25–46.

Arrow, Kenneth J. "The Economics of Agency." In *Principals and Agents: The Structure of Business*, ed. John W. Pratt and Richard J. Zeckhauser, 37–51. Boston: Harvard Business School, 1985.

Aufhauser, Keith. "Slavery and Scientific Management." *Journal of Economic History* 33 (December 1973): 811–24.

Bacot, Barbara SoRelle. "The Plantation." In *Louisiana Buildings, 1720–1940: The Historic American Buildings Survey*, ed. Jessie Poesch and Barbara SoRelle Bacot, 107–25. Baton Rouge: Louisiana State University Press, 1997.

Barton, Keith C. "'Good Cooks and Washers': Slave Hiring, Domestic Labor, and the Market in Bourbon County, Kentucky." *Journal of American History* 84 (September 1997): 436–60.

Bergad, Laird W. "The Economic Viability of Sugar Production Based on Slave Labor in Cuba, 1859–1878." *Latin American Research Review* 24 (1989): 95–113.

Berlin, Ira, and Philip D. Morgan. "Labor and the Shaping of Slave Life in the Americas." In *Cultivation and Culture: Labor and the Shaping of Slave Life in the Americas*, ed. Berlin and Morgan, 1–45. Charlottesville: University Press of Virginia, 1993.

Bernstein, Barton J. "Southern Politics and Attempts to Reopen the African Slave Trade." *Journal of Negro History* 51 (January 1966): 16–35.

Bezís-Selfa, John. "A Tale of Two Ironworks: Slavery, Free Labor, and Resistance in the Early Republic." *William and Mary Quarterly* 56 (October 1999): 677–700.

Blassingame, John W. "Status and Social Structure in the Slave Community: Evidence from New Sources." In *Perspectives and Irony in American Slavery*, ed. Harry P. Owens, 137–51. Jackson: University of Mississippi Press, 1976.

Bolland, O. Nigel. "Proto-Proletarians? Slave Wages in the Americas Between Slave Labour and Free Labour." In *From Chattel Slaves to Wage Slaves: The Dynamics of Labour Bargaining in the Americas*, ed. Mary Turner, 123–47. London: James Currey, 1995.

Bowles, Samuel, and Herbert Gintis. "Contested Exchange: New Microfoundations for the Political Economy of Capitalism." *Politics and Society* 18 (June 1990): 165–222.

Bowman, Shearer Davis. "Antebellum Planters and Vormarz Junkers in Comparative Perspective," *American Historical Review* 85 (October 1980): 779–808.

Briggs, Charles L. "Introduction." In *Disorderly Discourse: Narrative, Conflict, and Inequality*, ed. Briggs, 4–13. New York: Oxford University Press, 1996.

Bruegel, Martin. "'Time That Can Be Relied Upon': The Evolution of Time Consciousness in the Mid-Hudson Valley, 1790–1860." *Journal of Social History* 28 (spring 1995): 547–64.

Burrell, R. J. W., M. J. R. Healy, and J. M. Tanner, "Age at Menarche in South African Bantu Schoolgirls Living in the Transkei Reserve." *Human Biology* 33 (September 1961): 250–61.

Bush, Barbara. "Hard Labor: Women, Childbirth, and Resistance in British Caribbean Slave Societies." In *More Than Chattel: Black Women and Slavery in the Americas,* ed. David Barry Gaspar and Darlene Clark Hine, 193–217. Bloomington: Indiana University Press, 1996.

Calderhead, William. "The Role of the Professional Slave Trader in a Slave Economy: Austin Woolfolk, A Case Study." *Civil War History* 23 (September 1977): 195–211.

Camp, Stephanie M. H. "The Pleasures of Resistance: Enslaved Women and Body Politics in the Plantation South, 1830–1861." *Journal of Southern History* 68 (August 2002): 533–72.

Campbell, John. "As 'A Kind of Freeman'? Slaves' Market-Related Activities in the South Carolina Up Country, 1800–1860." In *Cultivation and Culture: Labor and the Shaping of Slave Life in the Americas,* ed. Ira Berlin and Philip D. Morgan, 243–74. Charlottesville: University Press of Virginia, 1993.

———. "Work, Pregnancy, and Infant Mortality among Southern Slaves." *Journal of Interdisciplinary History* 14 (spring 1984): 793–812.

Campbell, Randolph B. "Slave Hiring in Texas." *American Historical Review* 93 (February 1988): 107–14.

Caron, Peter. "'Of a Nation Which Others Do Not Understand': Bambara Slaves and African Ethnicity in Colonial Louisiana, 1718–60." In *Routes to Slavery: Direction, Ethnicity, and Mortality in the Atlantic Slave Trade,* ed. David Eltis and David Richardson, 98–121. London: Frank Cass, 1997.

Cashin, Joan E. "Introduction: Culture of Resignation." In *Our Common Affairs: Texts from Women in the Old South,* ed. Cashin, 1–41. Baltimore: Johns Hopkins University Press, 1996.

———. "The Structure of Antebellum Planter Families: 'The Ties That Bound Us Was Strong.'" *Journal of Southern History* 56 (February 1990): 55–75.

Castañeda, Digna. "The Female Slave in Cuba during the First Half of the Nineteenth Century." In *Engendering History: Caribbean Women in Historical Perspective,* ed. Verene Shepherd, Bridget Brereton, and Barbara Bailey, 141–54. Kingston: Ian Randle, 1995.

"A Century of Progress in Louisiana, 1852–1952." *Southern Pacific Bulletin* (October 1952): 1–55.

Clinton, Catherine. "Caught in the Web of the Big House: Women and Slavery." In *The Web of Southern Social Relations: Women, Family, and Education,* ed. Walter J. Fraser Jr., R. Frank Saunders Jr., and Jon L. Wakelyn, 19–34. Athens: University of Georgia Press, 1985.

———. "'Southern Dishonor': Flesh, Blood, Race, and Bondage." In *In Joy and Sorrow: Women, Family, and Marriage in the Victorian South, 1830–1900,* ed. Carol Bleser, 52–68. New York: Oxford University Press, 1991.

Coclanis, Peter A. "How the Low Country Was Taken to Task: Slave-Labor Organization in Coastal South Carolina and Georgia." In *Slavery, Secession, and Southern History,* ed. Robert Louis Paquette and Louis A. Ferleger, 59–78. Charlottesville: University Press of Virginia, 2000.

Cody, Cheryll Ann. "Cycles of Work and Childbearing: Seasonality in Women's Lives on Low Country Plantations." In *More Than Chattel: Black Women and Slavery in the Americas,* ed. David Barry Gaspar and Darlene Clark Hine, 61–78. Bloomington: Indiana University Press, 1996.

———. "Sale and Separation: Four Crises for Enslaved Women on the Ball Plantations, 1764–1864." In *Working toward Freedom: Slave Society and Domestic Economy in the American South,* ed. Larry E. Hudson Jr., 119–42. Rochester, N.Y.: University of Rochester Press, 1994.

Coles, Harry L. "Some Notes on Slaveownership and Landownership in Louisiana, 1850–60." *Journal of Southern History* 9 (August 1943): 381–94.

Conrad, Alfred, and John Meyer. "The Economics of Slavery in the Antebellum South." *Journal of Political Economy* 66 (April 1958): 95–123.

Crawford, Stephen C. "Punishments and Rewards." In *Without Consent or Contract: The Rise and Fall of American Slavery (Technical Papers).* Vol. 2, *Conditions of Slave Life and the Transition to Freedom,* ed. Robert W. Fogel and Stanley L. Engerman, 536–50. New York: W. W. Norton, 1992.

Davis, David Brion. "Looking at Slavery from Broader Perspectives." *American Historical Review* 105 (April 2000): 452–66.

DeFrantz, Thomas F. "Black Bodies Dancing Black Culture—Black Atlantic Transformations." In *EmBODYing Liberation: The Black Body in American Dance,* ed. Dorothea Fischer-Hornung and Alison D. Goeller, 11–16. Hamburg: Lit Verlag, 2001.

Dew, Charles B. "Disciplining Slave Ironworkers in the Antebellum South: Coercion, Conciliation, and Accommodation." *American Historical Review* 79 (April 1974): 393–418.

———. "Slavery and Technology in the Antebellum Southern Iron Industry: The Case of Buffalo Forge." In *Science and Medicine in the Old South,* ed. Ronald L. Numbers and Todd L. Savitt, 107–26. Baton Rouge: Louisiana State University Press, 1989.

Donaldson, Gary A. "A Window on Slave Culture: Dances at Congo Square in New Orleans, 1800–1862." *Journal of Negro History* 69 (spring 1984): 63–72.

Downey, Tom. "Riparian Rights in Antebellum South Carolina: William Gregg and the Origins of the 'Industrial Mind.'" *Journal of Southern History* 65 (February 1999): 78–108.

Dunn, Richard S. "'Dreadful Idlers in the Cane Fields': The Slave Labor Pattern on a Jamaican Sugar Estate, 1762–1831." *Journal of Interdisciplinary History* 17 (spring 1987): 795–822.

———. "Sugar Production and Slave Women in Jamaica." In *Cultivation and Culture: Labor and the Shaping of Slave Life in the Americas,* ed. Ira Berlin and Philip D. Morgan, 49–72. Charlottesville: University Press of Virginia, 1993.

Dupre, Daniel. "Ambivalent Capitalists on the Cotton Frontier: Settlement and Development in the Tennessee Valley of Alabama." *Journal of Southern History* 56 (May 1990): 215–40.

Durrill, Wayne K. "Routines of Seasons: Labor Regimes and Social Ritual in an Antebellum Plantation Community." *Slavery and Abolition* 16 (August 1995): 161–87.

Earle, Carville. "The Price of Precocity: Technical Change and Ecological Constraint in the Cotton South, 1840–1890." *Agricultural History* 66 (summer 1992): 25–60.

———. "To Enslave or Not to Enslave: Crop Seasonality, Labor Choice, and the Urgency of the Civil War." In *Geographical Inquiry and American Historical Problems*, 226–57. Stanford: Stanford University Press, 1992.

Egerton, Douglas R. "Markets without a Market Revolution: Southern Planters and Capitalism." *Journal of the Early Republic* 16 (summer 1996): 207–21.

Engerman, Stanley L. "Contract Labor, Sugar, and Technology in the Nineteenth Century." *Journal of Economic History* 43 (September 1983): 635–59.

———. "Slavery, Serfdom, and Other Forms of Coerced Labour: Similarities and Differences." In *Serfdom and Slavery: Studies in Legal Bondage*, ed. M. L. Bush, 18–41. London: Longman, 1996.

Fenoaltea, Stefano. "Slavery and Supervision in Comparative Perspective: A Model." *Journal of Economic History* 44 (September 1984): 635–68.

Ferlerger, Louis. "Cutting the Cane: Harvesting in the Louisiana Cane Industry." *Southern Studies* 23 (spring 1984): 42–59.

———. "Farm Mechanization in the Southern Sugar Sector after the Civil War." *Louisiana History* 23 (winter 1982): 21–34.

———. "Productivity Change in the Post-Bellum Louisiana Sugar Industry." In *Time Series Analysis*, ed. O. D. and M. R. Perryman, 147–71. New York: North Holland Press, 1981.

Fiehrer, Thomas M. "The African Presence in Colonial Louisiana: An Essay on the Continuity of Caribbean Culture." In *Louisiana's Black Heritage*, ed. Robert R. McDonald, John R. Kemp, and Edward F. Haas, 3–31. New Orleans: Louisiana State Museum, 1979.

Fildes, Valerie. "Historical Changes in Patterns of Breast Feeding." In *Natural Human Fertility: Social and Biological Determinants*, ed. Peter Diggory, Malcolm Potts, and Sue Teper, 119–29. Basingstoke: Macmillan, 1988.

Finley, Ronald. "Slavery, Incentives, and Manumission: A Theoretical Model." *Journal of Political Economy* 83 (October 1975): 923–33.

Fischer, Kirsten. "'False, Feigned, and Scandalous Words': Sexual Slander and Racial Ideology among Whites in Colonial North Carolina." In *The Devil's Lane: Sex and Race in the Early South*, ed. Catherine Clinton and Michele Gillespie, 139–53. New York: Oxford University Press, 1997.

Flanders, Ralph B. "An Experiment in Louisiana Sugar, 1829–1833." *North Carolina Historical Review* 9 (April 1932): 153–62.

Fogel, Robert, and Stanley L. Engerman. "The Slave Breeding Thesis." In *Without Consent or Contract: The Rise and Fall of American Slavery (Technical Papers)*. Vol. 2, *Conditions of Slave Life and the Transition to Freedom*, ed. Fogel and Engerman, 455–72. New York: W. W. Norton, 1992.

Follett, Richard. "Heat, Sex, and Sugar: Pregnancy and Childbearing in the Slave Quarters." *Journal of Family History* 28 (October 2003): 510–39.

———. "On the Edge of Modernity: Louisiana's Landed Elites in the Nineteenth-Century Sugar Country." In *The American South and the Italian Mezzogiorno: Essays in Comparative History*, ed. Enrico Dal Lago and Rick Halpern, 73–94. Basingstoke: Palgrave, 2002.

———. "Slavery and Plantation Capitalism in Louisiana's Sugar Country." *American Nineteenth Century History* 1 (autumn 2000): 1–27.

———, and Rick Halpern. "From Slavery to Freedom in Louisiana's Sugar Country: Changing Labor Systems and Workers' Power, 1861–1913." In *Sugar, Slavery, and Society*, ed. Bernard Moitt, forthcoming. Gainesville: University Press of Florida.

Foner, Eric. "Free Labor and Nineteenth-Century Political Thought." In *The Market Revolution in America: Social, Political, and Religious Expressions, 1800–1880*, ed. Melvyn Stokes and Stephen Conway, 99–127. Charlottesville: University Press of Virginia, 1996.

Foshee, Andrew W. "Slave Hiring in Rural Louisiana." *Louisiana History* 26 (winter 1985): 63–73.

Freudenberger, Herman, and Jonathan Pritchett. "The Domestic United States Slave Trade: New Evidence." *Journal of Interdisciplinary History* 21 (winter 1991): 447–77.

Frisch, Rose E. "Demographic Implications of Female Fecundity." *Social Biology* 22 (spring 1975): 17–22.

Gallay, Allan. "The Origins of Slaveholders' Paternalism: George Whitefield, the Bryan Family, and the Great Awakening in the South." *Journal of Southern History* 53 (August 1987): 369–94.

Gaspar, David Barry. "Sugar Cultivation and Slave Life in Antigua before 1800." In *Cultivation and Culture: Labor and the Shaping of Slave Life in the Americas*, ed. Ira Berlin and Philip D. Morgan, 101–23. Charlottesville: University Press of Virginia, 1993.

Geggus, David P. "Slave and Free Colored Women in Saint Domingue." In *More Than Chattel: Black Women and Slavery in the Americas*, ed. David Barry Gaspar and Darlene Clark Hine, 259–78. Bloomington: Indiana University Press, 1996.

———. "Sugar and Coffee Cultivation in Saint Domingue and the Shaping of the Slave Labor Force." In *Cultivation and Culture: Labor and the Shaping of Slave Life in the Americas*, ed. Ira Berlin and Philip D. Morgan, 73–98. Charlottesville: University Press of Virginia, 1993.

Genovese, Eugene D. "The Medical and Insurance Costs of Slaveholding in the Cotton Belt." *Journal of Negro History* 45 (July 1960): 141–55.

———. "'Our Family, White and Black': Family and Household in the Southern Slaveholders' World View." In *In Joy and Sorrow: Women, Family, and Marriage in the Victorian South, 1830–1900,* ed. Carol Bleser, 68–87. New York: Oxford University Press, 1991.

Gibbons, Robert. "Incentives in Organizations." *Journal of Economic Perspectives* 12 (fall 1998): 115–32.

Gould, Jeffrey. "The Strike of 1887: Louisiana Sugar War." *Southern Exposure* 12 (November–December 1984): 45–55.

Gundersen, Joan Ezner. "The Double Bonds of Race and Sex: Black and White Women in a Colonial Virginia Parish." *Journal of Southern History* 52 (August 1986): 351–72.

Gutman, Herbert G. "Enslaved Afro-Americans and the 'Protestant' Work Ethic." In *Power and Culture: Essays on the American Working Class,* 298–325. New York: Pantheon Books, 1987.

———, and Richard Sutch. "Sambo Makes Good, or Were Slaves Imbued with the Protestant Work Ethic?" In *Reckoning with Slavery: A Critical Study in the Quantitative History of American Negro Slavery,* ed. Paul A. David et al., 55–93. New York: Oxford University Press, 1976.

———. "Victorians All? The Sexual Mores and Conduct of Slaves and Their Masters." In *Reckoning with Slavery: A Critical Study in the Quantitative History of American Negro Slavery,* ed. Paul A. David et al., 134–64. New York: Oxford University Press, 1976.

Hall, Gwendolyn Midlo. "African Women in French and Spanish Louisiana: Origins, Roles, Family, Work, Treatment." In *The Devil's Lane: Sex and Race in the Early South,* ed. Catherine Clinton and Michele Gillespie, 247–62. New York: Oxford University Press, 1997.

Halpern, Rick. "The Iron Fist and the Velvet Glove: Welfare Capitalism in Chicago's Packinghouses, 1921–1933." *Journal of American Studies* 26 (August 1992): 159–83.

Hanger, Kimberly S. "Greedy French Masters and Color-Conscious, Legal Minded Spaniards in Colonial Louisiana." In *Slavery in the Caribbean Francophone World: Distant Voices, Forgotten Acts, Forged Identities,* ed. Doris Y. Kadish, 106–21. Athens: University of Georgia Press, 2000.

Hareven, Tamara K., and Maris A. Vinovskis. "Introduction." In *Family and Population in Nineteenth-Century America,* ed. Hareven and Vinovskis, 3–21. Princeton, N.J.: Princeton University Press, 1978.

Hendrix, James Paisley, Jr. "The Efforts to Reopen the African Slave Trade in Louisiana." *Louisiana History* 10 (spring 1969): 97–123.

Hensley, Paul B. "Time, Work, and Social Context in New England." *New England Quarterly* 65 (December 1992): 531–59.

Hilliard, Sam B. "Site Characteristics and Spatial Stability of the Louisiana Sugarcane Industry." *Agricultural History* 53 (January 1979): 254–69.

Hine, Darlene Clark. "Rape and the Inner Lives of Black Women: Thoughts on the Culture of Dissemblance." *Signs* 14 (summer 1989): 912–20.

Hudson, Larry E., Jr. "'All That Cash': Work and Status in the Slave Quarters." In *Working*

Toward Freedom: Slave Society and Domestic Economy in the American South, ed. Hudson, 77–94. Rochester, N.Y.: University of Rochester Press, 1994.

Hunt, Patricia K. "The Struggle to Achieve Individual Expression through Clothing and Adornment: African American Women under and after Slavery." In *Discovering the Women in Slavery: Emancipating Perspectives on the American Past,* ed. Patricia Morton, 227–40. Athens: University of Georgia Press, 1996.

Ingersoll, Thomas. "The Slave Trade and the Ethnic Diversity of Louisiana's Slave Community." *LH* (spring 1996): 133–61.

Irwin, James R. "Exploring the Affinity of Wheat and Slavery in the Virginia Piedmont." *Explorations in Economic History* 25 (July 1988): 295–322.

Jacoby, Karl. "Slaves By Nature? Domestic Animals and Human Slaves." *Slavery and Abolition* 15 (April 1994): 89–99.

Jasieńska, G., and P. T. Ellison. "Physical Work Causes Suppression of Ovarian Function in Women." *Proceedings of the Royal Society of London,* ser. B, 265 (1998): 1847–51.

Jennings, Thelma. "'Us Colored Women Had to Go through a Plenty': Sexual Exploitation of African-American Slave Women." *Journal of Women's History* 1 (winter 1990): 45–74.

Johnson, Michael P. "Smothered Slave Infants: Were Slave Mothers At Fault?" *Journal of Southern History* 47 (November 1981): 493–520.

———. "Work, Culture, and the Slave Community: Slave Occupations in the Cotton Belt in 1860." *Labor History* (1984): 325–55.

Kahn, Charles. "An Agency Approach to Slave Punishments and Rewards." In *Without Consent or Contract: The Rise and Fall of American Slavery (Technical Papers).* Vol. 2, *Conditions of Slave Life and the Transition to Freedom,* ed. Robert W. Fogel and Stanley L. Engerman, 551–65. New York: W. W. Norton, 1992.

Kaye, Anthony E. "Neighbourhoods and Solidarity in the Natchez District of Mississippi: Rethinking the Antebellum Slave Community." *Slavery and Abolition* 23 (April 2002): 1–24.

Kelley, Jennifer Olsen, and J. Lawrence Angel. "Life Stresses of Slavery." *American Journal of Physical Anthropology* 74 (1987): 199–211.

Kelley, Robin D. G. "Rethinking Black Working-Class Opposition in the Jim Crow South." *Journal of American History* 80 (June 1993): 75–112.

Kendall, John S. "New Orleans' 'Peculiar Institution.'" *Louisiana Historical Quarterly* 23 (July 1940): 3–25.

Kerber, Linda K. "Separate Spheres, Female Worlds, Woman's Place: The Rhetoric of Women's History." In *Towards an Intellectual History of Women: Essays,* 159–99. Chapel Hill: University of North Carolina Press 1997.

King, Wilma, "The Mistress and Her Maids: White and Black Women in a Louisiana Household, 1858–1868." In *Discovering the Women in Slavery: Emancipating Perspectives on the American Past,* ed. Patricia Morton, 82–106. Athens: University of Georgia Press, 1996.

Kiple, Kenneth F., and Virginia Kiple. "Slave Child Mortality: Some Nutritional Answers to a Perennial Puzzle." *Journal of Social History* 10 (March 1977): 284–309.

Klein, Herbert S., and Stanley L. Engerman. "Fertility Differentials between Slaves in the United States and the British West Indies: A Note on Lactation Practices and Their Possible Implications." *William and Mary Quarterly* 35 (April 1978): 357–74.

Kotlikoff, Laurence J. "The Structure of Slave Prices in New Orleans, 1804 to 1862." *Economic Inquiry* 17 (October 1979): 496–517.

Lachance, Paul F. "The Politics of Fear: French Louisianans and the Slave Trade, 1786–1809." *Plantation Society* 1 (June 1979): 164–73.

Levine, Bruce. "Modernity, Backwardness, and Capitalism in the Two Souths." In *The American South and the Italian Mezzogiorno: Essays in Comparative History*, ed. Enrico Dal Lago and Rick Halpern, 231–37. Basingstoke: Palgrave, 2002.

Levine, Lawrence W. "Slave Songs and Slave Consciousness: An Exploration in Neglected Stories." In *African-American Religion: Interpretive Essays in History and Culture*, ed. Timothy E. Fulop and Albert J. Raboteau, 57–89. London: Routledge, 1997.

Lichtenstein, Alex. "'That Disposition to Theft, with Which They Have Been Branded': Moral Economy, Slave Management, and the Law." *Journal of Social History* 21 (1988): 413–40.

Lockley, Timothy James. "Partners in Crime: African-Americans and Non-Slaveholding Whites in Antebellum Georgia." In *White Trash: Race and Class in America*, ed. Matt Wray and Annalee Newitz, 57–72. London: Routledge, 1997.

———. "Trading Encounters between Non-Elite Whites and African Americans in Savannah, 1790–1860." *Journal of Southern History* 66 (February 2000): 25–48.

Margo, Robert A. "Wages and Prices during the Antebellum Period." In *American Economic Growth and Standards of Living before the Civil War*, ed. Robert E. Gallman and John Joseph Wallis, 173–210. Chicago: University of Chicago Press, 1992.

———, and Richard H. Steckel. "The Heights of American Slaves: New Evidence on Slave Nutrition and Health." *Social Science History* 6 (fall 1982): 516–38.

Marshall, Woodville K. "Provision Ground and Plantation Labor in Four Windward Islands: Competition for Resources in Slavery." In *Cultivation and Culture: Labor and the Shaping of Slave Life in the Americas*, ed. Ira Berlin and Philip D. Morgan, 203–20. Charlottesville: University Press of Virginia, 1993.

McDonald, Roderick A. "Independent Economic Production by Slaves on Antebellum Louisiana Sugar Plantations." In *Cultivation and Culture: Labor and the Shaping of Slave Life in the Americas*, ed. Ira Berlin and Philip D. Morgan, 275–302. Charlottesville: University Press of Virginia, 1993.

McDonnell, Lawrence T. "Money Knows No Master: Market Relations and the American Slave Community." In *Developing Dixie: Modernization in a Traditional Society*, ed. Winfred B. Moore Jr., Joseph F. Tripp, and Lyon G. Tyler Jr., 31–44. Westport, Conn.: Greenwood Press, 1988.

———. "Work, Culture, and Society in the Slave South, 1790–1861." In *Black and White Cultural Interaction in the Antebellum South*, ed. Ted Ownby, 125–148. Jackson: University of Mississippi Press, 1993.

McKenzie, Earl. "Time in European and African Philosophy: A Comparison." *Caribbean Quarterly* 19 (September 1973): 77–85.

McMillen, Sally G. "Mother's Sacred Duty: Breast-Feeding Patterns among Middle- and Upper-Class Women in the Antebellum South." *Journal of Southern History* 52 (August 1985): 333–56.

McNeilly, Alan S. "Breastfeeding and Fertility." In *Biomedical and Demographic Determinants of Reproduction*, ed. Ronald Gray, Henri Leridon, and Alfred Spira, 391–412. Oxford: Oxford University Press, 1993.

Metzer, Jacob. "Rational Management, Modern Business Practices, and Economies of Scale in the Antebellum Southern Plantations." *Explorations in Economic History* 12 (April 1975): 123–50.

Milgrom, Paul, and John Roberts. "An Economic Approach to Influence Activities in Organizations." *American Journal of Sociology* 94 (1988): S154–S179.

Miller, Randall M. "Slaves and Southern Catholicism." In *Masters and Slaves in the House of the Lord: Race and Religion in the American South, 1740–1860*, ed. John B. Boles, 127–52. Lexington: University of Kentucky Press, 1988.

Miller, Steven F. "Plantation Labor Organization and Slave Life on the Cotton Frontier: The Alabama-Mississippi Black Belt, 1815–1840." In *Cultivation and Culture: Labor and the Shaping of Slave Life in the Americas*, ed. Ira Berlin and Philip D. Morgan, 155–69. Charlottesville: University Press of Virginia, 1993.

Millet, Donald J. "The Saga of Water Transportation into Southwest Louisiana to 1900." *Louisiana History* 15 (fall 1974): 339–56.

Mintz, Sidney. "Was the Plantation Slave a Proletarian?" In *An Expanding World: The European Impact on World History, 1450–1800*, Vol. 18, *Plantation Societies in the Era of European Expansion*, ed. Judy Bieber, 305–22. Aldershot: Variorum, 1997.

Moitt, Bernard. "Slave Women and Resistance in the French Caribbean." In *More Than Chattel: Black Women and Slavery in the Americas*, ed. David Barry Gaspar and Darlene Clark Hine, 239–58. Bloomington: Indiana University Press, 1996.

———. "Women, Work, and Resistance in the French Caribbean during Slavery, 1700–1848." In *Engendering History: Caribbean Women in Historical Perspective*, ed. Verene Shepherd, Bridget Brereton, and Barbara Bailey, 155–75. Kingston: Ian Randle, 1995.

Moreno Fraginals, Manuel, Herbert S. Klein, and Stanley L. Engerman, "The Level and Structure of Slave Prices on Cuban Plantations in the Mid-Nineteenth Century: Some Comparative Perspectives." *American Historical Review* 88 (December 1983): 1185–213.

Morgan, Philip D. "The Ownership of Property by Slaves in the Mid-Nineteenth Century Low Country." *Journal of Southern History* 49 (August 1983): 399–426.

———. "Task and Gang Systems: The Organization of Labor on New World Plantations."

In *Work and Labor in Early America,* ed. Stephen Innes, 189–220. Chapel Hill: University of North Carolina Press, 1988.

Morris, Christopher. "The Articulation of Two Worlds: The Master-Slave Relationship Reconsidered." *Journal of American History* 85 (December 1998): 982–1007.

Nettle, Daniel. "Women's Height, Reproductive Success, and the Evolution of Sexual Dimorphism in Modern Humans." *Proceedings of the Royal Society of London,* ser. B, 269 (2002): 1919–23.

Newby, Howard. "Paternalism and Capitalism." In *Industrial Society: Class, Cleavage, and Control,* ed. Richard Scase, 59–73. London: George Allen and Unwin, 1977.

Niemi, Albert W., Jr. "Inequality in the Distribution of Slave Wealth: The Cotton South and Other Southern Agricultural Regions." *Journal of Economic History* 37 (September 1977): 747–53.

O'Brien, John T. "Factory, Church, and Community: Blacks in Antebellum Richmond." *Journal of Southern History* 44 (November 1978): 509–36.

Olwell, Robert. "'Loose, Idle, and Disorderly': Slave Women in the Eighteenth-Century Charleston Marketplace." In *More Than Chattel: Black Women and Slavery in the Americas,* ed. David Barry Gaspar and Darlene Clark Hine, 97–110. Bloomington: Indiana University Press, 1996.

———. "'A Reckoning of Accounts': Patriarchy, Market Relations, and Control on Henry Lauren's Lowcountry Plantations, 1762–1785." In *Working toward Freedom: Slave Society and Domestic Economy in the American South,* ed. Larry E. Hudson Jr., 33–52. Rochester, N.Y.: University of Rochester Press, 1994.

O'Malley, Michael. "Time, Work, and Task Orientation: A Critique of American Historiography." *Time and Society* 1 (September 1992): 341–58.

Painter, Nell Irvin. "Soul Murder and Slavery: Toward a Fully Loaded Cost Accounting." In *U.S. History as Women's History: New Feminist Essays,* ed. Linda K. Kerber, Alice Kessler-Harris, and Kathryn Kish-Sklar, 125–46. Chapel Hill: University of North Carolina Press, 1995.

Paquette, Robert L. "The Drivers Shall Lead Them: Image and Reality in Slave Resistance," in *Slavery, Secession, and Southern History,* ed. Robert Louis Paquette and Louis A. Ferleger, 31–58. Charlottesville: University Press of Virginia, 2000.

Paskoff, Paul F. "Hazard Removal on the Western Rivers as a Problem of Public Policy." *Louisiana History* 40 (summer 1999): 261–82.

Pawlowski, B., R. I. M. Dunbar, and A. Lipowicz. "Tall Men Have More Reproductive Success." *Nature* 403 (2000): 156.

Perrin, Liese M. "Resisting Reproduction: Reconsidering Slave Contraception in the Old South." *Journal of American Studies* 35 (2001): 255–74.

Postell, Paul Everett. "John Hampton Randolph, A Louisiana Planter." *Louisiana Historical Quarterly* 25 (January 1942): 149–223.

Pratt, John W., and Richard J. Zeckhauser. "Principals and Agents: An Overview." In *Principals and Agents: The Structure of Business,* ed. Pratt and Zeckhauser, 1–35. Boston: Harvard Business School, 1985.

Prichard, Walter, ed. "A Forgotten Louisiana Engineer: G. W. R. Bayley and His 'History of the Railroads of Louisiana.'" *Louisiana Historical Quarterly* 30 (October 1947): 1065–85.

———. "Routine on a Louisiana Sugar Plantation under the Slavery Regime." *Mississippi Valley Historical Quarterly* 14 (September 1927): 168–78.

Pritchett, Jonathan B. "The Interregional Slave Trade and the Selection of Slaves for the New Orleans Market." *Journal of Interdisciplinary History* 28 (summer 1997): 57–85.

———, and Richard M. Chamberlain. "Selection in the Market for Slaves: New Orleans, 1830–1860." *Quarterly Journal of Economics* 108 (May 1993): 469–70.

———, and Herman Freudenberger. "A Peculiar Sample: The Selection of Slaves for the New Orleans Market." *Journal of Economic History* 52 (March 1992): 109–27.

Prude, Jonathan D. "To Look Upon the 'Lower Sort': Runaway Ads and the Appearance of Unfree Laborers." *Journal of American History* 78 (June 1991): 124–59.

Raboteau, Albert J. "The Black Experience in American Evangelicalism: The Meaning of Slavery." In *African-American Religion: Interpretive Essays in History and Culture,* ed. Timothy E. Fulop and Albert J. Raboteau, 90–106. London: Routledge, 1997.

Radano, Ronald. "Denoting Difference: The Writing of Slave Spirituals." *Critical Inquiry* 22 (spring 1996): 506–44.

Rankin, David C. "Black Slaveholders: The Case of Andrew Durnford." *Southern Studies* 21 (fall 1982): 343–47.

Ransom, Roger, and Richard Sutch. "Capitalists without Capital: The Burden of Slavery and the Impact of Emancipation." *Agricultural History* 62 (summer 1988): 133–60.

Reddock, Rhoda. "Women and the Slave Plantation Economy in the Caribbean." In *Retrieving Women's History: Changing Perceptions of the Role of Women in Politics and Society,* ed. S. Jay Kleinberg, 105–19. Oxford: Berg, 1988.

Reed, Merl E. "Boom or Bust—Louisiana's Economy during the 1830s." *Louisiana History* 4 (winter 1963): 35–54.

———. "Government Investment and Economic Growth: Louisiana's Ante-Bellum Railroads." *Journal of Southern History* 28 (May 1962): 183–201.

Reidy, Joseph P. "Obligation and Right: Patterns of Labor, Subsistence, and Exchange in the Cotton Belt of Georgia, 1790–1860." In *Cultivation and Culture: Labor and the Shaping of Slave Life in the Americas,* ed. Ira Berlin and Philip D. Morgan, 138–54. Charlottesville: University Press of Virginia, 1993.

Richter, William L. "Slavery in Baton Rouge, 1820–1860." In *Plantation, Town, and County: Essays on the Local History of American Slave Society,* ed. Elinor Miller and Eugene D. Genovese, 381–95. Urbana: University of Illinois Press, 1974.

Robertson, Claire. "Africa into the Americas? Slavery and Women, the Family, and the Gender Division of Labor." In *More Than Chattel: Black Women and Slavery in the Americas,* ed. David Barry Gaspar and Darlene Clark Hine, 3–40. Bloomington: Indiana University Press, 1996.

Rodrigue, John C. "'The Great Law of Demand and Supply': The Contest over Wages in Louisiana's Sugar Region, 1870–1880." *Agricultural History* 72 (spring 1998): 159–82.

Rose, Willie Lee. "The Domestication of Domestic Slavery." In *Slavery and Freedom,* ed. William W. Freehling, 18–36. New York: Oxford University Press, 1982.

Ross, Loretta J. "African-American Women and Abortion: A Neglected History." *Journal of Health Care for the Poor and Undeserved* 3 (fall 1992): 274–84.

Russell, Sarah. "Ethnicity, Commerce, and Community on Lower Louisiana's Plantation Frontier, 1803–1828." *Louisiana History* 40 (fall 1999): 389–405.

Sappington, David E. M. "Incentives in Principal-Agent Relationships." *Journal of Economic Perspectives* 5 (spring 1991): 45–66.

Schafer, Judith Kelleher. "New Orleans Slavery in 1850 as Seen in Advertisements." *Journal of Southern History* 47 (February 1981): 33–56.

Schmitz, Mark D. "Economies of Scale and Farm Size in the Antebellum Sugar Sector." *Journal of Economic History* 37 (December 1977): 959–80.

———. "Farm Interdependence in the Antebellum Sugar Sector." *Agricultural History* 53 (January 1979): 254–69.

Scott, Jim. "Everyday Forms of Peasant Resistance." *Journal of Peasant Studies* 13 (January 1986): 5–35.

Scott, Rebecca J. "Defining the Boundaries of Freedom in the World of Cane: Cuba, Brazil, and Louisiana after Emancipation." *American Historical Review* 99 (February 1994): 70–102.

———. "'Stubborn and Disposed to Stand their Ground': Black Militia, Sugar Workers, and the Dynamics of Collective Action in the Louisiana Sugar Bowl, 1863–1887." In *From Slavery to Emancipation in the Atlantic World,* ed. Sylvia R. Frey and Betty Wood, 103–26. London: Frank Cass, 1999.

Scranton, Philip. "Varieties of Paternalism: Industrial Structures and the Social Relations of Production in American Textiles." *American Quarterly* 36 (1985): 235–57.

Shammas, Carole. "Black Women's Work and the Evolution of Plantation Society in Virginia." *Labor History* 26 (winter 1985): 5–28.

Sheridan, Richard B. "Strategies of Slave Subsistence: The Jamaican Case Reconsidered." In *From Chattel Slaves to Wage Slaves: The Dynamics of Labour Bargaining in the America,* ed. Mary Turner, 48–67. London: James Currey, 1995.

Shlomowitz, Ralph. "Team Work and Incentives: The Origins and Development of the Butty Gang System in Queensland's Sugar Industry, 1891–1913." *Journal of Comparative Economics* 3 (1979): 41–55.

Short, R. V. "Lactation—The Central Control of Reproduction." In *Breast-Feeding and the Mother,* Ciba Foundation Symposium 45, 73–86. Amsterdam: Ciba Foundation, 1976.

Sitterson, J. Carlyle. "Antebellum Sugar Culture in the South Atlantic States." *Journal of Southern History* 3 (May 1937): 175–87.

———. "Financing and Marketing the Sugar Crop of the Old South." *Journal of Southern History* 10 (May 1944): 188–99.

———. "Hired Labor on Sugar Plantations of the Ante-Bellum South." *Journal of Southern History* 14 (May 1948): 192–205.

———. "Magnolia Plantation, 1852–1862: A Decade of a Louisiana Sugar Estate." *Mississippi Valley Historical Quarterly* 25 (September 1938): 197–210.

———. "The McCollams: A Planter Family of the Old and New South." *Journal of Southern History* 6 (August 1940): 347–67.

———. "The Transition from Slave to Free Economy on the William J. Minor Plantations." *Agricultural History* 17 (October 1943): 216–24.

———. "The William J. Minor Plantations: A Study in Antebellum Absentee Ownership." *Journal of Southern History* 9 (February 1943): 59–74.

Smith, Mark M. "Old South Time in Comparative Perspective," *American Historical Review* 101 (December 1996): 1432–69.

———. "Time, Slavery, and Plantation Capitalism in the Ante-Bellum American South." *Past and Present* 150 (February 1996): 142–68.

Smith-Rosenberg, Carroll, and Charles Rosenberg. "The Female Animal: Medical and Biological Views of Woman and Her Role in Nineteenth-Century America." *Journal of American History* 60 (September 1973): 332–56.

Spear, Jennifer M. "Colonial Intimacies: Legislating Sex in French Louisiana." *William and Mary Quarterly* 60 (January 2003): 75–98.

Starobin, Robert S. "Disciplining Industrial Slaves in the Old South." *Journal of Negro History* 53 (April 1968): 115–26.

Steckel, Richard H. "Birth Weights and Infant Mortality among American Slaves." *Explorations in Economic History* 23 (April 1986): 173–98.

———. "A Dreadful Childhood: The Excess Mortality of American Slaves." *Social Science History* 10 (winter 1986): 427–65.

———. "A Peculiar Population: The Nutrition, Health, and Mortality of American Slaves from Childhood to Maturity." *Journal of Economic History* 46 (September 1986): 721–41.

———. "Slave Height Profiles from the Coastwise Manifests." *Explorations in Economic History* 16 (October 1979): 363–80.

Steffen, Charles G. "The Pre-Industrial Slave: Northampton Iron Works, 1780–1820." *Labor History* 20 (winter 1979): 89–110.

Stephens, Carlene. "'The Most Reliable Time': William Bond, the New England Railroads, and Time Awareness in Nineteenth-Century America." *Technology and Culture* 30 (January 1989): 1–24.

Stevenson, Brenda. "Distress and Discord in Virginia Slave Families, 1830–1860." In *In Joy and Sorrow: Women, Family, and Marriage in the Victorian South, 1830–1900,* ed. Carol Bleser, 103–24. New York: Oxford University Press, 1991.

———. "Gender Conventions, Ideals, and Identity among Antebellum Virginia Slave Women." In *More Than Chattel: Black Women and Slavery in the Americas,* ed. David Barry Gaspar and Darlene Clark Hine, 169–90. Bloomington: Indiana University Press, 1996.

Suarez, Raleigh A. "Bargains, Bills, and Bankruptcies: Business Activity in Rural Antebellum Louisiana." *Louisiana History* 7 (summer 1976): 189–206.

Sutch, Richard. "The Breeding of Slaves for Sale and the Western Expansion of Slavery, 1850–1860." In *Race and Slavery in the Western Hemisphere: Quantitative Studies,* ed. Stanley L. Engerman and Eugene D. Genovese, 173–210. Princeton, N.J.: Princeton University Press, 1975.

———. "The Care and Feeding of Slaves." In *Reckoning with Slavery: A Critical Study in the Quantitative History of American Negro Slavery,* ed. Paul A. David et al., 231–301. New York: Oxford University Press, 1976.

Sweig, Donald M. "Reassessing the Human Dimensions of the Slave Trade." *Prologue* 12 (spring 1980): 5–21.

Tadman, Michael. "The Demographic Cost of Sugar: Debates on Societies and Natural Increase in the Americas." *American Historical Review* 105 (December 2000): 1534–75.

———. "The Persistent Myth of Paternalism: Historians and the Nature of Master-Slave Relations in the American South." *Sage Race Relations Abstracts* 23 (February 1998): 7–23.

Tafari, N., R. L. Naeye, and A. Gobezie. "Effects of Maternal Undernutrition and Heavy Physical Work during Pregnancy on Birth Weight." *British Journal of Obstetrics and Gynecology* 87 (1980): 222–26.

Tandberg, Gerilyn. "Field Hand Clothing in Louisiana and Mississippi during the Antebellum Period." *Dress* 6 (1980): 89–103.

Tansey, Richard. "Bernard Kendig and the New Orleans Slave Trade." *Louisiana History* 23 (spring 1982): 159–78.

Thapa, Shyam, Roger V. Short, and Malcolm Potts. "Breast Feeding, Birth Spacing, and Their Effects on Child Survival." *Nature* 335 (October 1988): 679–82.

Thompson, E. P. "Time, Work-Discipline, and Industrial Capitalism." *Past and Present* 38 (December 1967): 56–97.

Thompson, Robert Farris. "An Aesthetic of the Cool: West African Dance." *African Forum* 2 (fall 1966): 85–102.

Toledano, Roulhac B. "Louisiana's Golden Age: Valcour Aime in St. James Parish." *Louisiana History* 10 (summer 1969): 211–24.

Tomich, Dale. "*Une Petite Guinée:* Provision Ground and Plantation in Martinique, 1830–1848." In *Cultivation and Culture: Labor and the Shaping of Slave Life in the Americas,* ed. Ira Berlin and Philip D. Morgan, 221–42. Charlottesville: University Press of Virginia, 1993.

Touchstone, Blake. "Planters and Slave Religion in the Deep South." In *Masters and Slaves in the House of the Lord: Race and Religion in the American South, 1740–1860*, ed. John B. Boles, 99–126. Lexington: University of Kentucky Press, 1988.

Tregle, Joseph G. "Creoles and Americans." In *Creole New Orleans: Race and Americanization*, ed. Arnold R. Hirsch and Joseph Logsdon, 131–85. Baton Rouge: Louisiana State University Press, 1992.

———. "Louisiana and the Tariff, 1816–1846." *Louisiana Historical Quarterly* 25 (January 1942): 24–148.

Trussell, James, and Richard Steckel. "The Age of Slaves at Menarche and Their First Birth." *Journal of Interdisciplinary History* 8 (winter 1978): 477–505.

Tucker, Barbara M. "The Family and Industrial Discipline in Ante-Bellum New England." *Labor History* 21 (winter 1979–1980): 55–74.

Turner, Mary. "Introduction." In *From Chattel Slaves to Wage Slaves: The Dynamics of Labour Bargaining in the Americas*, ed. Turner, 1–8. London: James Currey, 1995.

———. "Slave Workers, Subsistence, and Labour Bargaining: Amity Hall, Jamaica, 1805–1832." In *The Slaves' Economy: Independent Production by Slaves in the Americas*, ed. Ira Berlin and Philip D. Morgan, 92–105. London: Frank Cass, 1991.

Usner, Daniel H., Jr. "Indian-Black Relations in Colonial and Antebellum Louisiana." In *Slave Cultures and the Cultures of Slavery*, ed. Stephan Palmié, 145–61. Knoxville: University of Tennessee Press, 1995.

Vlach, John M. "Not Mansions . . . But Good Enough: Slave Quarters as Bi-Cultural Expression." In *Black and White Cultural Interaction in the Antebellum South*, ed. Ted Ownby, 89–114. Jackson: University of Mississippi Press, 1993.

Wallerstein, Immanuel. "What Can One Mean by Southern Culture?" In *The Evolution of Southern Culture*, ed. Numan V. Bartley, 1–13. Athens: University of Georgia Press, 1988.

Way, Peter. "Labour's Love Lost: Observations on the Historiography of Class and Ethnicity in the Nineteenth Century." *Journal of American Studies* 28 (April 1994): 1–22.

White, Deborah Gray. "Female Slaves: Sex Roles and Status in the Antebellum Plantation South." *Journal of Family History* 8 (fall 1983): 248–61.

White, Shane, and Graham White. "Slave Hair and African-American Culture in the Eighteenth and Nineteenth Centuries." *Journal of Southern History* 61 (February 1995): 45–76.

———. "'Us Likes a Mixtery': Listening to African American Slave Music." *Slavery and Abolition* 20 (December 1999): 22–48.

Whitman, T. Stephen. "Industrial Slavery at the Margin: The Maryland Chemical Works." *Journal of Southern History* 59 (February 1993): 31–62.

Whitten, David O. "Medical Care of Slaves: Louisiana Sugar Region and South Carolina Rice District." *Southern Studies* 16 (summer 1977): 153–80.

——. "Slave Buying in Virginia as Revealed by Letters of a Louisiana Negro Sugar Planter." *Louisiana History* 11 (summer 1970): 231–44.

——. "Sugar Slavery: A Profitability Model for Slave Investments in the Antebellum Louisiana Sugar Industry." *Louisiana Studies* 12 (summer 1973): 423–42.

——. "Tariff and Profit in the Antebellum Louisiana Sugar Industry." *Business History Review* 44 (summer 1970): 226–33.

Wilks, Ivor. "On Mentally Mapping Greater Asante: A Study of Time and Motion." *Journal of African History* 33 (spring 1992): 175–90.

Williams, Robert. "Thomas Affleck: Missionary to the Planter, the Farmer, and the Gardener." *Agricultural History* 31 (July 1957): 40–48.

Wilson, Semvel, Jr. "Architecture of Early Sugar Plantations." In *Greenfields: Two Hundred Years of Louisiana Sugar,* 51–82. Lafayette: Center for Louisiana Studies, University of Southwestern Louisiana Press, 1980.

Wood, Betty. "'Never on a Sunday?': Slaves and the Sabbath in Low-Country Georgia, 1750–1830." In *From Chattel Slaves to Wage Slaves: The Dynamics of Labour Bargaining in the Americas,* ed. Mary Turner, 79–96. London: James Currey, 1995.

Wood, Peter H. "'Gimme de Kneebone Bent': African Body Language and the Evolution of American Dance Forms." In *The Black Tradition in American Dance,* ed. Gerald E. Myers, 1–15. Durham: Duke University Press, 1988.

Wyatt-Brown, Bertram. "The Mask of Obedience: Male Slave Psychology in the Old South." *American Historical Review* 93 (December 1988): 1228–52.

Young, Jeffrey. "Ideology and Death on a Savannah Rice River Plantation, 1833–1867: Paternalism amidst a 'Good Supply of Disease and Pain.'" *Journal of Southern History* 59 (November 1993): 673–706.

DISSERTATIONS, THESES, REPORTS, AND UNPUBLISHED PAPERS

Dal Lago, Enrico. "Southern Elites: A Comparative Study of the Landed Aristocracies of the American South and the Italian South, 1815–1860." Ph.D diss., University College, London, 2000.

Gudmestad, Robert H. "A Troublesome Commerce: The Interstate Slave Trade, 1808–1840." Ph.D. diss., Louisiana State University, 1999.

Jung, Moon-Ho. "'Coolies' and Cane: Race, Labor, and Sugar Production in Louisiana, 1852–1877." Ph.D. diss., Cornell University, 2000.

Lathrop, Barnes Fletcher. "The Pugh Plantations, 1860–1865: A Study of Life in Lower Louisiana." Ph.D. diss., University of Texas, 1945.

Martin, Jonathan D. "Divided Mastery: Slave Hiring in the Colonial and Antebellum South." Ph.D. diss., New York University, 2000.

McGowan, James T. "Creation of a Slave Society: Louisiana Plantations in the Eighteenth Century." Ph.D. diss., University of Rochester, 1976.

Paton, Diana. "Enslaved Women's Bodies and Gendered Languages of Insult in Jamaica during Late Slavery and Apprenticeship." Paper presented at the Fourth Avignon Conference on Slavery and Forced Labor, Women in Slavery, in Honor of Suzanne Miers, October 2002.

Russell, Sarah Paradise. "Cultural Conflicts and Common Interests: The Making of the Sugar Planter Class in Louisiana, 1795–1853." Ph.D. diss., University of Maryland, 2000.

Sacher, John M. "'A Perfect War': Politics and Parties in Louisiana, 1824–1861." Ph.D. diss., Louisiana State University, 1999.

Schmitz, Mark D. "Economic Analysis of Antebellum Sugar Plantations in Louisiana." Ph.D. diss., University of North Carolina, 1974.

Shea, Philip. "The Spatial Impact of Governmental Decisions on the Production and Distribution of Louisiana Sugar Cane, 1751–1972." Ph.D. diss, Michigan State University, 1974.

Suarez, Raleigh A. "Rural Life in Louisiana, 1850–1860." Ph.D. diss., Louisiana State University, 1954.

Twiss, Miranda C. "Slave Hiring in Louisiana, 1719–1861." M.A. thesis, University of London, 2002.

Whitten, David O. "Antebellum Sugar and Rice Plantations, Louisiana and South Carolina: A Profitability Study." Ph.D. diss., Tulane University, 1970.

Wingfield, Charles L. "The Sugar Plantations of William J. Minor, 1830–1860." M.A. thesis, Louisiana State University, 1950.